Irfan Moeen Khan

Re-centering the Sufi Shrine

Islamkundliche Untersuchungen

Band 348

Irfan Moeen Khan

Re-centering the Sufi Shrine

A Metaphysics of Presence

DE GRUYTER

ISBN 978-3-11-078102-1
e-ISBN (PDF) 978-3-11-078149-6
e-ISBN (EPUB) 978-3-11-078155-7
ISSN 0939-1940

Library of Congress Control Number: 2022945396

Bibliographic information published by the Deutsche Nationalbibliothek
The Deutsche Nationalbibliothek lists this publication in the Deutsche Nationalbibliografie; detailed bibliographic data are available on the Internet at http://dnb.dnb.de.

Cover image: *Construction Drawing 3*, marble sculpture, inlaid (Pietra dura) with granite
© Hamra Abbas
Printing and Binding: LSC Communications, United States

www.degruyter.com

To the memory of my father Muhammad Shafiullah Khan
and mother Anjum Shafiullah

Acknowledgements

I submitted this work as my Ph.D. dissertation to The Committee on the Study of Religion (GSAS) at Harvard University. It is a culmination of my interest in studying religious authority, especially in reference to ritual practice and mysticism. The study of ritual spaces of Sufi shrines requires the study of texts and anthropological fieldwork. Hence, I visited several Sufi *dargāhs* in Lahore, Qasur, Multan, Bahawalpur, Karachi, Faisalabad, Islamabad, and remote villages such as Fazilpur (in South Punjab) and other lesser-known rural vicinities. During these trips, I met several *dervish* figures (men and women) whose warm hospitality and words have been sources of wisdom.

In this journeying, my utmost gratitude is reserved for Manzūr Saīn, an unlettered expert on *wujūdī* philosophy and Persian poetry, who accompanied me during a visit to the tombs of Khawaja Ghulām Farīd in Kot Mithan on foot (at times, we used public transport, packed with passengers). During these pilgrimages, Manzūr Saīn spoke to me about the Sufi lore of friendship and the metaphysics of presence. I must also acknowledge that it was incredibly advantageous for me to be the great-grandson of Shaykh Muḥammad Chishtī (Gharīb Nawāz), whose *dargāh* is in a village outside of Lahore. It was at Shaykh Muḥammad Chishti's shrine that a chance encounter in observing a mystic during his *zikr* (Ar. *Dhikr*) on a late wintry night, and the words spoken in a state of ecstasy, left me bewildered, despite my life-long familiarity with Sufi spaces. There was something very different about the mystic's voice and the penetrating power of his words that I cannot explain. So, my question was, how could I write about what I had witnessed or experienced at the *dargah* in an academic language? I do not claim to have solved this problem in this book, but it does address the importance of understanding the theopoetics of ritual space beyond texts.

I was fortunate to have the friendship of several people whose kindness and advice helped me at different points of my academic journey. Nabeel Sarwar's guidance some twenty-three years ago will never be forgotten. Dr. Jamil Rashid pointed me to McGill University as a suitable place for my MA in Islamic Studies (Professor Sajida Alvi's expertise in Islamic intellectual history proved especially useful for my work). Dr. Warnock Davies for showing me the art of the application process and sharing his wisdom about what it means to devote a life to learning during the early stages of my career. I was fortunate to be a TA for the late Dr. Linda Walbridge of Indiana University, who was teaching at the Lahore University of Management Sciences, to work on her project about Christian communities in Pakistan. Linda thought I would make an excellent anthropologist and encouraged me to pursue my graduate studies. I am indebted to the late

https://doi.org/10.1515/9783110781496-001

Dr. Zafar Ishaq Ansari for the incredible affection showered on me when I showed up at his office one afternoon at the Islamic Research Institute, Islamabad, after finishing my master's degree from McGill, looking for potential employment. I wish I could have seen him one last time after completing my Ph.D. to offer my gratitude in person.

Dr. William A. Graham (Harvard University) taught me how compassion and scholarship could coexist. Each draft of the chapter was returned to me with extensive comments and edits that helped me improve the work. I was fortunate to have his guidance at every stage. If anyone whose presence was a source of comfort during the rigors of writing, it was Dr. Kimberley C. Patton. The current project emerged from a paper presented in her seminar, "Relics, Tombs, and Transcendence." I learned about the merits of comparative inquiry in religion from her. Thanks to Dr. Ali S. Asani for his insights into Sufism and the piping hot cups of tea at his office in the Barker Center. I also want to mention scholars who have contributed to my learning at Harvard University: Dr. David Lamberth (I still have his copy of John McDermott's monograph), Dr. Francis X. Clooney, Dr. Stephenie Paulsell, Dr. Davíd Carrasco, Dr. Steve Caton, Dr. Arthur Kleinman, Dr. Parimal Patil, Dr. Rob Robson, Dr. Luis Manuel Girón-Negrón, and the late Dr. Shahab Ahmad.

Thanks to Katharina Zülhke at the De Gruyter Press, whose expert help made the process of finalizing the manuscript go so well. Any shortcomings are solely my responsibility. Finally, I owe a mountain of thanks to my lovely wife, Hamra Abbas, for her incredible support and patience. The artwork on the cover (collection of Mr. Taimur Hassan) is by her, entitled *Construction Drawing*. My two boys Shaaf Khan and Vali Khan, have brought light into our lives that has (at least in my opinion) made me a better scholar.

Contents

Introduction

The title of this dissertation, "Re-centering the Sufi Shrine," might sound somewhat inappropriate, or rather ordinary, given the voluminous scholarship on Sufi shrine visitation (*ziyāra*) since the 19th century. However, this re-centering is to be done concerning the lived experience of Indus Sufism in Pakistan.[1] It refers specifically to religious (and ritual) worship, or *'ibāda*, at a Sufi shrine (*dargāh* or *darbār*). Therefore, the study aims to present a new scholarly perspective on understanding the religious significance of the Sufi shrine and uncover the strategies of *the 'ulamā* and the state-directed against shrine worship as these have long been practiced. The focus will be on a specific ritual action or gesture performed in visiting the *dargāh* of the Sufi saints in Indus Sufism, namely that which is known as *ḥāẓirī* (being present to the sacred gaze of the saint). This ritual is centered on the worship of the sacred, "resurrected," body of a Sufi saint, whom devotees experience as *present* (*mawjūd*) at the *dargāh*.[2]

---

1 Specifically, the call for "re-centering" the Sufi tomb here refers to taking into account the living presence of Sufi saints, its sacred efficacy, and agency as the reason why visitors travel to visit saints' tombs. Therefore, the notion of re-centering also entails describing and analyzing the ritual aspects of *ziyāra* that embody the metaphysics of presence. The contention is that the study of Sufi devotional metaphysics associated with shrines, aiming to highlight the ontological relationship between the poetics of ritual practice and the shrine, requires an approach that extends well beyond textual descriptions of *ziyāra*, mainly as discussed in legal discourses on the issue. The metaphysics of presence at Sufi shrines need to be examined as a *lived* religious reality in which practice refers to the use or performance of the body in experiencing that presence. Clive Cazeaux, "Metaphor and the Categorization of the Senses," *Metaphor and Symbol* 17.1 (2002), 9: "The body, Merleau-Ponty (1945/1962) asserted, is not an empirical object, one thing among others in the world, but an ontological condition, a "body schema" (p. 206): a framework of intentionality whereby consciousness and the world are opened up for each other. Not only is it the medium that organizes a coherent sensory picture of the world for us, but also something active and prospective that is able to play this constructive, objective role because it moves around the world and maps it out for us."

2 Traditionally, Indus Sufism is distinguished by its religious culture of pluralism and the performative arts of the *samā'*, best cultivated in the form of the *kāfī* verses attributed to its saints such as Bābā Farīd of PakPattan, Bābā Bulleh Shāh of Qasūr, Khawāja Ghulām Farīd of Mithankot, and 'Abdul Laṭīf Bhiṭ'āī of Sindh. The creative confluence between the mystical poetry of the Sufi saint and the metaphysics of union with God has been a point of theological contestation for many centuries, but the most vociferous criticism of devotional observances was formulated by the 13th-century theologian Ibn Taymiyya and carried forward by his student, Ibn Qayyim al-Jawziyya (d. 751/1350). Such virulent polemics became catalysts for attacks on the devotional culture of Sufi shrines in 19th century India by scholars of the 18th-century reformist school associated with the name of Shāh Walīullāh of Delhi (1703-1762).

https://doi.org/10.1515/9783110781496-002

In writing about the ritual of *ḥāẓirī* and distinguishing it from the orthodox view of pilgrimage known as *ziyāra* in Islamic law, the study identifies a case of ritual incongruity between the prescriptive (official) and popular mode of performance missed by other studies on Sufism.[3] This is not to suggest that scholarship has been oblivious about the ritual recitation of *fātiḥa* at graves or about the juristic arguments for and against, as discussed abundantly in Christopher Taylor's *In the Vicinity of the Righteous* and other studies.[4] The only claim is that scholars have not raised the question about its theological or metaphysical relevance in the ritual use for sacred or immortal bodies of Sufi saints. Scholars of Sufism have amply recognized that the entombed Sufi saints are considered alive by their devotees.[5] Nevertheless, the matter of their immortality has not been ade-

---

3 Dale F. Eickelman, *Morrocan Islam: Tradition and Society in a Pilgrimage Centre* (Austin: University of Texas Press, 1976), C. Bawa Yamba, *Permanent Pilgrims: The Role of Pilgrimage in the Lives of West African Muslims in Sudan* (Washington D. C.: Smithsonian Institution Press, 1995), Josef W. Meri, *The Cult of Saints Among Muslims and Jews in Medieval Syria* (Oxford: Oxford University Press, 2002).

4 Some important works on *ziyāra* which discuss the prayer for the dead are Christopher Taylor, *In the Vicinity of the Righteous: "Ziyāra" and the Veneration of Muslim Saints in Late Medieval Egypt* (Leiden: Brill, 1999), Werner Diem and Marco Scholler, *The Living and the Dead in Islam. Studies in Arabic Epitaphs* (Wiesbaden: Harrassowitz Verlag, 2004), and also, Carl W. Ernst, "An Indo-Persian Guide to Sufi Shrine Pilgrimage," edited by Grace Martin Smith and Carl W. Ernst, *Manifestations of Sainthood in Islam* (Istanbul: Isis, 1993), 43 – 67. F. De Jong, "Cairene Ziyāra-Days: A Contribution to the Study of Saint Veneration in Islam," *Die Welt des Islams* 17.1/4 (1976 – 1977), 27: "The noun *ziyāra*, meaning literally visit, when used to mean a visit to the tomb of a saint (*walī*), stands for a whole set of prescribed ritual behavior such as greeting the saint when entering his shrine, reading *al-fātiḥa*, touching the maqṣūra around the *tābut*, or the *tābut* itself, in order to obtain some baraka of the saint, presenting a votive offering (naḏhr, pl. nuḏhur), praying two *rak'as*, and leaving the shrine moving backwards while saying "that God's peace and blessing may be upon you", while performing all this in the proper state of ritual purity."

5 An excellent essay on the ritual space of a Sufi shrine in Karachi is by Richard Kurin, "The Structure of Blessedness at a Muslim Shrine in Pakistan," *Middle Eastern Studies* 19.3 (1983), 315: "According to the devotees (*malang*) at the shrine of Abdullah Shah, the saint is not dead. Rather, he lives in spirit form behind a curtain (*purdah*)...All those at the shrine are bathed in this light as they come within the spiritual gaze of the saint. This gaze focuses the blessedness flowing from Allah through the saint to those supplicants prepared to receive it. Because of this, the shrine is a place of power, a locus of actions and objects within which concentrated blessedness may be dispersed." Also, Carl W. Ernst, " An Indo-Persian Guide to Sufi Shrine Pilgrimage," in *It's Not Just Academic!: Essays on Sufism and Islamic Studies* (New Delhi: Sage, 2018), 177 – 78: "The ziyarat pilgrimage is not merely a journey to a place of burial, but is literally a visit to a living saint; one of the most common Persian terms for a saint's shrine is mazar, a place that is visited, indicating that the act of personal encounter takes priority over the structure's reliquary function. Pilgrimage to Sufi saints' shrines is, temporally, a search for union with God

quately linked with ritual practices at Sufi shrines. Nor has any study discerned the incongruity of reciting *fātiḥa* at their tombs. This question about the juristic use of scripture in signifying immortal bodies of popular devotion to be dead requires an explanation because that is not how devotees interact with their saints in practice.[6] The central question of this study is why legalists recommend reciting the "prayer for the dead" for the immortal bodies of Sufi saints. How can we understand its relevance or salience with other rituals that are contrary to its (*fātiḥa* recitation) juristic aim in the celebration of sacred bodies' transcendence of death?[7] Therefore, on the one hand, it is in the non-recognition in the contemporary scholarship of incongruity between rituals that designate the departed Sufi saints as dead, and on the other, practices that are at odds with this scriptural signification (in the use of *fātiḥa*) where the Sufi saint's body is treated as the locus of Divine Presence.

The legal theology of intercession in so far as its use is concerned to reverse or rupture the Sufi metaphysics of Presence has remained unexamined in studies on *ziyāra*, the ritual politics that *Recentering the Sufi Shrine* addresses.[8] The tex-

---

through synchronicity with the saint's death anniversary; physically it is an approach to the divine presence over the threshold of the saint's tomb."

6 Richard Kurin, "The Structure of Blessedness," 316: "Supplicants typically touch, kiss and caress the coffin. Many eat flower petals previously placed on the coffin. Some take handfuls of petals for redistribution to friends and relatives who could not attend. After finishing prayers, supplicants may 'wash' their faces with the saint's blessedness through appropriate gestures. Such actions, done with a clear heart would appear to allow for the penetration of the saint's gaze and the reception of blessedness. According to devotees, if such actions are performed correctly and properly, truly in the name of Allah or the saint and without trace of selfish or devious purpose, then the blessedness of the saint would fall upon the supplicant." Chapter I discusses the matter of the saint's sacred gaze (known as *naẓar-i karam*).

7 *Ibid.*, 315: "offerings may take many forms – the most common being flowers or garlands of flowers, usually rose, marigold and jasmine. Other items used as offering are thin cloths (*chaddar*), simple hard while sweets (*batasha*) and incense (*agarbhatti*). Money, oil, copies of the Qoran, and food (e.g. lamb, rice) may also be brought as offerings, although they seem less common than flowers and sweets." Carl W. Ernst, " An Indo-Persian Guide," 175: "One should especially avoid turning one's back on the saint's tomb or turning one's feet disrespectfully in that direction. The pilgrim is also instructed to perform a deep psychological self-examination while visiting tombs, for receptivity to supernatural communications is then greatly increased and one may hope for spiritual guidance by this means."

8 Christopher Taylor (*In the Vicinity of the Righteous*) studied the legal arguments between scholars in Medieval Egypt on *ziyāra*. It is a valuable contribution to understanding the legal perspectives on pilgrimage to Sufi shrines but does not discuss the Sufi metaphysics of immortality. Although Taylor does draw on Turner's work on pilgrimage in Christianity but only sees it in terms of social dynamics, 77: "Not only was the Qarāfa liminal space, but the ziyāra was what Victor Turner would describe as a "liminoid" ritual process." For example, Taylor writes, 78: "The

tual explanation by Sunni legalists of what pilgrims must do to avoid idolatry (*shirk*) and how the devotees of Friends of God perform the ritual when visiting their shrines is representative of underlying tension between the epistemology of Islamic law—the domain of textual Islam—and the performative realm of the vernacular mysticism at Sufi shrines. Identifying this ritual difference has consequences for understanding the meaning of Sufi tombs or shrines as well for rethinking the relationship between Islamic legalism and Sufi bodies—that is, between textual knowledge of the *'ulamā* and "direct experience" of the divine reported by the mystics. This work analyzes how theopoetics of vernacular mysticism represents an alternative to the economy of pious merit (*ṣawāb*).

The question, therefore, is why these earlier works of scholarship miss the theological and legal implications of a ritual ambiguity at the heart of pilgrimage to tombs. Several excellent studies on the shrines have examined their histories and diverse ritual practices.[9] However, it would not be possible to review each of these contributions to the study of Sufism. Therefore, I have focused on scholarship representing paradigmatic approaches that have influenced the direction of contemporary Sufi studies.[10] In particular, the study of Sufi shrines or Sufism is shared between anthropologists and Islamicists, albeit there has been a disciplinary divide traditionally between the two.[11] The critique of Clifford Geertz by

ziyāra seems to have served as an important medium through which the spontaneous communitas Turner describes. For it was in the ziyāra, as we have seen, that powerful sultans donned simple Ṣūfī garments, the rich mixed with the poor, men mixed with women, and the living mixed with the dead as a broad cross-section for medieval Cairene society journeyed out from the "mundane center" of the city to the "sacred periphery" of al-Qarāfa that "great medium of divine blessing."

9 Imtiaz Ahmad (editor), *Ritual and Religion among Muslims in India* (New Delhi: Manohar, 1981), Christian W. Troll (editor), *Muslim Shrines in India: Their Character, History and Significance* (Oxford: Oxford University Press, 1989), and some recent publications are Snehi Yogish, *Spatializing Popular Sufi Shrines in Punjab: Dreams, Memories, Territoriality* (New York: Routledge, 2019), and Chauhan, Abha (editor), *Understanding Culture and Society in India. A Study of Sufis, Saints and Deities in Jammu Region* (Singapore: Springer, 2021), Surinder Singh and Ishwar Dayal Gaur (editors), *Sufism in Punjab: Mystics, Literature and Shrines* (Delhi: Aakar Books, 2009).

10 Dale F. Eickelman: "The Study of Islam in Local Contexts," in *Defining Islam* (New York: Routledge, 2014), 60: "Because the published work of Clifford Geertz now spans more than two decades and has had a pervasive influence on scholars of many disciplines, it provides the best point of departure for tracing shifts in how anthropologists have sought to contribute to the study of Islam." Talal Asad is another such figure whose work has influenced a generation of scholars (discussed below).

11 John R. Bowen, *Muslims Through Discourse: Religion and Ritual in Gayo Society* (Princeton: Princeton University Press, 1993), 5: "But if anthropologists by and large, focused on the locally

Talal Asad and in another essay, "The Idea of Anthropology of Islam," he dismissed the methodological claims of interpretive anthropology.[12] Asad's critique made it difficult for anthropologists to ignore the textual sources and the interpretative tradition of religious scholars, that is, the "discursive tradition."[13] The implication of this turn to disciplinary structures is has become a dominant paradigm for the study of Sufism. A major premise of this disciplinary approach is the emphasis on religiopolitical negotiations whereby political patronage is offered in return for spiritual blessings (*baraka*).[14] Hence, Chapter 2 looks at three critical studies that deal with the politics of Sufism: Pnina Werbner, Omid Safi, and a third publication that concerns specifically the ritual of *īṣāl-i ṣawāb* (known as *Slametan* in Indonesia) by Mark Woodward.[15] In the work of each of these scholars, the focus is on the reciprocity between power and Sufism.

---

distinctive at the expense of the religious shared, Islamicists focused on texts and their interpretation at the expense of everyday religious understandings and practices."

12 Dale F. Eickelman, "The Study of Islam in Local Contexts," 59: "The main challenge for the study of Islam in local contexts is to describe and analyze how the universalistic principles of Islam have been realized in various social and historical contexts without representing Islam as a seamless essence on the one hand or as a plastic congeries of beliefs and practices on the other." See Talal Asad, The Idea of an Anthropology of Islam (Washington, DC: Georgetown University Center for Contemporary Arab Studies, 1986) and his "Anthropological Conceptions of Religion: Reflections on Geertz," *Man* 18.2 (1983), 237-259.

13 John R. Bowen, *A New Anthropology of Islam* (Cambridge: Cambridge University Press, 2012), 7 – 8: Some may see these two approaches – Geertz's cultural emphasis, Asad's political one – as irreconcilable opposites; I prefer to note that, taken together, they pointed toward something like what I have sketched above: an approach to Islamic traditions that takes seriously both religious thinking and social frameworks. I find that the analytical tensions and dissensions emerging from contemporary debates within anthropology – Is religion symbols or discipline? Is it coherence or fracture? – have themselves generated an impressive set of new work on the processes and practices surrounding the Islamic tradition. Over the past twenty-odd years, many of us have taken the twin tensions – one between cultural specificity and a shared Islamic tradition; the other between an Islam of individual creativity and one authorized by religious leaders – as generating our empirical and analytical framework."

14 Nile Green, *Sufism*, 7: "Since no form of powerful knowledge can exist in isolation from material forms of power, the economic strength of Sufism was crucial to the overall profile it acquired and so ultimately the discursive, miraculous and economic power of the Sufis were each interdependent...If in relation to many earlier presentations of Sufism, the shift in emphasis here towards power and tradition marks something of a Hobbesian turn, then this is perhaps a necessary corrective."

15 Omid Safi, *The Politics of Knowledge in Premodern Islam: Negotiating Ideology and Religious Inquiry* (Chapel Hill: The University of North Carolina Press, 2006), Pnina Werbner, *Pilgrims of Love: The Anthropology of a Global Sufi Cult* (Bloomington: Indiana University Press, 2003), and Mark Woodward, "The "Slametan": Textual Knowledge and Ritual Performance in Central Javanese Islam," *History of Religions* 28.1 (1988), 54-89.

This understanding is an outcome of the methodological approach that privileges literary narratives over sensory experiences.[16] As a consequence, Chapter 2 also questions Geertzian anthropology.[17] Recent studies have begun to correct this tilt to take interest in the ritual body in the study of Islam.[18] David Howes offers a compelling argument of what we can do to improve our methods in the study of experience:

> Dialogical anthropologists *do* appear to have exchanged an ear for an eye. But this is only because of the demands of textualization which, "from beginning to end," remain their most abiding concern. For this reason, they have not been able to escape from the (metaphorical) prison-house of literacy, the foundations of which were laid by Geertz in the Balinese cockfight article...But were the secondary textualists able to repress their textuality for just a moment, they would recognize that culture is not *created* out of an "interplay of voices: but *constructed* out of the interplay of all the senses.[19]

My thesis is that *ḥāẓirī*, as a form of religious practice or worship (Urdu *'ibādat; Ar. 'ibāda*), is the ritual of communion with the living presence of the Sufi saint, which means that *ḥāẓirī* is effectively equivalent to what is referred to in scriptural Islam as the act of ritual prayer (Pers. *namāz*; Ar. *ṣalāt*). However, describing the ritual devotion at the Sufi shrine as *'ibādat* is classified as idolatry (*shirk*) by traditionalist *'ulamā'* (including the pro-Sufi variety). Since the 19th century,

---

16 David Howes, "Introduction: To Summon all the Senses," in David Howes (editor), *The Varieties of Sensory Experience: A Sourcebook in the Anthropology of the Senses* (Toronto: University of Toronto Press, 1991), 8: "This [sensory] dimension cannot be comprehended within the framework of either interpretive or dialogical anthropology because both remain, in effect, 'verbocentric' (text-centered in the first case, and speech-centred in the second). Thus, the shift from the ocular to the oral must be accompanied by a further shift, which takes in the gustatory, olfactory, and tactile modalities as well."

17 David Howes, "Controlling Textuality: A Call for a Return to the Senses," *Anthropologica* 32.1 (1990), 64 - 65: "For the interpretative ethnographer is not only a member of a literate (or text-producing) culture, he also uses the model of the text as a grid for the modulation of his perception-interpretation of *other* cultures. In short, what the adoption of a Geertzian approach imports is a double interiorization of reading-writing. The effect of this double interiorization is apparent in the research agenda of those in the vanguard of contemporary "culture theory."

18 Yogish Snehi. *Spatializing Popular Sufi Shrines in Punjab*, 77: "Sacred shrines and sacrality have predominantly been understood through the agency of texts and the textual notion of community. Therefore, they are either construed as belonging to the realm of dominant religious identities or, if this does not explain the aspects of belonging, they are reduced to metonyms like syncretism or tolerance. Historians, do not pay much attention to the social spaces of sacrality and structures of power that explain the embeddedness of sacred shrines in the realm of a quotidian lived practice."

19 David Howes, "Controlling Textuality," 68.

Islamic reformism has concentrated on suppressing such "idolatry" and reversing the lived metaphysics of Sufism. In addition, the Islamic state of Pakistan in the 1960s joined them and entered fully into the sacred space of Sufi shrines to rupture the perceived "idolatrous" relationship between the devotees and the Sufi saints of the Indus.

There has been strong interest among scholars working in different disciplines in understanding the religious role of Sufi shrines, not simply in Pakistan but across the Islamic world. This focus has been strongly affected by a major fault line that has developed in the study of Islam, especially following the work of Clifford Geertz.[20] This is the rift between those scholars emphasizing textual tradition on one hand and those interested in oral culture, practices, and customs on the other. Central to both approaches is an interest in the historical relevance of power as a dialectic between orthodoxy and heterodoxy. This dissertation argues that these approaches have been misplaced in their theoretical assumptions when used in examining the religious significance of the Sufi *dargāh*. Instead, I contend that the ritual space of the Sufi shrines cannot be classified as a site of some form of "syncretic" religious practice that fundamentally differs from orthodox rituals. Nor should shrine-centered (Indus) Sufism be categorized primarily as an "intercessory" (*tawassul*) form of normative, or traditional, Islam, as it has been by Islamicists seeking to emphasize the orthodox status of Sufism (over against the Wahhabī Muslim treatment of ritual devotion at Sufi shrines as idolatry or *shirk*).

The dissertation argues that to gain a clearer perspective on the religious significance of the Sufi shrine, we have to look at ritual "on its own terms."[21] This means examining ritual practice at shrines not as it is described and mandated in textual sources nor as viewed through the lens of anthropological theory, but as a mode or embodiment of an authentic human encounter with the Divine, which is necessarily a "sensual experience."[22] What studies have missed in their characterization of Sufi shrines is the full scope of religious worship at these places. My contention is that *ḥāẓirī* is effectively an act of prayer, *namāz* (Ar. *ṣalāt*), but consisting in the worship of the sacred body of the Sufi saint.[23]

---

20 Clifford Geertz, *The Religion of Java* (Chicago: The University of Chicago, 1976).

21 See William A. Graham, "Islam in the Mirror of Ritual," in *Islam and Comparative Religious Studies: Selected Writings* (Burlington: Ashgate, 2010), 103.

22 See William A. Graham, *Beyond the Written Word: Oral Aspects of Scripture in the History of Religion* (Cambridge: Cambridge University Press, 1987), 6-7.

23 The proposal to study Sufism or mysticism as a lived religious practice associated with Indus shrines is thus essentially a proposal to study the instrumentality of the body in this practice. Experiencing the presence of the Sufi saint, and indeed engaging with the social environment

Central to this dimension of religious experience or worship is communion with the living presence of the Sufi saint. In other words, the devotees travel to the *dargāh* to find the Sufi saint *present* (*mawjūd*), and this presence is realized in *ḥāẓirī*. However, defining the ritual devotion of *ziyāra/ḥāẓirī* as an act of *'ibādat* is not acceptable in "scriptural" Islam among the *'ulamā,'* who would see such an interpretation as signifying outright *shirk* (idolatry, association of something else with God).

Nonetheless, *ziyāra* in the *lived* experience of the shrine devotees is very much performed as an act of *'ibāda*, and as such its practice is analogous to the ritual of *ṣalāt* (Pers., *namāz*). However, the suppression of this lived experience in Pakistan has been carried out through the rationalization of its practice in the following ways: (1) in the ideology of intercession (*tawassul*)[24], which is carried out through (2) the use of the scriptural *fātiḥa* to regularize ritual devotion at the Sufi *dargāh*. [25] In legal interpretation of the *'ulamā'*, *ziyāra* receives

---

of its space is dependent upon the perceptual abilities of the body. The body is the corporeal medium of experience; it is the concrete locus of recognizing the metaphysics of presence as experienced in devotional performance. Such an understanding of the role of the body promises to tell us much about the nature of sainthood in Islam. In the presence of the Sufi saints, gestures of praying, sitting, singing, and dancing, all become meaningful modes of *being* and *acting*. Studies of *ziyāra* that have sought to examine religious experience at Sufi shrines have not concentrated on the body's participation to see how the meaning of *ziyāra* arises in the devotional interaction of the visitors with the visited saint.

24 For the technical meaning of the terms used for intercession, such as *waṣila* and *shafā'a*, see Constance E. Padwick, *Muslim Devotions: A Study of Prayer-Manuals in Common Use* (London: S. P. C. K, 1961). In traditional Sunni Islam, the *'ulamā* are in general agreement over the permissibility of sending pious merit (*ṣawāb*) to the ordinary dead, and since this is accepted to be a legitimate way for the living to aid the dead, its practice is considered to be equally applicable to the Sufi saints as well. According to the Sunni (orthodox) interpretation of *tawassul*, the Sufi saint, like any other pious person, in the grave, after receiving merit offered by the living, prays (*du'ā*) to God in favor of that person—discussed in Chapter 1. This ritual act of supplication to God and its beneficial effect in the exchange is described as *tawassul* (intercession). Despite the differences between the Deobandī and the Barelwī, this doctrine of intercession ultimately prevents the identification of saints with the creative power to help the ordinary directly. Nevertheless, the *karama* of saints is well known among the devotees of shrines. As I have argued, reciting *fātiḥa* at their graves is a scriptural intervention that attempts to regulate the relationship between the devotees and the Sufi saints. The point is that the doctrine of "intercession" helps regulate the ritual of *ziyāra*. For a comparative perspective, see J. W. Bowker, "Intercession in the Qur'ān and the Jewish Tradition," *Journal of Semitic Studies* 11 (1966): 69-84.

25 Offering supplications (*du'ā*) for the dead is popularly known by the first surah of the Qur'ān, namely, *al-Fātiḥa*. Despite its brevity, it is considered widely, in popular faith and practice, to be representative of or even equivalent to the entire Qur'ān in its power when recited devotionally. The Deobandīs reject the practice of pilgrimage to Sufi shrines for any purpose other

legitimation only if performed for the purpose of reciting *fātiḥa* as a prayer *for* (not *to*) the dead saint (*īṣāl-i ṣawāb*). This mode of performing *ziyāra* as recognized by the *'ulamā* and the Pakistani state, is in practice used as a coercive ideology for the suppression of *ḥāẓirī* as a legitimate form of religious devotion. It entails the use of scripture (Qur'ān) to reverse the lived metaphysics of Indus Sufism in which the body of the Sufi has transcended death to be fully present and alive to the devotees. *Fātiḥa* at the *dargāh* is given here a textual signification (as scripture) to rupture the traditional, active communion between the devotees and the Sufi saints, which is considered aberrant and dangerous by both the religious and the political establishments in Pakistan.

The ritual space of the Sufi shrine is, however, demonstrably a place of communion (prayer, *ṣalāt* and *du'ā*) with the divine, and understanding this vital dimension of religious experience allows us to understand the cosmological significance of the sacred body of the Sufi saint. This dissertation is thus especially concerned with the sacred *body* of Sufi saints in the context of Indus Sufism. Specifically, it is a study of how the "resurrected" body of the Sufi saint at his tomb is experienced by the worshipper through what I describe as the theopoetics of popular ritual devotion, especially and specifically of the alternative ritual form of communion (prayer) known as *ḥāẓirī*. Practices of ritual devotion at the Sufi *dargāhs* are a means for gaining direct, yet ineffable, religious experience of God, a form of communication that the mainstream *'ulamā*, in conjunction with Department of *Awqāf* of the state of Pakistan, have sought to silence.[26]

The first chapter is an ethnographic comparison of these two different modes of ritual practice in the performance of *ziyāra* in Pakistan. The aim is to understand the ontological status of ritual incongruity, that is, between a devotee's ritual performance as a *ḥāẓirī*, on the one hand, and the performance of *fātiḥa*, on

---

than offering a gift to the soul of the dead. Meanwhile, the Barelwīs accept the central logic of *īṣāl-i ṣawāb* [prayer for the dead; see below] as a means for sending a pious gift of merit (*ṣawāb*) but consider its practice in the case of Sufi saints as being particularly efficacious for the supplicants because of the saint's friendship with God. A significant issue for the Deobandīs is that the feasts associated with *īṣāl-i ṣawāb* are meant to be *ṣadaqa* for the poor, and therefore criticized its consumption by the followers, which is seen as a violation of its primary aim. Eating the food as a form of *taburrak* is considered un-Islamic by the Deobandīs and also by the neo-traditionalist Ahl-ḥadīs and the modernist Jama'āt-i Islāmī, two of the groups that share a greater affinity with rigorist Wahhābī interpretations of Islam. The chapter also looks at the work of Muḥammad Ṭahir-ul-Qādirī (1952-present), a Ḥanafī-Barelwī scholar who leads his international organization known as the Minhaj al-Qur'ān. The debate sheds light on the nature of the disagreement among the *'ulamā* on ritual practice at Sufi shrines and the ontological status in the Islamic law of the Sufi saint at his tomb.

26 See Muhammad Arkoun, *Islam: To Reform or To Subvert?* (London: Saqi Books, 2006).

the other. The ritual logic of offering *fātiḥa* at the graves of Sufi saints is at odds with the ways in which devotees perceive and experience the power of sacred ("resurrected") bodies.[27] Chapter One will establish how the ritual of *ḥāẓirī* at the *dargāh* of saints, in the lived experience of devotees, is a form of worship (*'ibāda)—in spite of this being* a classification of this practice unacceptable to Sunni orthodoxy. It also argues that the scriptural intervention of *fātiḥa* is specifically intended to displace the ritual practice of *ḥāẓirī*, to prevent it from becoming an act of worship, so it does not come to undermine or subvert the formal hierarchy of ritual acts established in Islamic law.

Despite the scholarly interest in Sufism, the study of Sufi rituals has received very little attention (as explained above). Neither has there been a comparative study that seeks to see how orthodox rituals of Islam measure up to devotional practices performed at the *dargāh* of saints (the only study to provide a kind of a detailed spectrum of Islamic ritual practices that includes both is William Graham's "Islam in the Ritual of Mirror,"). Studies on Sufism have rarely taken ritual as a conceptual starting point for examining its larger place in the prescriptions of Islamic law.[28] In this, I will revisit the debate between Islamic Studies and anthropology in a line of inquiry suggested by Graham's study. Hence, chapter two seeks to explain how the neglect of ritual has to do with methodological (and theoretical) assumptions that have guided the scholar's gaze in framing Sufism.[29]

---

27 In this dissertation, and specifically in our case study, we find that the aforementioned legal discourse regarding *ziyāra* does not fit the lived practice of people visiting the shrine to commune directly with their saint. There is a significant disparity between what scholarship has presented as constituting the discourse on *ziyāra* and how the followers of saints regard and perform their devotions.

28 The first in-depth anthropology of ritual feasts known as *Slametan* was conducted by Clifford Geertz and published as his *The Religion of Java* (Chicago: University of Chicago Press, 1976). The aim was to pay attention to local practices and account for religious diversity. For Geertz, the *Slametan* was a distinctly Javanese form of Islamic practice and, as a complex of religious practices, removed from the *santri* version of Islam represented by the learned class of *'ulamā*. Mark R. Woodward (also see Mitsuo Nakamura), in turn, criticized Geertz for overlooking textual Islam and identified the *Slametan* as a legally and politically sanctioned form of mystical Islam in Java, as well as a ritual practice pervasive in other regions of the Muslim world. Although the issue of *Slametan* ritual feasts was initially concerned with Southeast Asia, it has relevance for South Asia as well.

29 Earlier Arabists and Islamicists distinguished scripturalist Islam and Sufism. The former was seen as part of the socio-political and religious order, and the latter was seen as an apolitical phenomenon primarily manifest in music, poetry, and contemplative practices. The work of anthropologists in the second half of the 20th century increased interest in studying the periphery and local practices in the Islamic world. In the ensuing debate over "local" versus "universal" dimensions of Islam, recent approaches have also been influenced by the work of Foucault and

Such approaches have privileged textual Islam, failing in the process to take seriously *lived* experience and to validate the Sufi experience as an embodied mystical experience.[30] This omission has not only kept scholars, I argue, from properly assessing the underlying rift between Sufism and legalism in Islam but also from adequately grasping the popular dimensions of devotion to Sufi shrines. By taking account of the "sensual dimensions" of ritual, in this case, the ritual of *ḥāẓirī*, this work offers to "re-center" the Sufi shrine. In Indus Sufism, visiting the tombs of Sufi saints is essential to religious practice aimed at gaining divine favor. In other words, the centrality of the living presence of Sufi saints at their tombs cannot be understated in assessing the nature of religious experience in Indus Muslim religious practice. Still, ironically, in much of the best scholarship on both Indus and other regional expressions of Sufism, the tomb has not figured prominently, despite the centrality of its sacred space in the religious experience of the devotees. In drawing attention to the role of authority in studying Sufism, even the most influential studies have neglected the role of living presence in understanding the meaning and significance of Sufi tombs and the associated practices.[31]

A major aim of this dissertation is also to expose the state's efforts to delineate in theologically acceptable terms the function of the recitation of the *fātiḥa* in a veiled attempt to regulate the "inner" faith and overt practices of shrine visitors. In addition, I argue that despite the official status of the *fātiḥa* promoted in legal discourse about *ziyāra*, for the followers of the saints, the living presence of Sufi saints at their tombs continues to be a powerful source of religious experience and a continuing part of the devotional imagination in ways counter to the

---

his emphasis on knowledge/power as a significant factor in the social episteme. Studies now examine Sufism through the historical study of discourses, traditions, and texts in analyzing the power of its appeal to its followers and how this translates into the social and political order. However, in the emphasis on discourse and history, there seems to be a dearth of studies investigating devotees' symbolic meaning and "lived" religious experience at the Sufi tomb.

30 For its part, secondary scholarship in Sufism has misled us by dwelling on important, but not defining, characteristics of these shrines, such as their role as distributive repositories of blessing, *barakah*, or their socio-political dimensions as centers of Sufi practice. These studies have inadvertently marginalized the worship of the saints at their shrines (*ziyāra*) by portraying this phenomenon as an ethnographically rich yet still epiphenomenal entity in the history of Sufism and Islam, a view of the tomb-shrine tradition that sees it much more as an interpretable "discourse" rather than a lived experience.

31 For example, Carl W. Ernst, *Eternal Garden: Mysticism, History, and Politics at a South Asian Sufi Center* (Albany: State University of New York Press, 1992); Vincent J. Cornell, *Realm of the Saint: Power and Authority in Moroccan Sufism* (Austin: University of Texas Press, 1998).

governmental and religious establishments' desires and reform efforts.[32] In Pakistan, the state has carried out a systematic program of reform to suppress and rechannel the devotional energy generated at the Sufi shrines of the Indus valley. In line with a particular trajectory of religious thinking among more rigorist *'ulama'*/religious scholars, and to curtail this potent strand of the popular religious imagination, the state has sought to refashion the hagiographic identities of Indus Sufi saints as, essentially, normative models of piety as opposed to sources of intercession and response to prayer.[33] The state has been involved in directly regulating Sufi shrines and promoting an image of the saint's body that detracts from its recognition and veneration as that of a saint who has realized union in God. *'Urs* (marriage with God) traditionally signifies the transcendence of death granted to the Sufi saint. This immortality is precisely what his followers celebrate in bringing their love and desires to the threshold of the *dargāh*. This has made for a direct conflict between state and rigorist religious reform on the one hand and a long tradition of popular Sufi piety on the other (see Chapter 3). The Department of *Awqāf* manages day-to-day operations at Sufi shrines and collects its revenues. It distributes pamphlets and treatises to the public recounting

---

32 In 1947, Pakistan inherited the extensive network of Sufi shrines in the Indus Valley regions of Panjāb and Sindh. Since the 19th century, reformist groups such as the Deobandī and the Barelwī have been theologically at odds with one other regarding the rituals of supplication (*du'ā*) at Sufi shrines. Despite the disagreements between the Deobandī and the Barelwī, both schools believe that all mystical experiences must be subservient to matters of Islamic law. The legal polemics between factions of the *fuqahā'* regarding popular practices were transported to Pakistan, where religious scholars interpret du'ā at Sufi shrines as legitimate if it is undertaken as a prayer for the dead. In recent decades, the state and religious scholars have encouraged the recitation of the *fātiḥa*, a Qur'ānic surah commonly employed in gaining merit for the dead, as the correct way of observing *ziyāra* at Sufi shrines and graves. Such a ritual intervention, then, is aimed at reversing the popular acceptance of the metaphysics of Divine union achieved by the saint by emphasizing the status of the Sufi saint as a pious dead person instead. Through legal reasoning, religious scholars have also sought to negate other ritual practices that celebrate the power of Sufi saints to communicate directly with their followers even in death.

33 The reality, however, is that in Indus Sufi piety the saints had been far from model citizens in the "prehistory" of the nation-state of Pakistan. Instead, they were often antinomian, ecstatic teachers, healers, poets, and visionaries whose followers' devotion potentially threatened the order above. After their death, they were, and have remained, far from dead in the religious consciousness of the faithful who visit them at their *dargāhs*. In managing *ziyāra* to Sufi shrines today, the Ministry of *Awqāf* promotes the offering of *Fātiḥa* (*du'ā*) to the soul of the saints as the correct ritual performance to be observed at the shrines, thereby signifying the agency of the living over the dead rather than vice-versa. This is a major incongruity because each of the Sufi saints have, in the view of their faithful devotees, in fact transcended their deaths, a transcendence that represents an act of tremendous grace on God's part, one that is celebrated in popular imagination and practice in the ritual of *'urs* at each saint's shrine.

its version of the pious lives of the Sufi saints. Furthermore, the shrines of Bābā Bulleh Shāh (ca. 1680-1757), ʻAlī ibn ʻUsmān Hujwīrī (d. ca. 1072), and more recently, Barī Imām (1617-1705) in Islamabad have been thoroughly revamped physically. Through renovations of Sufi shrines, the state has marked boundaries around the shrines and, through its reconstruction efforts, reconstituted their physical landscapes, thus radically altering the sacred geography and the ecology of their premises. Chapter three also includes an exposition of *Ṭariqa-i* Muḥammadiyya's (a 19th-century movement) scripturalist reform of Sufism unleashed to counter the *wujūdī* metaphysics of presence.

Chapter 4 will present the case of Bābā Bulleh Shāh (1680-1757), an 18th-century Panjābī Sufi poet whose shrine in Qasūr, Pakistan, is a prominent center of Indus pilgrimage.[34] In researching this chapter, I visited and studied this shrine as a participant-observer during the annual ʻurs of the saint in August 2015. (In this discussion, I want to show how popular devotion to Bulleh Shāh manifests the increasingly acute tensions in representing all Indus saints.)

Bulleh Shāh's mysticism is suffused with notions of *ʻishq*, a "passionate desire or longing [for God]" rooted in the inner heart, and a rejection of political identities in favor of the Divine self. It is not without reason that followers of Bulleh Shāh see in him an ecstatic mystic, ablaze with Divine love. Similarly, the *kalām* of Bulleh Shāh has a powerful effect on his followers, and his sacred shrine is a center for all types of Sufi and religious communities who gather to experience the ecstasy of sacred abandon associated with Bulleh Shāh. By all accounts, Bulleh Shāh's mysticism represents a rejection of the dominant order of the day and clear defiance of the *fuqahā'* of Qasūr. He cast his lot with the singers and the musicians instead. Consequently, to this day, Bulleh Shāh remains a patron saint of all the musicians in the Indus region, just as his *dargāh* attracts people from the lower rungs of the social ladder across the Panjāb. Above everything, Bulleh Shāh's poetry is popular for hard-hitting verses

---

34 Bulleh Shāh's Sufism represents one of the major strands of Indus Sufism. This saint is known for his antinomianism regarding formal rituals such as *ṣalāt*, *ḥajj*, and *ṣawm*. Moreover, Bulleh Shāh wrote verses criticizing ascetic practices and affiliated modes of ritual piety and valorized the efforts to appease one's *pīr* as a concrete encounter with the Divine, the *pīr* in his case being Shāh 'Ināyat (1643-1728). Bulleh Shāh's hagiographical identity is based on his rejection of his own *sayyid* social and religious status to join the ranks of a lowly class of musicians (*mirasis* and *kanjars*) often associated directly or indirectly with the courtesan culture (*bazār-i ḥusn*). Bulleh Shāh himself was a skilled musician, and it was through his ecstatic performances he was reunited with his *pir* or *murshid*. Bulleh Shāh was a mystic whose experiential tasting of Divine truths was affected through composing *kāfī* verses and performing them.

in the critique of textual (*kitābī*, trans. "bookish") knowledge of religious scholars.

The vernacular *kāfī* verses, which devotees believe to be the spoken word of the Sufi saints of the Indus, are a regular feature of the ritual devotion at Sufi shrines. Chapter 5 examines the notion of a festival, described as a *mela*, celebrating what is commonly known as *'urs*: marriage union with God. The chapter looks at the ritual meaning of the festival and its theopoetics (through the *kāfī*) as a means of understanding the centrality of the mystical body of each saint. The theopoetic meaning of the ritual of *'urs* as a festival (*mela*) offers a compelling contrast and alternative to the economy of *īṣāl-i ṣawāb* in orthodox interpretation and its disciplinary constraints on the performance of rituals at the shrine. Moreover, the festival of *'urs* celebrates the presence of the resurrected body of the saint by which the Divine bestows its gifts of grace on the world. Hence, the festival of *'urs* is open to all religious communities. In this metaphysics of hospitality, the devotional culture of Sufi shrines transcends any monolithic notion of religiopolitical identity. This was a primary reason that a group like *Ṭariqa-i Muḥammadiyya* (a 19th-century reformist movement) saw Sufi shrines as places where Hindus and Muslims interacted with each other in devotion to the Sufi saints as a threat to their agenda of reform.

The dissertation's last chapter (6) focuses on the ritual meaning of *samā'* and its performative salience as a significant element of aural/oral culture in Indus Sufism. The metaphysics of *ḥāẓirī* in its devotion (*'ibāda*) to the sacred body of saints is evident in the scriptural function of the performance of *samā'*. The chapter explains the relationship between ritual and poetry in the context of Indus Sufism. It argues that poetry performed orally at the *dargāh* derives its sacral quality from the living presence of the Sufi saints at their tombs. This devotional connection (*nisbat*) serves as a means for the "interiorization" of the divine word.

Nisbat does not denote a single type of experience but encompasses varied religious experiences.[35] It is the devotional cultivation of a personal relationship, *nisbat*, of each devotee with the Sufi saint(s).[36] As noted, *ḥāẓirī* refers, specifically, to the ritual of visiting the *dargāh*; it is an embodied gesture of both *nisbat*

35 No matter how sophisticated our concepts become, they fail to do justice to what William James called the "plenum of existence—full range of human experience, intransitive and transitive, fixed and fluid, rational and emotional, coherent and wild, real and symbolic," Michael Jackson and Albert Piette, *What is Existential Anthropology?* (Oxford: Berghahn Books, 2017): 7.

36 For this reason alone, the importance of the experience of the continued presence of a saint's grace in the lives of the devotees is hard to underestimate in the study of Sufism—at least from my emic/etic perspective.

and personal devotion. It refers to the experience of being in the living presence of a Sufi saint. This is the realm of embodied practices, feelings, and experiences. The performative art at Sufi shrines speaks, often through metaphor and allegory, to this dimension of personal connectedness (*nisbat*) with saints. Furthermore, in considering *nisbat* with saints to study religious devotion, we must recognize that the name of the Sufi saint and its invocation carry sacred power.[37]

In my experience of involvement with Sufism and my field observations for the current project, the Sufi *dargāh* is a sacred space that facilitates communion with the living presence of the saint of the tomb.[38] The direct experience at a grave of a Sufi saint's living presence, as a transcendence of death, is a form of religious knowing to which the scholar has little access. Still, its reality as an experience for the devotee must be recognized and acknowledged by the scholar if s/he is to begin to understand the role of *dargāh*s in Indus Islam. Despite political interventions, religious, and political involvements, the existential significance of Sufism is rooted in the way it continues to offer a timeless path for personal journeying toward (or searching for) —and tasting (*dhawq*) of—divine truths. Through the years of visiting Indus Sufi shrines and being bodily attuned to their ritual space, I have experienced how shrines are a place of mysterious encounters, happenings, and sightings that open into a world of ineffable experiences for the visitors. As experiential theosophy of practice, the power of Sufism is best understood through the devotional engagement of the heart. In my experience, Sufism is existential theosophy that is centrally concerned with the sacrality of all human existence and the means of experiencing the living presence of the Divine.

Centered on and oriented towards the *dargāh*, Indus Sufism has been a site of remarkable effervescence in mysticism. Despite its affinity with Persian metaphysics and symbolism, it has included a distinct complex of practices and Sufi folklore that have survived the travails of modernity. Indus Sufism is a complex

---

37 As it is a refrain in a popular *qawwalī* "*dam dam karo Farīd, ākho ḥaq Farīd, yā Farīd*" (invoke with your breath: *Farīd*, say *Farīd* is the truth, Oh Farīd!). Similarly, there is an especially sweet, and unique, resonance to *yā Gharīb-nawāz*. Another definition of *samā'*, is to see it as the performative art of reciting the name of Sufi saints. Devotees attend the *samā'* to listen to the ritual invocation of the name of their saints.

38 Jorge N. Ferrer and Jacob H. Sherman (editors), *The Participatory Turn: Spirituality, Mysticism, Religious Studies* (Albany: State University of New York Press,2008), 20: "Participatory approaches to language see human poiesis as a creative manifestation of life or the spirit in the human realm—a swallowing the heart of the world—and they thus radically overcome the modern split between language and ontology. We suggest that it is in the particularities and constraints of nature, culture, and history that language becomes truly revelatory, a stance that affirms the immanence of the mystery without in any way repudiating its transcendence."

phenomenon that comprises several mystical strands. Its *dargāh*s are sacred spaces celebrating the living presence of its Sufi saints and a vital source of experiential wisdom that is part of its living heritage. Nevertheless, this heritage has been threatened by the rising forces of religious reformism and state attempts to manipulate its practices for political ends. An extensive network of musicians and singers is dedicated to its devotional culture and performance of its mystical verses in Panjābī and Sindhī. These musicians are one of the most important resources for preserving its mystical folklore and modes of experience. Predictably, the state has actively sought to interpret this form of devotion only as an element of folk culture to negate its legitimacy and religious significance. In addition, the *Awqāf* ministry is actively involved in large-scale architectural plans that would fracture the local ecology of Sufi shrines and transform their structures of intimacy into large, impersonal spaces. This has already affected the erection of colossal mosques that envelop and functionally eclipse the shrine's landscape. This building program is ideologically linked to the state program of Islamization of local Sufism.

Indus Sufism, in its most basic form, is about the relationship between the self and the Divine, expressed in the symbolism, metaphysics, and devotional features connected to the *dargah*s and ritually celebrated in the bodies of the Sufi saints they entomb.[39] Its form of devotionalism is a part of the everyday religiosity of the region and has much to tell us about religion in the Indus region. Whatever meaning has been threatened by the religiopolitical moves to de-center the tombs of the saints, and whatever deeper understanding may have been lost by their superficial treatment in Islamic studies scholarship, can, I suggest here, begin to be repaired by placing the focus of inquiry on the authentic nature, history, and motivations of Indus Sufism. Using a multidimensional approach that brings political-historical, ethnographic, theological, literary, aesthetic, and phenomenological vectors in a robust analysis of *dargāh* worship in the Indus Valley, I want to offer a new taxonomy of the sacralized tomb. This work aims to show how latent, subtle tensions can permeate contemporary external religious observances without necessarily being explicitly noticed or challenged. It tries to consider seriously what is at stake in the efforts to de-center shrine worship by religious reformers and the government, as well as, intentionally or unintentionally, by scholars in the academy, and what we gain by re-centering it in the inquiry into Indus Sufism.

---

39 Richard Kurin, "The Structure of Blessedness," 322: "Relationships enacted at the shrine of Abdullah Shah [Karachi, Pakistan] are paradigmatic of those thought to exist between humans and God as well as within persons themselves."

Chapter One
A Case of Ritual Incongruity in the Practice of *Ziyāra: Ḥāẓirī* and *Fātiḥa*

This first chapter will describe (and illustrate with photographic material collected during ethnographic fieldwork) the difference between two ritual modes of pilgrimage to Sufi shrines in the Pakistani context. Ordinary folk, devotees of saints, describe visiting a saint's shrine, or *dargāh,* as a *ḥāẓirī* (this is especially the case for vernacular Sufism). In *ḥāẓirī,* the visitors enter the shrine to find the saint present (*mawjūd*) and thus experience an unmediated or "direct" connection with the Divine. The saint's body is the locus of the Divine, hence the center of interest in the performance of pilgrimage. However, in the prescriptive (legal) texts on *ziyāra* written by the *'ulamā,* such visits are considered permissible only if they are performed not as *ḥāẓirī* but to transfer pious merit (known as *īṣāl-i ṣawāb*) to the souls of the dead, which involves the recitation of the first Qur'ānic *sūra, al-Fātiḥa.* In the case of *ḥāẓirī,* the devotees "pray to" the saints, and in the recitation of the *Fātiḥa,* the visitors "pray for" the saint.[1]

While reciting the *fātiḥa* to confer/transfer merit is acceptable in cases of the ordinary dead, in the lived experience of devotees, the Sufi saints have transcended their deaths or been "resurrected" by the grace of God. Therefore, such an offering (as a transfer of merit) is an unnecessary and inappropriate ritual. The saints require no assistance whatsoever, no intercession, from the living; instead, devotees visit their thresholds because the *sacred bodies* of the Sufi saints buried in their dargahs are sites of Divine presence. The pilgrim's focus, in other words, is the "resurrected" body of the friends of God (*awliyā' Allāh*). On the other hand, the purpose of having devotees focus on reciting *fātiḥa* is to prevent pilgrimage to shrines from becoming an act of worship (*'ibāda*), which according to Islamic law as interpreted by the *fuqahā,* would be idolatry (*shirk*). In Indus Sufism, devotees do not perceive the saints as dead nor bound by the boundaries of the *barzakh. In this intermediate zone,* all the dead reside

---

1 Jane I. Smith and Yvonne Y. Haddad, *The Islamic Understanding of Death and Resurrection* (Oxford: Oxford University Press, 2002), 183–191: a short but very useful discussion on the case of Sufism. The question of difference between "praying to" and "praying for" is further connected to gaining an understanding of what it means "to pray." So, to pray for is to request God to help or aid someone, and in terms of orthodoxy, "pray to" is only acceptable or legitimate as an act of worship *to* God. In ritual performance, *ṣalāt* or *namāz* (in Persian) translates as "praying to" God, and *du'ā* or supplication is an extension of it where the person petitions God, and thus supplication is also in the category of "praying to."

https://doi.org/10.1515/9783110781496-003

until the final Day of Resurrection (*yawm al-qiyāma*). In fact, it is not just a matter of their mobility in the afterlife, but rather that their having transcended death signifies the great transformation in which Sufi saints achieve union with God.

Thus here, we have two different and contradictory or competing forms of religious experience at work in the performance of *ziyāra:* one accepted by the Sunni orthodox *'ulamā* (*fātiḥa-khwanī*), and the other (*ḥāẓirī*) prevalent among the people. So, what is the reason for this ritual incongruity? Why does the interpretation of *ziyāra* offered by the *'ulamā* differ from its lived practice? The answer to these interrelated questions and the key to understanding the metaphysical and ontological significance of this ritual incongruity (one that has been missed in previous analytical studies on Sufism or *ziyāra* to shrines) is gained by attending to the phenomenological dimensions of *ḥāẓirī.*

Ḥaẓirī as a ritual experience is a means of cultivating a direct and personal connection on the part of the devotee with the *sahib-i mazār* (master of the shrine) in which lighting a lamp, laying a *chado*r, or standing at the threshold of the saint's grave becomes a channel of communication with the sacred presence of the Sufi saint. *Ḥāẓirī* prepares the body for hearing the Divine word and having it enter the heart, and this direct "experiencing" of the word is what results in ecstasy. (In traditional accounts, ecstasy is understood as a kind of disconnection with one's surroundings, an intensely personal experience—which it is, but it is also a mode of seeing fully what is before us). Intentionality here is critical to religious experience. The ritual of *ḥaẓirī* involves, in essence, a theopoetics of the sacred gaze (*naẓar-i karam*).

For the traditional scholars and the governmental interventionists, the use of the *fātiḥa* as a prayer for the dead is a scriptural intervention to prevent the devotees from the act of communion with the saint and its *interiorization* of the saint's presence in the ritual act of *ḥāẓirī.* The *Dargāh* is primarily a sacred place for prostration (*sajda*), contested by the legal guardians of Islam, who find this ritual act of submission unacceptable. *Ḥaẓirī* is a bodily orientation toward the resurrected body of the saint. In this embodied interaction, in which listening to the spoken word of the Sufis in *samā'* is ritually inseparable from its performance, the sacred space of the shrine offers worshippers a site for a lived experience of the Divine presence. *Ḥāẓirī* and listening to the *samā'* constitute the sensual dimension of the ritual of visiting the tombs of "friends of God" (*awliyā' Allāh*). As a material manifestation (*nishanī*) of the Divine, Sufi shrines have functioned historically in the eyes of devotees as sites where the ordinary person can have a direct connection with God. In *ḥāẓirī,* there is a connection between the body of the devotees and the mystical body of the saint, which is what the politics of the (scriptural) *fātiḥa* wants to discipline and channel in more acceptable channels, according to the socio-moral vision of *sharīʿa* as interpreted by the *ʿulamā*.

In what follows, I will first describe the ritual practice of *ḥāẓirī* to show that its performance at the *dargāh* is an *act of communion* with the saint—this being the main point of the section (and of this study of Indus mysticism). The latter part of the chapter then examines the Barelwī-Sufi interpretation of how *īṣāl-i ṣawāb* is a channel for mediating the relationship between the living and "dead" saints and how it serves an intercessory function meant to benefit all Muslims collectively. This interpretation maintains the primacy of *fātiḥa* as an equally valid ritual directed to the ordinary dead and the saints alike. Despite their defense of *ziyāra*, as far as ritual is concerned, the legal-minded Sufis of the Barelwī are hostile to any theopoetics of *ḥāẓirī*, and scholars accept their defense of Sufi shrines as proof of the traditional synchronicity between *sharīʿa* and *ṭarīqa*. Neither does the Barelwī version of Sufism support *ziyāra* as an act of communion; since the dargah is not accepted as a place of Divine presence, any such communion qualifies in their view as *shirk*. In Islamic law, the only valid ritual that establishes a "direct" contact by worship (*ʿibāda*) of God is that of *ṣalāt*. Additionally, supplications (*duʿā*) are offered legitimately after *ṣalāt* in which the believer petitions God. In terms of ritual space, it also means that the mosque for Muslims, according to orthodoxy, is the only designated sacred place for entering into the Divine presence.[2]

1.1 *Ḥāẓirī*

The religious character of Indus Sufism is marked by devotional love for its Sufi saints, which is expressed primarily through encounter with their living presence at their tombs. As a consequence of this spiritual (*ruhanī*) connection (*nisbat*) between devotees and the saints, or "friends of God" (*awliyā' Allāh*), Sufi Tombs and shrines play a central role in the religious life of the Indus regions of Panjāb and Sindh.[3] Shrines are hubs of ritual devotion, performative arts (espec. variet-

2 The mosque (*masjid*, from the root in Arabic: *s j d*, translated as place of prostration: *sajada*) as a place of worship and social action is symbolically central to representing the religiopolitical power of Islam.

3 For a good topography of shrines, including brief accounts of the rituals and devotional practices at these places, see Annemarie Schimmel, *Islam in the Indian Subcontinent* (Leiden: E. J. Brill, 1980). Ucch in South Panjāb, and Maklī Hill in Sindh are two of the largest shrine-cities (necropolises), along with, in India, Khuldābād and Gulbarga in the Deccan. A more recent pub-

ies of *samā'*), meditative practices (*zikr*), feasts (*langar*), and at some of the major shrines, the *'urs* ("marriage" with God, celebrated on the death anniversary) of the saint, which also features carnivals, or *melas*.[4] Shrines of Sufi saints such as Bābā Farīd, Bābā Bulleh Shāh, Shāh Hussaīn, Sachal Sarmast, 'Abdul Latif Bhiṭ'āī, or Khawāja Ghulām Farīd continue to receive visitors who travel from far and near to be in their sacred presence and to seek their intercession in their lives.[5]

In its performative or lived dimension, the religious culture of Sufi shrines in the Indus region of Pakistan arguably derives its rich theopoetic character from the relationship of mutual love and compassion between the devotees and their immortal saints. The devotional connection of visitors with the living presence of the Sufi saints is the key to understanding the efficacy of the visitation/veneration of Sufi saints at their tombs.

The sacred efficacy of Sufi-shrine devotions subverts the usual distinctions between life and death. This efficacy extends beyond the rational order of things because in the eyes of the devotees the living presence of an entombed Sufi saint is the reason a shrine possesses a special agency that can miraculously help those who petition there. This presence is not abstract or ethereal, but very much corporeal. The saint is present to the visitor in the body in the tomb. Furthermore, the saint's sacred power as a blessed immortal transcends the visible manifestations of socio-political structures, and other embedded hierarchies, whether in the form of the state bureaucracy's regulation of a shrine or in the organization of whatever Sufi order is conventionally and/or historically associated with the shrine.[6] So while these social structures and relations around the shrine play a large part in the organization of daily activities and directly affect

lication on the Indus Sufis is by Samina Qureishi, *Sacred Spaces: A Journey with Sufis of the Indus* (Cambridge: Peabody Museum Press, 2009).

4 These carnivals are purely for entertainment purposes, with joyrides and games for children; some kind of daredevil stunts are also a popular feature, along with theatrical performances. The *'urs* of Shāh Hussaīn in Lahore is known for its *mela*, and the shrine of Barī Imām in Islamabad is a venue for such events. However, the more conservative minded do not consider the *mela/carnival* a part of the *'urs* activities, and one may speculate that the occurrence of such carnivals is on the decrease currently. We shall discuss the poetic and metaphoric significance of the carnival below, Ch. 5.

5 The mystical verses of these Sufi saints are an important aspect of shrine devotionalism and represent the vernacular strand of Sufism in the wider Indus region. All these poet-saints have shrines in different parts of the Panjāb and Sindh, which are centers of pilgrimage for the devotees. Their poetry plays a significant role in shaping the religious sensibility of the shrine devotional culture.

6 See Spencer J. S. Trimingham, *The Sufi Orders of Islam* (Oxford: Oxford University Press, 1998).

the devotional life of its space, ultimately the power of the living presence of the Sufi saint associated with each shrine is the prime locus of the lived experience of the faithful.

Pilgrimage to Sufi shrines, or tombs, is widely known in the Islamic world as *ziyāra*, but in contemporary, everyday parlance in Pakistan and North India, the visit is typically described as *ḥāẓirī*. This term, which might best be rendered as "being present"/"presenting oneself to," conveys the intentionality of presenting oneself in the service of a superior personage.[7] The Sufi saints are unequivocally and by consensus (*'ijmā*), seen as "friends of God" (*awliyā' Allāh*) in the popular imagination and thus have an exalted status.[8] Used in this sense for attending upon a saint at his/her tomb, *ḥāẓirī* is also a speech act, something that is physically done, far from passive, and involving a volitional, vocal performance.[9]

Moreover, *ḥāẓirī* as a ritual action encompasses a wide variety of practices and gestures that have more to do with the inner orientation of the visitor than they do with a formalized ritual act. The visitor offers her *salām* (greetings) to the Sufi saint in genuflection with hands clasped in front to mark obeisance in the presence of the saint. Laying a shawl (*chador*), along with garlands and rose petals, on the grave (*maqbara*), burning incense in the courtyard or near the

---

7 *Ziyāra* is derived from the Arabic root زار which simply means to visit. Hans Wehr, *Arabic-English Dictionary*, edited by Cowan 4th edition (Ithaca: 1994). The word *ḥāẓirī* is derived from the Arabic root *h-d-r*, *"to be present"*. It denotes action, and willing submission, on the part of the one visiting because it means to make oneself available to the will of the other. It is also used in polite, or formal, conversation to describe the willingness of one person to make an appearance at the behest of the other, (implied) more respectable person. The idea of being fortunate to receive the benevolent and powerful gaze of the Sufi saint is celebrated in devotional verse. John T. Platts (1830 – 1904), " *'Ḥāẓirī'' A Dictionary of Urdu, Classical Hindi, and English* (New Delhi: Munshiram Manoharlal Publishers, 2000): "Presence, attendance, appearance in a Court, also "an offering of food to a saint or a departed spirit, a funeral feast (sent to the household of the deceased), p. 473. Platts' meaning of the term reflects the ambiguity in the matter because offering food *in the name* of the saint is very different from offering it *to* them.
8 For a comprehensive treatment of the "friends of God" genre, see John Renard, *Friends of God: Islamic Images of Piety, Commitment, and Servanthood* (Berkeley: University of California Press, 2008). It is rare to find *'ijmā* invoked for Sufism, yet I think its use does have merit, that is, in the light of the popularity of saints. *'Ijmā* is a legal term in *usūl al-fiqh*, where it is considered to be one of the means by which the truth of religion is preserved. *Fiqh* scholars after al-Shāfi'ī overturned his interpretation of *'ijmā* as the consensus of the entire community (*umma*). The later *fuqahā'* were inclined to interpret *'ijmā* as the consensus only of scholars—in this function, it is equivalent to the two sacred sources of Islam. On the issue of its circularity in Ḥadīth, see George F. Hourani, "The Basis of Authority of Consensus of Sunnite Islam," *Studia Islamica* 21 (1964), 13 – 60, but also the response of Wael B. Hallaq, "On the Authoritativeness of Sunni Consensus," *IJMES* 18 4 (1986), 427 – 454.
9 John L. Austin, *How To Do Things With Words* (Cambridge: Harvard University Press, 1975).

Figure 1: *Ḥāẓirī* at the shrine of Shāh Hussaīn, Lahore. Photo by I. M. Khan.

tomb, lighting an oil lamp (Urdu: *chiragh*, Panjābī: *diya*), and distributing sweets are all practices that are part of *ḥāẓirī* at a Sufi tomb. These are modes of embodied performance in devotional communication with the living presence of the Sufi saint at the tomb, after which the visitor brings her petition to the notice of the saint orally through *du'ā*. In its ritual sense, then, *ḥāẓirī* is also a form of *adā*[10] to find favor in the eyes of the Sufi saint, and in Panjābī, a near equivalent to the meaning of *adā* is the word "*chajh*."[11] *Chajh* implies a kind of sincere

10 See John T. Platts, "*adā*," grace, beauty, elegance, charm, blandishment, coquetry, etc.
11 "*sikh koī chajh yār manāwan dā*" (learn, if you will, some tact to appease the friend) was performed by the great Panjābī Qawwāl, Nuṣrat Fateḥ 'Alī Khān (d. 1997) in his inimitable style. The verse belongs to Bābā Bulleh Shāh (now searchable on *YouTube*). The ritual of *ḥāẓirī* as a mode of communication with the living presence of the saint is also a world of creative imagining, a kind of religious poesis. There are many *kāfī*s (songs) that celebrate journeying

alertness or an awareness born of devotional love that the visitor brings to the living presence of the Sufi saint at the tomb. It is a bodily attention, to know that one is standing in a sacred presence. It is primarily about the spiritual bond between the devotee and the Sufi saint, and a mindfulness of the divine that is embodied in the human saint. Some devotees bow at first sight of the *dargāh*, and always before entering the main tomb. All the senses focus on the shrine, which initiates the communion during *ḥāẓirī*, and all are directed towards the saint. Shrines have different landscapes and architecture, where the devotees have to ascend stairs to reach the tomb (as in 'Abdullah Shāh in Karachi, or at the Kamāl Chishtī in Qasūr), while many others may be located merely a few feet from the public street. The smell of roses and incense, sight and sounds are equally part of the experience. Upon entering the door of the main structure, visitors usually touch, kiss, or prostrate before the tomb grating, or perform all three activities, to greet the saint. During the *'urs* festival, it is harder to move and stay next to the holy grave than on ordinary days. For devotees, *ḥāẓirī* is a *felt* experience; a warming up of the heart for many, but it is fair to say that *ḥāẓirī* is inclusive of a range of emotions. Many shed tears, sharing their innermost secrets, issues, and problems with the saint: women ask for children, a man prays for the health of his mother, a daughter asks for help against her tyrannical in-laws, a young man pleads with the saint to help him find employment, or a mystic seeks spiritual training. Whatever the nature of the problem, the *dargāh* is a place where untold numbers visit to meet the saint personally and converse, to seek guidance or other help, or merely to show devotion to the saint. A number of shrines have musicians, who usually sit in the central courtyard and may periodically recite a few couplets. Shrines under state management do not allow musicians to sit too close to the grave, but those in private custody may have musicians sitting close beside it. Pilgrims return to perform *ḥāẓirī* periodically, and those living closer may perform it every morning before leaving for work.

Furthermore, to visit the sacred site of the *dargāh* or *darbār* as an act of *ḥāẓirī* is to be present to the *gaze* of the Sufi saint.[12] Specifically, it is journeying

---

to see the saint—or in less favorable circumstances, searching for an opportunity to do so—to please or entice the saint for a favorable glance on the devotee. As a desire, it is also related to other moods and motivations; for example, in one *kāfī* by Shāh Ḥussain (1538–1599), Paṭhānā Khān (d. 2000), a mystic folk singer, sings about relinquishing the rings on his fingers (*hath dey chaley*) *if anyone* brings news (*khabar*) about his beloved *Ranjhān*.

12 For Christianity, see David Morgan, *The Sacred Gaze: Religious Visual Culture in Theory and Practice* (Berkeley: University of California Press, 2005). I am not sure if the sacred gaze works well for *naẓar-i karam* because more than a gaze, it is a momentarily look or glance since that is

Figure 2: Oil lamps at the *dargāh* of Bābā Bulleh Shāh, Qasūr, Pakistan.
Photo by I. M. Khan.

to the Sufi tomb with the desire and expectation of benefiting from the sacred gaze (*naẓar-i karam*) of the saint.[13] Pilgrimage to a Sufi tomb (*ziyāra*) can thus also be defined as carrying out the intention of *ḥāẓirī* to obtain the *naẓar-i karam* of the saint.[14] This journeying for *naẓar-i karam* at a Sufi tomb underscores the saint's direct and active agency (or sacred power) that, if received by the fortunate visitor, is capable of bestowing a profound grace on them. One only needs to look at the vast array of songs and recitations in praise of the saints to realize the prevalent belief in the spiritual potency in the gaze (*naẓar)* of the "friends of God." There is a steadfast faith among pilgrims that loving and overwhelming compassion emanates from a direct encounter with those absorbed in God. Shrines are sacred spaces that give access for ordinary folk to Divine presence, and although people approach these sites with tremendous care and respect, there is much more that can also be said about the informal,

---

all that is needed for the realization of grace (*karam*). The visitor's desire is to have that momentarily glance fall upon her, even if by chance.

13 The perfect (*kamil*) Sufi, the one who is blessed with special sight, can see the divine face or manifestation of the saint. This vision is described as *tajjalī*—a Sufi term—and in the vernacular verses of Bulleh Shāh, people are unable to see because they do not have the eyes, or the esoteric vision, as a result of which they can *truly* see: "*weykhān walī akh nahi.*"

14 Cf. On seeing and being seen by a god in Hinduism, see Diana L. Eck, *Darsan: Seeing the Divine Image in India* (New York: Columbia University Press, 1998).

almost casual interaction of the devotees with saints in the way one would interact with a close, intimate friend. *Ḥāẓirī* is the internalization of communication with saints, a process of dialogue that initiates the visitors into what mystics may describe as the mysteries of Divine presence. *Ḥāẓirī* requires no form of literacy or formal ritual obligations that must be performed, or without which it could potentially be declared null and void—as is the case with orthodox rituals, other than the necessity of bringing loving attention to the living presence of the saint.

Kursī

The photograph (figure 3) is of a local shrine, that of Bābā Sam'ay Shāh, near Lahore in what was not too long ago primarily a rural area, and which of late is undergoing rapid transformation through construction of urban infrastructure.[15] Villages in the area (such as *Ḥallokī*, *Janjhatey* etc.) are now being absorbed in the city of Lahore. The *'urs* here is observed on the first day of the monsoon season (Ur. "*sāwan kī phelī tarikh*"), as mentioned in the epigraphic inscription.

At the present time, the shrine of Bābā Sam'ay Shāh is not under the supervision of the governmental *Awqāf* authorities, and the resulting lack of intervention is clear from the iconography of the shrine. The Qur'ānic verse one encounters is not the *fātiḥa*—nor is there any instruction about reciting that particular verse, as there is in the case of the shrine of Madho Lāl (see figure 1.5). In fact, the marble tombstone has the *āyat al-kursī* (the verse of the throne, Qur'an 2.255), commonly understood as the verse of Divine sovereignty, etched in arabesque style on the ornate, border area of the epitaph in black ink (more on this below).[16] What this iconography conveys is that the shrine *is* the

15 I had the opportunity to visit one such shrine, photographed here, during ethnographic work in the autumn (October) of 2017. I made a *ḥāẓirī* at this shrine, requesting permission to take the photographs. A *faqīr* dressed in red robes sat on the doorstep, a sleepy rested dog nearby. The structure of the shrine is under the shade of large tree.

16 Since the Throne Verse is known for its talismanic qualities, Muslims often recite it for Divine protection. To have it printed on the epitaph of a saint's grave, which pilgrims visit daily, is much more than just a talismanic function. Feras Hamza, Sajjad Rizvi, and Farhana Mayer, *An Anthology of Qur'anic Commentaries, Volume I, On the Nature of the Divine* (Oxford: Oxford University Press, 2008), 131: "By most, the *kursī*, whether taken literally or purely symbolically, is understood as signifying God's sovereignty and authority: it is God's seat 'over' His creation." Moreover, the difference between the Sufi shrines and the graves of Muslim sovereigns is a common enough motif in the performance of Sufi poetry at shrines. For example, the Sufi *kalām* recited at

Figure 3: Shrine of Sam'ay Shāh in Lahore. Photo by I. M. Khan.

Chishtī *'urs* festivals in Qawallī performances is often replete with notions of divine power expressed through symbols of kingship and lordship. After all, were not sultans seeking proximity to Sufi tombs upon death? Suffice to say, seeing Sufi shrine architecture as symbols (*nishanī*) of

place for encountering the Divine presence: "*alā inna awliyā' Allāh lā khawfun alayhim walā hum yaḥzanūna*" (Q. 10.62: "Indeed, the friends of God will have nothing to fear and nor will they grieve"). There follows in Persian/Urdu, this couplet:

> *Jo ho guzar tū ā adhir ye hai qalandarī mazār*
> *Gar tuj pe paṛ ga'ī naẓar tu ho jā'ey beṛā pār*
>
> [If you should happen to pass-by here, know that this is the shrine of a *qalandar,* If the (saint's) gaze rests on you, your ferry will reach the safety of the riverbank]

These verses are evidence of the saint's ability to give spiritual help with a mere glance (*naẓar*) even to visitors who just happen to pass by (*guzar*). Such are the qualities of "this *qalandarī* shrine" (*ye hai qalandarī mazār*) says the hemistich of the first line.[17] As for the notion of a ferry (*beṛā*) successfully crossing the river, it is a common enough trope in South Asia to express the overcoming of difficulties, or misfortune, to be out of danger. In this case, in praise of the Sam'ay Shāh, the couplet says that the glance of the *qalandarī* saint carries such spiritual potency that should the visitor be fortunate enough to have it fall upon her, it will carry her across from a place of uncertainty and danger (i. e. river) to the realm of safety. The couplet promises the visitor relief or deliverance from whatever troubles she may be facing in the present.

---

divine friendship and power in popular imagination opens up a window into Islam's mythological narrative as well. The Sufi tomb is a glorification of the potentiality of humans to achieve divinity, claim to a mighty transformation of mythic proportions. It is believed that kings, and mighty pharaohs of the past, went to their graves as ordinary men, divested of all political power or claims of divinity. It is not difficult to see why the status of the dead and even death itself would involve issues of politics and governance. In my opinion, eschatology in Islam is intimately connected to questions of political consciousness and power, but very little has been written about these dimensions.

17 It is not difficult to further see or interpret this trope of saint's *naẓar* as being synonymous with divine light, to identify the vision of the saint with *nūr* of God, for example, see Gerhard Böwering, "The Light Verse: Qur'ānic Text and Ṣūfī Interpretation," *Oriens 36 Journal of the International Society for Oriental Research* (Leiden: Brill, 2001): 113 – 144. But for a more literal reading, there is also the case of a famous, and authoritative, *ḥadīth* "my servant does not cease to draw near to Me through his supererogatory prayers (*nawāfil*) until I love him. Now when I love [My servant], I am his ear (*sam'*) by which he hears, his eye (*baṣar*) by which he sees, his hand (*yad*) by which he grasps, and his foot (*rijl*) by which he walks. There is nothing more distasteful to Me than taking the life of a believer (*mu'min*)..." Cf. Merlin Swartz, *A Medieval Critique of Anthropomorphism: Ibn al-Jawzī*'s *Kitāb Akhbār aṣ-Ṣafāt* (Leiden: Brill, 2002), 269. For more on this divine saying, see William A. Graham, *Divine Word and Prophetic Word in Early Islam* (The Hague: Mouton & Co., 1977).

The second couplet, in praise of ʻAlī, is by a Sufi saint known as *Bū ʻAlī Qalandar* (Shaykh Sharīf al-Dīn Pānipatī):

> *ḥaydaram qalandaram ḥastam banda'-i murtaẓā ʻalī hastam*
> *payshawā-i tamām rindanam ke sag-i ko'i shair yazānam*

This translates, "I am a Haidarī (belonging to Haider); I am a qalandar; I belong to Murtaẓa ʻAlī. I am Imām (leader) of the drunkards (profligates), a dog at the court of *ʻAlī,* the lion of God."[18] *Mawlā* ʻAlī is venerated across shrines in India and Pakistan, seen as the inheritor of esoteric knowledge transmitted to him by the Prophet. In fact, no *samāʻ* ceremony (*qawwalī*) at the Sufi *dargāhs* is considered complete without the mention of his name (*manqabat-i ʻAlī* or the *qawl* of the Prophet that establishes ʻAlī's status or *maqām* as such: *man kuntu mawlā fā ʻAlī mawlā,* "whoever accepts me as master, ʻAlī is his master too").[19]

Karamāt

The *dargāh* is a place for miraculous happenings and occurrences (*karamāt*), healings, and, indeed, ecstasies that are experienced during devotees' experience of Sufi saints being *present* (*mawjūd*). Such presence connects ordinary folk with God; to visit the threshold of the *dargāh* is to stand in the presence of God. Devotees experience God's nearness personally, to feel a closeness to and intimacy with Him through *ḥāẓirī.* Thus devotees recognize and associate the names of their Sufi saints with divine power: hence the emission often of a spontaneous cry such as *yā ʻAlī*—sometimes just a drawn-out *ʻAlī*—or *yā Firūz ḥaq Firūz, yā Farīd ḥaq Farīd, mast qalandar mast* (referring to Lāl Shāhbaz of Sehwan), *yā Gharīb-nawāz.*[20] The name of each of these holy figures is an active force, spoken with devotion, gratitude, reverence, and joy. This mode of religious devotion that people enter into when visiting the Sufi saint of the *dargāh* I would describe as revealing and embodying the *lived* metaphysics of Indus Suf-

---

18 The title was bestowed upon (*mawlā*) ʻAlī by the Prophet, *asad Allāh.*

19 See Seyyed Hossein Nasr, "Shiʻism and Sufism" Their Relationship in Essence and in History," *Religious Studies* 6.3 (1970): 229–242, and Valerie J. Hoffman-Ladd, "Devotion to the Prophet and His Family in Egyptian Sufism," *IJMES* 24 (1992): 615–637.

20 In popular speech, the son-in-law of the Prophet is known as *Maulā* (master, lord) *ʻAlī,* considered to be the master of all Sufi saints, and therefore his name is invoked at all Sufi shrines. *Gharīb-nawāz* applies to one who is a "bounteous-giver to the poor." Any Sufi saint who displays such generosity is said to be like Khwāja Muʻīnuddīn, the revered leader of the Chishtī order, whose shrine is in Ajmer, India.

ism. Therefore, the main aim of this dissertation is to examine the religious experience of *ziyāra*, or specifically *ḥaẓirī*, to the *dargāh* as it is practiced in Indus Sufism, to grasp the meaning of this ritual as a *lived* experience, and thorough study of its ritual practice to arrive at a phenomenological understanding of the Sufi shrine as a site of hierophany and what it means for a Sufi saint to have a "resurrected" body.[21] The challenge is to understand how for the worshippers, the eternally "present" body (of the saint) is the *axis mundi* around which the universe revolves, the center of the cosmos.

1.2 Fātiḥā (du‘ā)

In contrast to *ḥaẓirī* at Sufi shrines, the rationalization of *ziyāra* in Islamic law focuses on the offering of *du‘ā* (personal prayer, supplication) at the *dargāhs*. In the Islamic textual tradition that treats *ziyāra*, both in legal writings and in polemics against the practice by religious scholars (*‘ulamā*), the overriding concern is the regulation of *du‘ā* (supplication), i. e., as to where the visitor stands or sits to engage in supplication near the grave of the saint. The "discourse" on *ziyāra* is entirely concerned with the "correct performance" of *du‘ā* at graves and tombs. In the eyes of the legal specialists, or *fuqahā'*, the central concern of Islamic law with regard to devotional practices is the avoidance of *shirk*, idolatry, and petitioning the Sufi saint directly is seen as a violation of *tawhīd* (Oneness) and hence a kind of *shirk*. At the shrine of Shāh Hussaīn (figure 3), Lahore, the management of *Awqāf* has engraved instructions on how to perform *ḥāẓirī* ("*darbār mein ḥāẓirī kā ṭariqa*"): recite *durūd sharīf* 11 times, *surat fātiḥa* 7

21 Chapter 2 discusses methodological implications of and notable contributions in the phenomenology of Islam. To my knowledge, there is no work in scholarship that takes note of *ḥaẓirī* as an experience at the *dargāh*. The one work that does so discusses the meaning of the word in the context of demonic possessions and speaks of *ḥaẓirī* as referring to the manifestation of this evil presence at the Sufi shrine because of the spiritual power of the saint is a book by Carla Bellamy, *The Powerful Ephemeral: Everyday Healing in Ambiguously Islamic Places* (Berkeley: University of California Press, 2011). Secondly, there has been no attempt to conduct a phenomenology of Sufi shrines. Annemarie Schimmel's classic work, *Deciphering the Signs of God: a Phenomenological Approach to Islam* (Albany: State University of New York, 1994), has very little on the Sufi shrine. Apart from an excellent collection of essays in W. C. Troll's *Muslim Shrines in India: Their Character, History and Significance* (New Delhi: Oxford University Press, 1989), but not much by way of phenomenology, the work by the late Samina Quraeshi, *Sacred Spaces: A Journey with the Sufis of the Indus* (Cambridge: Peabody Museum Press, 2009) contains an account of her experience with Sufi shrines in the Indus ("Storytelling as Imaginative History").

times, *surat ikhlās* 11 times. The aim of the ritual is to send pious merit to the Sufi saint (*īṣāl-i ṣawāb*), after which *du'ā* is offered, although it is not stated alongside, and neither do the pedagogical instructions explain the purpose or the theology of this method.

Figure 4: The grave (*maqbara*) of Mādho Lāl, adjacent to the tomb of Shāh Hussaīn, Lahore. Photo by I. M. Khan.

For the *'ulamā*, *du'ā* at Sufi shrines (or when visiting any grave) is a vehicle for sending a gift of religious merit to the dead, and it is directed to God, by whose power the gift is received on behalf of the dead. In theological and legal terms, *du'ā* at Sufi *dargāhs* is defined as *īṣāl-i ṣawāb* (transfer of merit). It helps ease the burdens of sinners and improves their spiritual condition in the *barzakh* (the state between death and the Resurrection). Still, those among the dead who were always pious, or those whom God has chosen as his friends, are said to raise their hands in supplication (*du'ā*) for the living as reciprocation for their gift. The logic is that the *du'ā* of pious servants and friends of God is always accepted. This, in principle, is known as *tawassul* (intercession) in Islamic discourse and is used by pro-Sufi *'ulamā* to legitimize offering *du'ā* at Sufi shrines. It is according to this principle that *'ulamā* such as the Barelwīs defend the sanctity of Sufi tombs, legitimize the offering of *du'ā* at *dargāhs*, and disagree with the Deobandīs in the latter's stricter interpretation of the practice, which denies *tawassul* as legitimate in Islam. In Pakistan, this disagreement among the religious scholars over *du'ā* constitutes the primary legal debate about *ziyāra*. As far as ritual practice is concerned, the Barelwīs believe in the efficacy of Sufi tombs and explain it as a form of reciprocity between the Sufi saints (who are in *barzakh*), and the living, while the Deobandīs only accept the mediation of the living at the shrines on behalf of the dead through *du'ā*. In the Deobandī view, even the pious dead do not have any agency in helping the living, such as that implied in the theology of *tawassul.*

Gesture

It is important first to understand the performance of *fātiḥa-khwānī* for the soul of the dead in the vicinity of the grave—which is specifically relevant for our issue—or for that matter, in any other location. The act itself involves the raising of the arms, and extending

the hands with open palms, to supplicate (*du'a*) God on behalf of the dead. Trimingham describes it in the following way: "The simplest form is to stand in front of the tomb and recite the Fātiḥa, which is caught by the symbolic act of raising the hands, palms upwards, during the recitation and then transferred by passing them down upon the face."[22] Although this bodily gesture is recognized as the Muslim way of communicating (supplicating) with God, it is not necessarily the case that this is the only reason why devotees of shrines raise their hands is to offer *fātiḥa*. Devotees also raise their hands in supplication to the saint of the shrine for his direct help.[23] Even the seventeenth-century *Farhang-i Jahāngīrī* contains a miniature painting showing the Mughal Emperor Akbar (d.1014/1605) making *ziyāra* to the shrine of the Chishtī Sufi Shaykh Farīd ad-Dīn Ganj-i Shakar (d. 664/1265). Akbar is shown kneeling next to the tomb of the Sufi with his hands raised in supplication (*du'a*).[24] However, in modern Pakistan such supplicatory actions seeking a departed saint's intercession are seen as a serious form of *shirk* by Deobandīs, ahl-i Ḥadīs, Jama'at-i Islāmī and other modernist Islamic movements.

22 Trimingham, *The Sufi orders*, 180.

23 Mattison Mines, "Islamization and Muslim Ethnicity in South India," in *Ritual and Religion Among Muslims in India*, edited by Imtiaz Ahmad (New Delhi: Manohar, 1981), 70: "Thus the orthodox stress that saints are great mean worthy of respect, but that prayers should never be offered to them, except, some argue, with the under-standing that the saints will pass them on to Allah."

24 The figure of the miniature is reproduced *inter alia* in John Renard, *Friends of God: Islamic Images of Piety, Commitment, and Servanthood* (Berkeley: University of California Press, 2008), 164. Cited as Mughal, *Farhang-i Jahāngīrī*, The Trustees of the Chester Beatty Library, Dublin, In 61.9.

Īṣāl-i Ṣawāb

The ritual of *īṣāl-i sawāb* "transfer of merit"—commonly also known as *fātiḥa khwānī* (recitation of the Fātiḥa as a prayer to the dead saint)—is a way to send *sawāb* (merit or reward, *thāwab* in Arabic), as "spiritual almsgiving," to the dead, but also to the living

according to some religious groups/schools of thought, such as the Barelwīs.[25] This is traditionally done through *du'ā*, as described above. *Īṣāl-i sawāb* (transfer of merit) has been formalized through a number of religious customs that pertain specifically to the dead, such as gatherings known as *qul* or *soyim* (on the third day after death), *daswān* (on the tenth), *chalīswān* (on the fortieth), *barsī* (on the anniversary and on *shab-i barāt* [Ar. *laylat al-barā'a*], the 15th Sha'bān).[26] The *fātiḥa* is one of the *suras* most commonly recited among the Muslims in *du'ā'* for the dead. Its brevity (a total of seven verses if the introductory *tasmiya/basmala* is counted as a verse) coupled with its synecdochal significance in the Qur'ān make it an efficacious offering.[27]

Although the recitation of Qur'ānic verses, and the legally obligatory acts of worship (*'ibadāt*)— prayer (*ṣalāt*), Fasting *(ṣawm)*, Pilgrimage (*ḥajj*), and charity (*zakāt*)—can be offered as pious means of transferring merit to the dead without

---

25 Constance E. Padwick, *Muslim Devotions: A Study of Prayer-Manuals in Common Use* (London: S.P.C.K, 1961), 111. As far as I know, Padwick is the only one to define it as spiritual almsgiving, which is an almost accurate description of the ritual activity, but it may not always hold in other ritual applications, for example, in the case of *īṣāl-i sawāb* for the esteemed Shaykh Abdul Qādir Gilānī (1077–1163 CE) by the *Qādirīyya silsila*. The Barelwīs would not interpret it as almsgiving in their celebration of *gyirawīn sharīf*, the 11th of Rabī' al-Thānī, and neither would the Chishtīs in their *chatī* (sixth) *sharīf* of *hind al-walī*, Khwāja Mu'īn al-Dīn *Gharīb Nawāz* of Ajmer *sharīf*. The reason for Padwick's use of almsgiving to describe the ritual has to do with the notion's (*īṣāl-i sawāb*) antecedents in *ṣadaqa*, as attested in a well-known Prophetic ḥadīth; this aspect is discussed later in the chapter.

26 On the third day, along with *the fātiḥa, surah ikhlās* is read in abundance hence it is called *qul*, and forty days (*chalīswān*) after burial of the dead body Qur'ān *khwānī* is organized, also known as *khatm al-Qur'ān* or just *khatm* for short. The Deobandī strictly oppose such a formalization of these ritual activities, and see them as an innovation, *bid'a*. Practice of such ritual activities is a major point of contention between the two reformist camps in Pakistan. For a classification of these rituals in Islam see William A. Graham's "Islam in the Mirror of Ritual" in *The Development of Islamic Ritual*, ed. Gerald Hawting (Burlington: Ashgate/Variorum, 2006).

27 See (William A. Graham, "Fātiḥa," in *Encyclopedia of the Qur'ān*, 1st ed.), M. Mahmoud Ayoub, "The Prayer of Islam: A Presentation of the Surat al-Fātiḥa in Muslim Exegesis," *Journal of the American Academy of Religion*, Thematic Issue S, XLVII 4 (1980), 636–47. However, unlike Ayoub, I do not think it is the "only link." Surah *Yasīn* is also seen as being extremely beneficial for this purpose, and there are other recommended combinations as well.

any additional rite, most *fātiḥa khwānī* rituals involve feasts, especially when observed for the prophet and the Sufi saints. *Fātiḥa* for Sufi saints, in contrast to that for the ordinary dead, is observed with special care and religious devotion, not only in South Asia, but other parts of the Islamic world—as in the *slametan* ritual in Indonesia, which offers an excellent comparative example for the purposes of the present study. A major point of religious contention is the formalization of such observances within the mainstream "orthodox" Islamic calendar.

In general, the notion of *īṣāl-i sawāb* pertains to the relationship between the living and the dead. It essentially applies to the pious act (*'amal*) of sending a gift of merit (*sawāb*), either through charity (*ṣadaqa* and *zakat*), religious rituals (*namāz, rozā, hajj*), or Qur'ānic recitation, to the soul of a dead—or a living—person in order to improve his or her spiritual condition in the *barzakh*, the interlude in the grave from the time of death to the Day of Judgment.[28] The dead can receive the gift in a tangible form in *barzakh*. The literal translation of the term *īṣāl-i sawāb* is the transmission of merit. Both the words are of Arabic origin, but the concoction is peculiar to Urdu, and from a linguistic angle can said to be an invention of 19th-century Islamic reformism in India. The operative principle of *īṣāl-i sawāb* is the intangible transfer of (sacred) merit, through the performance of non-obligatory prayers (*nawāfil*) and other pious activities that generate *sawāb* as a gift.

The theology of *īṣāl-i sawāb* is also important in the wider Islamic religious tradition with regard to the practice of visiting the dead. As mentioned above, it encompasses belief in the metaphysical transfer of the merit generated by legally sanctioned (*shar'ī*) religious actions dedicated to the dead. This notion also had a strong impact on socio-political life through the institution of the *waqf*, or pious endowment, to provide *īṣāl-i sawāb* for the departed souls by establishing welfare foundations as *ṣadaqa-jarīya* (charity in perpetuity). The importance of this foundational aspect of *īṣāl-i sawāb* is established in a Prophetic ḥadīth, and it has long been the subject of interpretive disagreement among the various religious factions.

As part of a larger narrative, *īṣāl-i sawāb* is embedded in the monotheistic cosmology of divine reward and punishment. By default, it involves a moral economy of virtuous deeds, clearly distinguishable from sinful actions that can bring shame, calamity, and divine wrath. It involves the religious belief that righteous deeds will bring personal salvation and that sinful deeds can de-

28 William Chittick, "Muslim Eschatology," in Jerry D. Walls (ed.), *Oxford Handbook on Eschatology* (Oxford: Oxford University Press, 2007). (Jane I. Smith has written about this as a paradox of time.)

crease chances of redemption, inviting divine punishment in the afterlife—the worst and most reprehensible sin in Islam being that of *shirk*. The Day of Judgment is the ultimate test when souls will be rewarded with heaven or punished in hell for what each has earned during their sojourn in the world. It is firmly believed that no one will be wronged on this day because of God's justice. "So intense is the Qur'ānic concern for and insistence on the day to come, when all will be held accountable for their faith and actions, that the ethical teachings contained in the Book must be understood in the light of this reality."[29] At the heart of the matter, *īṣāl-i sawāb* is about the topology of virtuous actions and deeds; it is about religious praxis and, therefore, inevitably about religious aesthetics as well.

In the introduction I noted that the ritual of *īṣāl-i sawāb* deals with the relationship between the living and the dead.[30] This relationship has many components, psychological, religious, economic, architectural, social, philosophical and political, and it can therefore be studied from a number of perspectives. For those scholars who consider it a legitimate practice in Islam, the ritual effectively places a certain responsibility on the shoulders of the living to care for the dead of their community. In this pious concern for the dead, then, Islam is similar to many other traditions around the world and also historically linked with much older ones that emphasize the need to cultivate the relationship with their dead by enacting rituals that would benefit and give them solace in the next world. The justification for such practices can vary tremendously across religious

---

29 Jane I. Smith and Yvonne Yazbeck Haddad, *Islamic Understanding of Death and Resurrection* (Oxford University Press, 2002), 2. In the discussion of God's divine omnipotence, the authors refer to the power of God to create simply by saying: BE! Interestingly, a *walī* of God too is given such a power by God in Sufi folklore. See Sara Sviri's article on the divine command. From personal experience I can cite the case of my grandmother, whose grandfather's shrine is located on the outskirts of Lahore and with whom she interacted as a young girl, bringing on occasion home cooked food to him in a secluded spot where he spent his time. I heard her say on several occasions that a *walī*—referring to Khawaja Shaikh Muḥammad (d. 1926)—of God has the power to create several worlds (*jahān*). Certainly, the ability to have such an imaginative hold at the popular level could be threatening to the religious orthodoxy and even more so for those directly in political power.

30 In the west, and other highly modernized societies, cemeteries are often organized and maintained very well, but rarely the focus of weekly or daily visitations. There tends to be a sharp divide, ritually and socially unbridgeable, between the dead and the living. We do know that during the Middle Ages in Europe, and also in the Muslim territories, cemeteries had a far more interactive function. In modernity, it is essential to maintain the fundamental difference between the dead and the living. In line with scientific biology today, there is no possibility that the dead can communicate with the living, and all that remains are the decomposed remains, nothing more.

divides, and also the role such practices play in organizing religious life, and how they relate to issues of authority and power in their respective contexts.[31]

In the case of Islam, the principle of *īṣāl-i sawāb* is connected with the theology of intercession (*tawassul*). Interpretations that legitimize its practice base its ontological validity in the intercessory privilege of the Prophet, by whose virtue God has allowed Muslims to benefit from the intercessory power of their pious deeds, be these supplicatory prayers (du‘ā) or even association with pious fellow Muslims. All of this stems ultimately from the Prophet's concern for the welfare of his community. Such intercessory acts have become a way to save the dead after they no longer have the ability to save themselves. In short, *īṣāl-i sawāb* is a way to send pleasing gifts to the dead, with the expectation that the same will be done for the present living once they too are in their graves. In such a way of thinking, the relationship between the dead and the living is easily sustained as a dynamic one.

Īṣāl-i sawāb as a concept or practice is easy to grasp, and on the surface of it, quite ordinary in its theological logic. It entails belief in the transfer of merit from virtuous acts of the living to the dead, a transfer that is grounded in the economy of divine reward and punishment in Islam. All those *‘amāl* (deeds or actions) before death defined as pious according to religious dictates automatically generate merit. Virtuous actions or deeds will be rewarded, and sinful or reprehensible actions will earn divine wrath (although they can also be forgiven). In the context of this Islamic moral economy, a major issue is that of personal responsibility, and *īṣāl-i sawāb* seems to undermine this central ethical dictate of Islam by allowing the living to reverse the status of the dead. Critics find this a major contradiction of Islam's ethical core and thus see intercession as *shirk*. The Deobandī *‘ulamā’* in Pakistan and the Tablighī Jama'at accept the *‘amal* of the living on behalf of the dead but vehemently reject any agency on the part of the dead, that is, any ability they might have to intercede on behalf of the living. The Jamā'at-i Islamī and the Ahl-i Ḥadīs do not consider *īṣāl-i sawāb* to be a valid ritual at all because, in their interpretation, it undermines the Qur'ānic law of justice according to which every believer stands helplessly alone in front of God and is accountable for their actions alone; there is a reward for pious deeds and punishment for sins. These groups also make a strict distinction between *du‘ā* and *īṣāl-i sawāb*. Therefore, the only means of mediation (*tawassul*) that will determine personal salvation or not is the personal balance of good deeds (worship, charity, acts of kindness, humility, etc.) over against sins and all transgressions against God. Once the person is dead, there is nothing that

31 For Christianity, see the work of Peter Brown and Douglas Davies.

can interfere with God's system of justice being fully implemented, and because God is a just God, humans have been given the freedom to shape or determine their ultimate fate.

Ritual Paradox

The premise of *īṣāl-i sawāb* is that interventions on behalf of the dead by the living can influence the soul's chances of gaining favor with God by negating the dead person's immoral or religiously unmindful actions during his or her lifetime, as a consequence of which the soul suffers in the grave. Moreover, pious souls, who have already gained favor with God, also benefit from *īṣāl-i sawāb* in the form of surplus gifts sent to them by their living relatives and friends. In essence, the main problem in the case of saints is that people visit their *dargahs* to ask for Divine help (in *du'ā*), to receive the saint's *faiz*, and for communion with the saint, who is perceived to be alive. "The attraction of a Muslim shrine is defined by the super-natural power of the saint's (Pir) spirit, who is buried in the tomb inside the *dargah*."[32] Yet, paradoxically, the practice of *īṣāl-i sawāb* is employed in order to invoke the help of the dead saint (who is in fact alive), and sending merit has become a feature in shrine rituals of *du'ā* as popularized by the Barelwī *'ulamā'*—see below.

The problematic issue is that the invocation of *fātiḥa* by the individual visitor at the grave of the saint, meant as a gift for the soul of the departed saint, is logically a superfluous ritual because it is the visitor who needs the saint's help, and therefore his pious act (*sawāb*/merit) of reciting *fātiḥa* contradicts the saint's cosmological status of being in the superior position of bestowing his *faiz*. Secondly, if the saint is *alive*, reciting *fātiḥa* for him/her, as is done for ordinary dead Muslims, is a cosmological contradiction in terms, and thereby even an affirmation that the saint is dead. However, despite such reservations, the practice of *īṣāl-i sawāb* is particularly pervasive among the followers of Barelwī Islam as part of their striving to access the saint's *baraka* or *faiz*.[33] Hence, such an econ-

32 Beatrix Plfleiderer, "Mira Datar Dargah: The Psychiatry of a Muslim Shrine," in *Ritual and Religion among Muslims in India* (New Delhi: Manohar, 1981), 198.

33 Pnina Werbner has is correct in describing the difference between the two (e.g *faiz* and *baraka*) in the context of South Asia because *baraka* is not used in everyday idiom for saints as is the case in the Middle East, rather *faiz* is commonly employed. Werbner does not delve into etymological details only noting *faiz* as a substitute for *baraka*, but it might be useful to look closer at its literary and customary usages to reveal important nuances between the two terms.

omy of exchange remains problematic in the context of popular Sufi cosmology, something that thus far has escaped notice in contemporary scholarship.

1.3 A Barelwī Defense of *Ziyāra:* 'Allamā Ṭahir al-Qādirī

Turning to a traditional Muslim scholarly interpretation, Ṭahir al-Qādirī's defense of *ziyāra*, in his treatise entitled *Īṣāl-i Sawāb Kī Shar'ī Ḥaisīyyāt*, rests on the traditional use of Qur'ān and Ḥadīth, accompanied by his commentary on the selected texts (*matn*), to extract principles through analogical reasoning.[34] Qādirī's argumentation follows the traditional method of referencing earlier authorities on Islamic law, such as Suyūtī, al-Jawziya, al-Ṭahṭāwī, ibn Ḥajar Makkī, al-Tabrizī, or Bayhaqī, to support his commentary. In building his case for the legitimacy of *īṣāl-i sawāb*, Qādirī does not restrict his selection of authorities to Ḥanafī *fiqh*, but draws as well from other *madhhabs*, such as the Shafi'ī and Ḥanbalī. This aspect of Qādirī's exposition in a region that is predominantly given to Ḥanafī *fiqh* also reflects his legal cosmopolitanism.[35] In comparison to many of the other Barelwī works on *īṣāl-i ṣawāb*, Qādirī's conclusions are unique to his analysis, and therefore also relatively modern.[36] It is noteworthy how Qādirī argues for the legitimacy of *īṣāl-i sawāb* as being essentially in the tradition of God's *sunna*. That is, according to Qādirī, the principle of reciprocity in *īṣāl-i sawāb*, and the notion of giving for the sake of benefitting others, is embedded in Divine reflexivity—discussed below.

Ṭahir al-Qādirī opens his short, albeit important, a treatise on the issue of *īṣāl-i sawāb* by emphasizing the salience of piety (Urdu. *naikī*) in Islam. Piety is among the numerous sources of honor that humans can readily have or

34 Qādirī is a follower of the Barelwī school of thought, a scholar of Ḥanafī law and also a Ph.D. in Islamic law from the University of Panjab (Lahore), and a spiritual leader in the Qadiriyya *ṭariqa*. However, Qādirī's writings deal with a variety of political and legal topics in a style that makes him far more modern-oriented than the senior Barelwī *'ulamā* who traditionally trained and ran their madrassahs. Qādirī is the founder of Minhaj al-Quran International, established in 1980, reflecting his modernist outlook in its organizational structure while remaining committed to the Barelwī ideology.

35 The appeal to sources outside of Ḥanafī *fiqh* also places the practice of *īṣāl-i sawāb* in a broader legal perspective, and therefore Qādirī's attempt to demonstrate a consensus (*'ijmā'*) over it in Islam.

36 A scholar such as Qādirī, it is fair to assume, would be familiar with the entire gamut of arguments, for and against, *īṣāl-i sawāb* and the innovative mode of his analysis in thinking about *īṣāl-i sawāb* is a conscious effort to address the most pressing criticisms of the day against its ritual use.

earn, mainly because of the nature of the religion (*dīn*)—Islam—for the guidance of all humanity by the Prophet Muhammad. In his view, the notion of piety in Islam is broad enough to encompass various acts.[37] To put it in another way, the daily opportunity to perform thousands of pious deeds is easily realizable in Islam. Actions (Urdu, Ar. *a'māl*) will subsequently determine the outcome for all individuals on the Day of Judgment (U. *ākhirat*), and just as those actions that are good will become a source of forgiveness (U. *bakhshish o maghfirat*) for oneself, they will also be of use in the salvation of others. According to Qādirī (henceforth), it is for this reason that the term (*iṣṭilāḥ*), *īṣāl-i sawāb*, has been a central part of Islamic teachings at all times. It is a rather sophisticated way of describing the unique merciful nature of Islam, that is, not only are pious actions easy to perform, but their merit also benefits others. *Īṣāl-i sawāb* epitomizes the compassionate spirit of Islam in that, given it, individual acts of piety come to have a broader, collective impact.

Therefore the meaning of *īṣāl-i sawāb* is the transfer of the merit of one's good actions (*'amal-i ṣaliḥ kā sawāb*) to another person, dead or living. This includes all virtuous deeds, along with acts of worship, such as *namāz, roza,* Qur'ān recitation (*tilāwat-i qur'ān*), *dhikr* (U. *zikr*), *ḥajj*, or *'umra*. In support of his view, Qādirī refers to a notable scholar of Ḥanafī *fiqh*, 'Allamā Ṭaḥṭāwī, who writes that in the teachings of *Ahl-i sunnat*, any person can transfer merit gained from their pious actions to another (*ghayr*), be it from *namāz, roza, ḥajj, ṣadaqa* or *khairāt* (charity) or recitation of the Qur'ān, or for that matter any other pious action. All the benefits from such activities reach and are beneficial to the dead.[38]

On the authority of Imām Ṭaḥṭāwī, Qādirī defines the role and meaning of *īṣāl-i sawāb* as being firmly grounded in *Sharī'ah*. The idea that one Muslim can be a means of intercession (*shafā'a*) for another and that the pious actions of one person can help in the spiritual elevation of another is legally valid. It is even stated in the Qur'ān, *Surat al-ḥashr* (59:10), that people supplicate (*du'ā*) God for the forgiveness of their sins and the sins of earlier generations, and Qādirī establishes the validity of the practice of calling upon God to forgive the sins of those who have died by quoting such verses.

---

37 Ṭāhir al-Qādirī's work entitled *Īṣāl-i Sawāb kī Sharī'ī ḥaisīyyat* has been published several times, its ninth edition was published in March 2005. All his writings, and various editions, are published from his own institution *Minhaj al-Qur'ān*, a religious network which is has come to spread around the world.

38 Ṭāhir al-Qādirī, *Īṣāl-i Sawāb*, cf from *Ḥāshiya al-Ṭaḥṭāwī 'alā Marāqi al-Falāḥ* (Miṣr: Muṣtafā al-Bābī), 15.

Qādirī's following reference is to al-Suyutī's *Sharḥ al-sudūr*, which narrates a Ḥadīth that says that the community of the Prophet will descend into their graves laden with sins but will arise on the Day of Judgment pure from any blemish because of the supplication (*istighfār*) of their living brethren.[39] As for those who have achieved God's grace, Qādirī explains in his commentary that the supplication for the forgiveness of sins (*du'ā-i maghfirat*) counts towards the elevation of their spiritual rank (*darajāt*). It was the practice of the Prophet, although he was infallible (*ma'ṣūm 'an al-khiṭā'*), still to ask for forgiveness for his sins a hundred times a day. Indeed, Qādirī clarifies that such a notion is unthinkable since it is because of the Prophet that past and future generations will receive, and have received, Divine favor. As mentioned in the Qur'ān, in *Surat al-fatḥ* (48:2), the Prophet's supplication served as a means of continual spiritual elevation. Similarly, Qādirī adds that supplications on behalf of all pious Muslims, prophets, and saints, who are already forgiven and have passed away, are for the elevation of their spiritual ranks (*darajāt kī bulandī*), which are unlimited. Therefore, through our acts of ṣadaqa and charity, these already pious souls achieve an even higher status, and in response to our actions (*'amal*), they send us a gift (*tuḥfa*) of their prayers, that is, they pray to God for us.

In short, the mechanism of *īṣāl-i sawāb* in reference to saints and prophets hinges on the principle of reciprocity between the dead and living. In the case of ordinary persons, the process helps to relieve their burden of sin, but for those who are already blessed, like saints and prophets and other pious souls, it translates into an exchange of gifts between the dead and the living.

The work does not categorically state the ontological difference between the saints and ordinary persons. Qādirī tactfully includes all those who have already succeeded in receiving Divine forgiveness and grace, and are close to God, such as the *muttaqīn, ṣaliḥīn barguzīda anbīyā', awlīyā'* and *ṣalihā,'* as capable of returning the favor. All of these terms are, in a way, synonymous because saints and prophets necessarily possess the virtues of being *muttaqīn* (God-fearing) and *ṣalihīn* (virtuous, pious). However, other pious souls can also be described in the same manner. However, there is a difference and a hierarchy in the spiritual stations of pious souls, which Qādirī does not mention. In practice, people visit the tombs of Sufi saints, and not just any other grave, expecting special blessings in return.

In addition to arguing for the legitimacy of the practice of *īṣāl-i sawāb* in sacred sources of Islam, such as Ḥadīth and Qur'ān, Qādirī turns to the Islamic

39 Qādirī, *Īṣāl-i Sawāb*, 17. Qādirī dedicates a section of his treatise to discuss how this does not detract from personal responsibility or violate Divine justice.

mode of greeting as an example of reciprocal giving. The example further confirms the essential nature of pious reciprocity rooted in the social fabric of Islam. For Qādirī, the recommendation to return goodness with double generosity is substantiated in *Surat al-nisā'*, which encourages Muslims to return greetings (*salām*) with a similar, or even better, more generous form of greeting. So in response to *assalām 'alaikum*, one answers with *wa 'alaikum assalām wa raḥmatu llāh wa barakātuhu.*[40] Using this analogy, Qādirī builds a case for the exchange of gifts between people and the pious deceased souls, such as the *anbiyā'*, *awliyā,'* and all those close to God. In fact, these noble souls can return superbly better gifts for our gifts of *ṣadaqa*, recitation of Qur'ānic verses, and pious acts. This also means that only the pious, those who have found favor with God, can reciprocate the gifts of the living.[41] In this capacity, he mentions Ḥaẓrat Dātā Ganj Bakhsh of Lahore, who, upon receiving a gift from anyone, would most certainly raise his hands in supplication (to God), to grant that person the benefit of the Prophet's intercession and Divine closeness.

In Qādirī's interpretation, intercession is a system (*nizām*) that has two components: *shafāt-i kubrā* and *shufāt-i ṣughrā*. The former (larger) is reserved for the Prophet, and the latter (smaller) for other prophets, saints, and pious folks. According to a *saḥīh* (sound) hadith, the Prophet was visited by an angel and given a choice of two things. First, the Prophet could be given the right to have half of his community forgiven on the Day of Judgment, without any interrogation (*ḥisab kitāb*), or, second, the Prophet could choose to have the right (*ḥaqq*) of interceding for anyone. Accordingly, those souls would be forgiven. In this matter, he then references Ibn Ḥanbal's *Musnad*, where God offers Muḥammad different options or modes of intercessions on behalf of Muslims. He wisely chooses the one that gives him unlimited intercessory privilege.[42]

Qādirī thus interprets intercession of the Prophet as a right granted to him by God. In this, he explains the context of a Qur'ānic verse that promises to bestow (from God) upon the Prophet favors in such abundance as to satisfy or please

---

40 Ṭāhir al-Qādirī, *īṣāl-i sawāb kī shari'ai ḥaisīyyat*, 16. Qādirī here does not mention how the exchange of pious greeting pertains only to the dead.

41 In this example, Qādirī grants action, or power, of the dead to deliver benefits to the living. In general, he does not make a distinction between the ordinary dead and the saints, other than by reference of the martyrs (*shuhada*), who were killed in the lesser *jihad*, and the saints who engage in a greater *jihad*, logically speaking must have a higher status. The martyrs are not to be believed as dead, therefore it would follow that those who have engaged in an even greater struggle (*jihad-i Akbar*) should be considered, comparatively speaking, more alive than the martyrs.

42 Qādirī, *īṣāl-i sawāb*, 19. God is the ultimate giver, and bestower of all honors, and gives the Prophet the choice to take unlimited advantage of his generosity towards him.

him.[43] Qādirī further comments on how the entire universe, the cosmos, has been created for the beloved Prophet in the first place, and as long as the Prophet keeps raising his hands for *du'ā*, the process of intercession will not cease.[44] Qādirī cites from *Saḥiḥ Bukhārī* a hadīth that narrates how the Prophet will not rest even after all of the members of his *umma* are retrieved from the confines of Hell; at that time, he will intercede even for those people who had only uttered the *kalima*—who were not practicing Muslims—, and God will certainly grant him that favor as well.[45]

As for the lesser *shafa'āt*, Qādirī believes that all good deeds and *'ibadāt* will intercede on behalf of their doers; even children who died in their infancy will plead on behalf of their parents, and those who have been the recipient of one's good actions will intercede for clemency towards one. He sees this as a net ("*shafāt kā jāl*") cast across the Muslim community, and if nothing else works for someone, the Prophet will use his power and privilege of intercession. In any case, Qādirī finds an undeniable attribute of Islam to be that it provides several paths by which intercession works for all Muslims. In this regard, Qādirī cites a hadith from *Jāmi' al-Tirmidhī* narrated on the authority of Muḥammad bin Ziyād al-Lahānī, who told Abū Imāma that he had heard the Prophet say God promised him that 70, 000 people from his community would enter paradise without any accountability. Then another 70,000 will be saved from punishment (on account of good works performed by a thousand believers).[46] Similarly, according to another saying of the Prophet, Muslims will receive the punishment due to them on the Day of Judgment in their graves, so when the day arrives, they will arise purified—another example of how Muslims will receive salvation despite their sinful deeds.

Metaphorically, then, Qādirī's law of intercession operates much like a dense network of cables transmitting *sawāb* to all Muslims, in some form or the other, to save them from facing the torments of hell. It is important to remember that this *nizām* is only possible because of the Prophet's ontological, intercessory sta-

43 Qādirī, *īṣāl-i sawāb*, 20: surah *al-ḍuhā* (5:93).

44 Qādirī's entire treatise is a commentary on this image or gesture of *du'ā* described as the raising of hands ("*hath du'ā kay li'ay uṭhtay rahain*"), 20. In my opinion, in the case of Sufi devotional culture, the visual imagery power of the Sufi saint's gaze (*naẓar*) constitutes a more potent form of gesture. This imagery is amply found in the performative Sufi culture. A good example is the popular Qawallī known as *chap tilak* by Amīr Khusrau that describes how Khawaja Nizamuddīn stole Khusrau's identity with through an exchange of a glance. It represents the magnetic and sacred *seeing* of the Sufi saint that devotee's desire to experience, and embodies a more direct mode of reciprocity between the Sufi saint and the lover.

45 Qādirī, *īṣāl-i sawāb*, 19.

46 Ṭāhir al-Qādirī, īṣāl-i sawāb kī shari'ai ḥaisīyyat, 22.

tus. Simultaneously, and at a more mundane level of *īṣāl-i sawāb* in its most elementary form, salvation hinges on the ability of one person to help another by pious deeds. Qādirī concludes the introductory section of his work on this point and, in his second section (chapter one), discusses the topic of transferability in the light of Qur'ān and Ḥadīth, the two primary sources of Islamic law.

The first chapter of the treatise is an exposition of the notion of *nisbat* (relationship or connection) and how it operates as a tangible connection between people separated in time and space. This relationship is based either on genealogy or devotional love (U. *'aqīdat o maḥabbat*) solely to win the acceptance of God.

Just as an individual can transfer her pious merit to another and also benefit the dead, as confirmed in the Qur'ān, the pious deeds of a *saliḥ* (pious), *mutaqī* (God-fearing), *momin* (believer), or *buzurg* (saint, sage) can yield merit for their offspring as well. Qādirī sees this principle confirmed in *Surat al-Tūr,* in which God says that persons of good faith and whose progeny follow them in their pious ways will be reunited in paradise.[47] In granting such a favor to the progeny of the pious, God promises not to take away from their reserve of merit that they have earned. It means that the pious acts of a community, or an individual, help in the salvation of the next generation. Qādirī explains that this verse refers to God forgiving the shortcomings of a community of believers, who, although they may remain steadfast in following the path of their ancestors, still do not fully measure up to their predecessor's standards of piety. In this case, the pious merit earned by their ancestors becomes a source of benefit for them. The point is that the pious acts of one person, or a community, serve as *īṣāl-i sawāb* for future generations, as affirmed in the Qur'ān. In this example, the *nisbat* (relationship or bond) is based on genealogy.

The second connection, in understanding the nature of *nisbat* and its importance for the principle of *īṣāl-i sawāb*, is forged based on mutual affection or love between two persons to please God.[48] Qādirī narrates a tradition from Tabrizī's (d. 1340) expanded *Mishkat al-masābih* by al-Baghawī, in which the Prophet explains a situation where two persons love each other for the sake of God but live poles apart—one in the east and the other in the west—but on the Day of Judgment God will reunite them on account of their mutual affection for the sake of God. According to Qādirī, the Ḥadith is a *tafsīr* of *Surat al-ṭūr.* It confirms the importance and efficacy of devotional love between two persons who lack physical proximity but share an emotional bond in piety.

---

47 Qādirī, *Īṣāl-i sawab*, 27.
48 Qādirī, *Īṣāl-i sawāb*, 28.

The second example, from *Ṣaḥīḥ Bukhārī*, reports a question by a man who wished to know the Prophet's views on holding deep adoration towards a nation or community (*qawm*) that existed in the past but no longer exists. The Prophet's response to the question is that the person will meet the same fate (*ḥashar*) as the object of her adoration, in this case, the nation. Here again, Qādirī claims to prove that having love or respect for a community or an individual one has never met serves as a metaphysical bond. Qadirī uses the ḥadīth to argue that an individual ends up receiving divine reward because of his connection of love towards another person or community for the sake of God.[49]

The principal of *nisbat*, mentioned above, either in terms of genealogy or by way of adoration or devotion, also establishes an intercessory link, as substantiated through *īṣāl-i sawāb.* It is not directly mentioned, but it is pretty evident that by these examples, Qādirī is countering criticism of the widespread adoration of saints whom they have not seen, saints who lived in much earlier historical periods. In Qādirī's view, the *aḥādīth* in *Saḥīḥ Bukharī proves* that the love for pious persons of bygone ages—in other words, the dead—is a sufficient criterion for God to accept the *iṣāl-i ṣawāb* for them from the living, and vice-versa.

In another example from *Ṣaḥiḥ Bukhārī*, Qādirī illustrates the power of intention (*nīyat*) by discussing a man who had killed a hundred men but was seeking forgiveness. After being told to join a particular group of pious people and sit in their company, the man sets out but dies en route to his destination. The angels descend to possess his soul, and a dispute ensues between the angels who wish to drag him to hell and the ones who want to save him (angels of mercy, *raḥma*). To reach a decision, they agree to measure the distance between his point of departure to that of the city of virtue; when this shows that he is closer to the town than his starting point, the angels of mercy immediately claim him as a saved soul. According to Qādirī, not only did his intentions count, but his proximity to the pious also determined his salvation. He quotes *from Sūrat al-nisā'* that any person who dies en route to a good deed will receive their recompense as if the act were done. Even a murderer can benefit from the pious actions of others he does not know but toward whom he looks to be forgiven for his heinous crimes.[50]

The point of the preceding discussion is to make clear that in practice, there exist different, competing modes of interpreting *ziyāra*. Whereas the notion of *ḥaẓirī* is linked to a ritual devotion addressed to the Sufi saint as one who has obtained union with the Divine and becomes a source and conduit for divine

---

49 *Ibid.*, 27–30.

50 Ṭāhir al-Qādirī, *īṣāl-i sawāb kī shariʻai ḥaisīyyat*, 29–33.

love, another mode of interpreting *ziyāra* stresses that the correct performance of *du'ā* is grounded in Islamic law. The former way of *ziyāra*, as *ḥaẓirī*, is characteristic of vernacular Sufism in the Indus region, both in Panjāb and Sindh (what here is described as Indus Sufism). In this differentiation, *ḥaẓirī* at Sufi *dargāhs* is the stuff of everyday *lived* experience and religious poetry, but *ziyāra* as an occasion for *du'ā* and *īṣāl-i ṣawāb* is something that the religious scholars argue about as regards the legitimacy or illegitimacy of the practice. Therefore, ritual practice at Sufi shrines in the Indus region of Pakistan conceals an underlying tension, or what can be bettered described as an incongruity, between two approaches to *ziyara* at these shrines. Second, this incongruity between the two modes of ziyāra can also be seen as symptomatic of more profound, epistemological rifts that run beneath the surface in the history of Sufism and Islam. This is the core of the debate over the relationship between rational knowledge, as exemplified in the legal discursive tradition (*usul al-fiqh*), and the experiential understanding of theosophical mysticism (*'irfān*), which is believed to be inexpressible in language. To understand the nature of this *incongruity* and how it specifically affects the practice and function of rituals at Sufi *dargāhs*, it is necessary to look at the role of Islamic reformism and state intervention in regulating Sufi shrines in Pakistan.

Chapter Two
Study of Sufism and Ritual

In the present study, I concentrate on Indus Sufism in Pakistan as an example of vernacular devotion to show the significance and meaning of the Sufi shrine in any adequate understanding of Sufism. Specifically, I focus on the ritual to unpack the meaning of *ziyāra*.[1] In its refusal to grant devotional acts such as *ziyāra* an epistemological validity of its own, scholarly work on Sufism has largely overlooked the centrality of the Sufi saint and his/her shrine as a sacred locus of religious experience. Understanding how the shrine works for the ordinary Muslim requires a phenomenological engagement that takes into account bodily experience. Therefore, the argument for "re-centering" the shrine, in addition to seeing the shrine as a locus of power beyond the Islamic textual tradition, is also an

1 In focusing on ritual as a way to study religious experience, the present work is indebted to the work of William A. Graham on the subject. In particular, I am guided first by Graham's recommendation for studying religious phenomena by paying more serious attention to their sensual dimensions (*Beyond the Written Word* [Cambridge Univ., 1987], pp. 162–5). Graham also devoted an earlier (1983) article entirely to Muslim rituals through the example of the Hajj. Here the fundamental importance lies in his candid discussion of the theoretical issue that has besieged Islamic studies from its inception as an academic subject. Most relevant here is his explanation of why in the particular case of ritual, both Arabists and social scientists have failed to reach a definitive basis upon which to build a more robust and meaningful academic consideration of Muslim ritual "*on its own terms*—namely those of Muslim interpretation and understanding themselves" ("Islam in the Mirror of Ritual," in *Islamic and Comparative Studies* [Burlington: Ashgate, 2010], 92). Amina Steinfels, in another brief assessment of the study of Islamic ritual, expresses surprise that "given the primacy of position of ritual discussion in the Islamic legal tradition, the lack of attention paid to it by Islamicists is striking" ("*Ritual,*" in *Key Themes for the Study of Islam* [Oxford: Oneworld, 2010], 304). Graham's essay on ritual, at the time of publication, can be seen as a response to this dilemma. Nancy Tapper and Richard Tapper, "The Birth of the Prophet: Ritual and Gender in Turkish Islam," *Man* 22 (1987), pp. 69–92, strongly criticized Graham for privileging orthodox Islam and marginalizing popular practices that, in their opinion, better represent Islam as lived religion. I think their critique failed to understand Graham's approach and, consequently, his argument. However, their misreading is symptomatic of the divide between anthropologists and Islamicists, which may have been one of the reasons Graham wrote his essay in the first place. Even if the social sciences "have taken the lead, producing ethnographic material about ritual performances in Muslim communities around the world. [And] by contrast text-based research—the province of Islamicists—has yielded much less scholarship on ritual" [306]; nevertheless, the theoretical issues that Graham brought to our attention still have not been addressed in any significant manner. This project seeks to develop an approach that may contribute, in a small way, to bridge the gap.

https://doi.org/10.1515/9783110781496-004

argument for seeing the emotive body of the devotee as quintessential to the meaning of religious experience.[2]

The ritual performance of *ḥāẓirī* at the *dargāh*, as a bodily devotional act, has received scant or no attention in scholarship on Sufism. Instead, scholars predominantly prefer to view Sufism through the prism of texts and narratives. On the basis of analytical study of literary tropes, scholars claim that Sufism is part of the ideological structure of knowledge/power relations: ordinary folk are unable to see behind this ideological veil and thus unable to question the logic of exchange between material interests and spirituality. In this chapter, I question discourse-centered approaches to argue that scholars have failed to grasp the significance of the ritual spatiality of the Sufi shrine as a medium of sensual religious experience.[3] Furthermore, this neglect has prevented scholars from observing the role of popular devotion in defining the locus of the Muslim faithful's most direct experience of God's sovereignty on earth. I argue that it is ritual devotion as an embodied practice at saints' shrines that tells us about the

2 There now exists a substantial corpus of literature on post/anti-Cartesian philosophy that argues for paying greater attention to the body in our inquiries of religious experience. Two of the notable studies in this field are Mary Douglas, *Purity and Danger: An Analysis of the Concept of Pollution and Taboo* (London: Routledge, 1966) and Carolyn Bynum, *Fragmentation and Redemption: Essays on Gender and the Human Body in Medieval Religion* (New York: Zone, 1991). It is important to clarify that the approach this study is concerned with is in understanding the body beyond its status as a mere social construct. Rather, the emphasis here is on the phenomenology of bodily experience in the direction of Merleau-Ponty's (1962) work, but also an earlier study that argued for taking sensory experiences seriously is by William James, *The Varieties of Religious Experience: A Study in Human Nature* [see "Religion and Neurology"] (New York: Longmans, 1902) that still stands as a convincing exposition of how to overcome Cartesian mind-body dualism perpetuated by scientific materialism. A good study on the topic is Sarah Coakley's (editor), *Religion and the Body* (Cambridge: Cambridge University Press, 1997). Later in this chapter, I discuss the contributions, and critique, of Talal Asad, Clifford Geertz, and Michel Foucault.

3 Therefore, the religious, oral culture of Indus Sufi shrines is not circumscribed by the Sunni textual tradition of Islam. For example, the ritual celebration (*'urs*) of the body of the saint as an eternal body fundamentally challenges the authority of the legal scholars and their texts. My argument is that Sufi shrine devotion offers an alternative interpretation of revelation in its emphasis on the sacred body of the Sufi saint as the true locus of divine communication. The experiential knowledge of the saint supersedes the discursive knowledge of the *fuqahā*, and the shrine as a devotional space is the symbolic manifestation of that supersession. The veneration of the body of the saint as divine archetype (*insān-i kāmil*) is also the sacralization of the human body itself as the ultimate locus for knowledge and experience of God. An excellent study on the sacrality of the human body is by Seyyed Hossein Nasr, *Religion & The Order of Nature. The 1994 Cadbury Lectures at the University of Birmingham* (New York: Oxford University Press, 1996) and also, *The Garden of Truth: The Vision and Promise of Sufism, Islam's Mystical Tradition* (New York: Harper Collins, 2007).

nature of the *living presence* of Sufi saints in Muslim religious life and experience.

The living presence of Sufi saints at their tombs (sing., *dargāh* or *darbār*) is the main reason for *ziyāra* in Indus Sufism.[4] Despite acknowledgment of this ontological dimension of shrine piety in previous scholarship on Sufism, the meaning, implications, and full significance of this aspect of devotional Sufism have received meager attention. What the current paradigm of scholarship on Sufism fundamentally misses in its investigation of religious experience is the *embodied* relationship between the devotee and the tomb of the saint, that is, the significance of *ḥāẓirī* as a direct, physical experience of the presence of the saint at the sacred shrine. It is this lack of scholarly attention to, or rather scholarly denial of the possibility of, an unmediated religious experience of holy presence that the present work seeks to redress by arguing for the *re-centering* of the Sufi shrines and their attendant rituals.

In the first chapter, I discussed the difference between two ritual modes of devotion in *ziyāra*. The first, *ḥāẓirī*, is representative of popular devotion, and the second, *fātiḥa*, has been legitimized by the religious scholars of Islamic law (and, in recent times, encouraged by the Pakistani government). Moreover, I argued that the former mode of visiting Sufi shrines is congruent with the cosmology of Indus Sufism, which considers the mystical body of the saint (as one who has transcended death) to be *present* (*mawjūd*) to the perception of his/her devotees at his/her tomb. In popular devotion, in addition to possessing a status or station (*maqām*) as a special immortal, a Sufi saint, who is a friend of God (*walīullāh*), is also described as an *insān-i kāmil*, a perfect(ed) human being.[5] In this self-realization as a perfect human being, then, the saint is said to be a manifestation of the image of God (*imago Dei*).[6]

---

4 See chapter 1 above for the meaning of *dargāh*.

5 ʻAbd al-Karīm Jīlī (d. 1428) wrote a treatise entitled *Insān al-Kāmil* see "al-Insān al-Kāmil" in the *Encyclopedia of Islam;* for its treatment in Islamic philosophy, see Seyyed Hossein Nasr, *An Introduction to Islamic Cosmological Doctrines: Conceptions of Nature and Methods Used for Its Study by the Ikhwān al-Ṣafā, al-Bīrūnī, and And Ibn Sīnā* (Albany: State University of New York Press, 1993).

6 See Yahya Michot, "The Image of God in Humanity From a Muslim Perspective," in *Abraham's Children: Jews, Christians and Muslims in Conversation,* edited by Norman Solomon, Richard Harries and Tim Winter (New York: T & T Clark, 2005), 163–174; Lloyd Ridgeon, " ʻAzīz Nasafī's Six Ontological Faces (Whenever You Turn There Is The Face of God)," *Iran: Journal of Persian Studies* 34 (1996): 85–99. See also Merlin L. Swartz, *A Medieval Critique of Anthropomorphism: Ibn al-Jawzī's Kitāb Akhbār aṣ-ṣifāt: A Critical Edition of the Arabic Text with Translation, Introduction and Notes* (Leiden: Brill, 2002); Wesley Williams, "Aspects of the Creed of Iman Ahmad Ibn Hanbal: A Study of Anthropomorphism in Early Islam," *IJMES* 34.3 (2002): 441–463; Binyamin Abra-

2.1 "Discourse" and Ritual Experience

In seeking to establish boundaries for Sufism in the modern period, the aim of the reformers was to realign actual devotional practice with the doctrine of the mystical body of the saint, as developed within the textual framework of Islam, and to align and to merge saints' sacred power with the envisioned religio-political order. Death is an important boundary for Islamic scholars in maintaining the sanctity of its moral parameters and a clear division between the living and the dead.[7] Between the debates of the *'ulamā* over Sufism in the subcontinent, first in the 19th century and then in the 20th-century national context of Pakistan, the *lived* dimensions of devotional practice at Sufi shrines were pushed to the periphery.[8] Second, scholarship has, in general, come to envision Sufism as congruent with political order in pre-modern Islam or even as a central constituent of its moral economy in maintaining public order. The rulers offered patronage in return for spiritual legitimacy for their power.[9] Sufism is considered to be an integral, if sometimes dangerous, part of the intellectual framework of Islamic knowledge maintained and mandated by its scholars. It is further held that rulers' patronage of Sufi shrines and the membership of Sunni religious scholars (*'ulamā*) in Sufi orders (*silsilas*) have fostered a mutual alliance between Sufism and power. Disputes and tensions are understood to have been the result of power struggles rather than as the result of ideological differences.

The popular construction of the mystical body of the saint as an immortal sacred being, especially in the case of Indus Sufism, is a source of deep consternation for Muslim religious scholars as well as scholars of religion/Islam of all

---

hamov, *Anthropomorphism and Interpretation of the Qur'an in the Theology of al-Qasim ibn Ibrahim* (Leiden: Brill, 1996).

7 See Leor Halevi, *Muhammad's Grave: Death Rites and the Making of Islamic Society* (New York: Columbia University Press, 2007).

8 The main purpose of the discussion in this section is to say that the transcendence of death by the saint challenges the orthodox interpretative tradition of the *sharī'a*-minded *'ulamā'*. The authority of the *'ulamā'* rests on their interpretive claim on the sacred texts of Islam for evaluating the morality and legitimacy of actions. In the modern period the *'ulamā* have had to contend with new challenges to their interpretive authority, especially in their fidelity to their schools of law in dealing with the sacred texts. Nevertheless, all the reformist movements derive their religious identities from the rational effort to discover the meaning of God's law from the text of the Qur'ān. In relation to textual Islam as a disciplinary mode of reasoning, the shrine represents an alternative, devotional mode of experience of scripture, that is, in the body of the immortal saint.

9 Omid Safi, *The Politics of Knowledge in Pre-modern Islam: Negotiating Ideology and Religious Inquiry* (Chapel Hill: University of North Carolina Press, 2006).

shades.[10] It challenges the latter's interpretative authority as intermediaries between sacred revelation, which includes in addition to the Qur'ān the Prophetic *Sunna* as recorded in the Ḥadīth traditions, and the Muslim faithful by extending the sphere of revelation to include the saintly bodies that in popular belief are held to have special efficacy. Thus, the popular veneration of the shrines of saints as divine embodiments challenges the textual authority of the *'ulamā*, both modernists and Islamists alike.

As I will discuss below in chapter 4 on Bābā Bulleh Shāh, the Indus Sufis mocked the discursive, textual knowledge of the *'ulamā*, and this certainly included a critique of their epistemological understanding of the Qur'ān as a fixed "text." For this reason, the important oral devotional culture of Indus Sufism is dedicated to the sacred body of the saint as a locus of divine presence and therefore derives a "scriptural" status from that connection. As an alternative to Qur'ān as "text" of God's speech, the Sufi shrine is a valorization of the human body as the true field of the human experience of the Divine, one superior to the textual knowledge of God. This experiential knowledge goes beyond the written word, beyond the text, and this accompanies a transformation of the body in the direct experience of divine truth (*al-ḥaqq*) in *ḥāẓirī*. The efficacy of bodily experience is encompassed in the lived metaphysics of *ḥāẓirī* at Sufi shrines, a ritual mode that the *fuqahā* seek to refute and suppress as an alternative means of directly experiencing the divine presence. The popularity of such practices has been widespread in Islamic history and a source of challenge to the juridical authority of the *'ulamā*, even as popular modes of devotion have influenced the learned class of scholars and created divisions over the status of these devotions—the Barelwī and Deobandī dispute in the South Asian/Pakistani case being an example of this tension within the ranks of the *'ulamā*.[11]

---

10 Thus far two studies on the significance of the body in Sufism have been published. The first is by Scott Alan Kugle, *Sufis & Saints' Bodies: Mysticism, Corporeality, & Sacred Power in Islam* (Chapel Hill, 2007), and the more recent is by Shahzad Bashir, *Sufi Bodies: Religion and Society in Medieval Islam* (New York: Columbia University Press, 2011). Both studies focus not on devotional ritual but on constructions of the body in the "discourses" about it, and as a result, what we get is the body as an abstraction, a purely theoretical construct. Even though Kugle offers a corrective on this same point in his review of Bashir's work, Kugle's own work is also driven entirely by literary narratives. Both studies assume the relationship between Sufism and law in terms of the body to be an unproblematic one in pre-modern Islam. Scott does discuss the threat of Wahhābī ideology to the idea of the sacred body in Sufism, which is a valuable contribution for the modern period and modernist attacks on the shrine, but he does not quite hit the mark in pinpointing exactly what kind of body the Wahhābīs were attacking.

11 Much has been written about the reform movements that erupted during the 18th and 19th centuries in different regions of the Muslim world. Typically, these diverse movements were in

In the following section we look at three important studies in the field of Sufism that view these phenomena from the angle of political power, following Foucault. The assumption behind such a perspective is that "discourse", or the textual tradition, is the real source that gives legitimacy to any religious or mystical experience. In seeing Sufism as an ideology of power, as a process of exchange between political power and spirituality, these studies of the "discourse" of Sufi texts have adopted an approach to Sufism that glosses over any incongruities that might indicate that other phenomena or other aspects of Sufi practice may be crucial to understanding the tradition. The final section of the chapter tries to show how our current scholarly approaches in their emphasis on language and literary narratives prevent our taking into account the *unsayable*, and thus keeps us from understanding the presence of incongruities, contradictions and paradoxes that complicate any truly adequate picture of Sufi thought and practice.

2.2 Mystical Body

The sacrality of the precincts of shrines (*dargāhs*) derives from the efficacious potency of the bodies of the Sufi saints entombed there, as we have stressed above. In the Indus Sufism of Pakistan, the operative cosmology is one in which humanity (as creation) and God (as creator-sustainer) are joined through the sacred body of the *insān-i kāmil* saint who subsists eternally in God and in turn is able to perpetuate divine creativity in the world (on which, see chapter

good part religious responses to declining political power, which was perceived to be the result of moral failings, especially the neglect of the "true" religious teachings of Islam. Muslim scholars held that the ills of the *umma* could only be corrected through a thorough reform of the reprehensible religious innovations (*bidā'*, sing. *bid'a*), mostly identified with practices associated with Sufi traditions, such as shrine rituals, by turning afresh to the sacred texts of Qur'ān and Ḥadīth. Chapter 3 of the dissertation will discuss how these reform movements brought about an epistemological shift in the process, one that would eventually undermine religious authority in Sunni Islam, specifically in the case of 19th-century India, and have an impact on the veneration of shrines through the polemical accusations of *shirk* (idolatry) flung at its practitioners. The reassertion of the primacy of sacred texts in reforming religious practices to bring about social reform needed for political revival of Islam is characteristic of all the modern reformist movements, even if there has remained also widespread disagreement over the hermeneutics of deriving God's law in a radically changed world. This as we shall see essentially led to diverging views among the religious scholars about the meaning of "scripture" as text and conflicting claims regarding which group who has the most authentic interpretive claim. See Muhammad Qasim Zaman, *The Ulama in Contemporary Islam: Custodians of Change* (Princeton: Princeton University Press, 2002).

5 below on the festival [*mela*] of saints).[12] The cosmological model of *insān-i kāmil* in the context of *wujūdī* and *shuhūdī* doctrines has received extensive attention in theoretical discussions of Islamicists and religious-studies scholars. The problem is that scholarship on Islam continues to rely exclusively upon such writings without much inclination to relate their ideas to the domain of actual practice.[13]

Consequently, the role of devotional practice focused upon the *insān-i kāmil* remains vague and largely unstudied. For the devotee of a given saint, establishing a proximate connection (*nisbat*) with the body of a Sufi saint at his/her tomb provides a gateway or link to the experience of the saint's living presence (as the *insān-i kāmil*)—in accordance with that particular devotee's own perceptual and spiritual capacities. In addition to its symbolic significance, the Sufi shrine is experientially an embodiment of the immortal, sacred body of the Sufi saint, present as the anthropomorphic image of God.[14]

The saint's body is thus the fullest experiential medium and metaphor for the divine presence in a human body, and it is also an important focus of per-

---

12 See Ahmet al-Bayrak, "Insan/Nas," in *the Qur'an: an encyclopedia* edited by Oliver Leaman (New York: Routledge, 2006): "in *tasawwuf* (Islamic mysticism), is the link between the *haqq* (the truth) and the people. *Al-insan al-Kamil* is the point where the *haqq* manifests itself," 302. There is a vast spectrum of scholarship on this subject, see Toshihiko Izutsu, *Sufism and Daoism: A Comparative Study of Key Philosophical Concepts* (Berkeley: University of California Press, 1983); William Chittick C., "The Perfect Man as the Prototype of the Self in the Sufism of Jami," *SI* 49 (1979): 71–101.

13 This indeed concerns the bigger question of how historians should contextualize their relationship with the texts, as being determinative how meanings and narratives are derived from them, a relationship with texts that is the central preoccupation of the field of hermeneutics. I will discuss this further below, but I want mention how important the "presence" and role of our bodies are in the way we conduct our inquiries. In this, I see my work as leaning towards what can be called as the anthropology of presence.

14 The question of anthropomorphism is mainly dealt as an issue of speculative theology, or as a subject in the theosophy of Sufism on *tanẓīh* (transcendence) and *tashbīh* (immanence) in the reception of Ibn 'Arabī's work. Its significance in devotional practice, especially in the role of mystical poetry, and possible relevance in the case of Sufi shrines has not been readily recognized. Studies discussing the role of Sufi devotional culture do so as an analytical study of these ideas that reduces aspects of religious experience to literary narratives of power. Such readings in the end offer little understanding of what might be really taking place on the ground in terms of the operative subtexts and subversions in ritual or gestures. What we get in the end are one-dimensional narratives without any sensitivity to the significance of embodied practices. A good display of such analytical treatment of the devotional significance of living presence of saints without concern for the insider's experience is Samuel Landell Mills, "The Hardware of Sanctity: Anthropomorphic Objects in Bangladeshi Sufism," in *Embodying Charisma: Modernity, Locality, and the Performance of Emotion in Sufi Cults* (New York: Routledge, 1998).

formances of mystical verse in the Sufi devotional practice of "audition" (*samāʿ*). The metaphor here offers an entry into knowing and experiencing what is absolutely real. To be clear, as a theoretical construct the notion of anthropomorphism has been controversial in Islamic intellectual thought, but it is not the aim of this dissertation to discuss the debates on this issue. What is relevant here is that the question of the *immanence* of God is closely related to matters of devotional practice, and the sacred space of the Sufi shrine provides for a devotional experience of God in the living presence of the saint. Our concern here is thus not with theological interpretation or theory so much as with understanding more adequately actual devotional practice in the veneration of saints at Sufi shrines. The emotive meaning of "celebration of death" as it pertains to Sufi saints, assessed in terms of its thematic, symbolic, or ritual relevance in studies on Sufism, has been passed over as unimportant in the history of Islam. What is it about the body of the saint, after all, that calls for a celebration? Is it not a moment of sobriety and sadness at a soul's departing this world and the perils of the grave turned upside down? The immortal body of the saint is at the center of pilgrimage (*ziyāra*), ritual devotion, and the performative arts dedicated to transmitting the saint's mystical verses. Existing scholarship in its hyper-focus on the textual tradition as the real driving force behind the social construction of Sufism has been incapable of understanding, let alone identifying, the saint's immortal body as a locus of sacred power. To be precise, this mystical body has been invisible to the scholarly lens that relies on "textuality" or "discourse" as the interpretive means of understanding the meaning of ritual action. In failing to take note of the centrality of the shrine and shrine devotional practice, studies have also failed to deal adequately with the phenomenological relationship between ritual devotion and devotional poetry.[15]

If we attempt an interpretation of the lived practice of *ḥāẓirī*, immortality in this case involves a *resurrection* of the sacred body of the saint at the time of death (*wisāl*). In the union with God at death, the body of the saint undergoes a transformation or transmutation: the saint is not just alive but has *risen*, so to speak. This interpretation is in contradiction to the orthodox belief that the resurrection of the body is to happen only on the Day of Judgment (*qiyāma*). This resurrection means that the Sufi saint, because of divine grace, has passed beyond the salvation that is applicable to the ordinary dead.

---

15 Due such a lapse on part of studies on Sufism, the performative space of Sufi shrines, at least in the Indus region of Pakistan, has yet to be examined as a source of experiential knowledge in Sufism. Later chapters will devote attention on this performative dimension.

The attempt by *'ulamā* and the government to promote *fātiḥa* for the dead in the textual traditions and to link this to the ritual of *ziyāra* to Sufi shrines, is aimed at an ideological suppression of belief in the immortal body of the saint. In the interpretation of *ziyāra* as focused on the *Fātiḥa* and its recitation as a pious gift (*īṣāl-i ṣawāb*) for the saint, the performance of *du'ā* places the body/soul of the saint in *barzakh*, the intermediary, liminal realm where all the souls of the dead await the time they are to be awakened or bodily "resurrected" on the Day of Judgment, which is the time for final recompense of each person after the period spent in the grave.[16] Thus the popular belief in the resurrection of the saint at death challenges the traditional interpretive scheme of eschatological doctrine in Islam.[17] To be sure, the most notable difference is that a saint's immortality in the form of a resurrection is an existential theophany or a sacred hierophany *in the body* of the saint as opposed to an orthodox resurrection of the body at an historical eschaton, on the Day of Judgment (*qiyāma*). Such an immortality runs counter to mainstream "orthodox" theological interpretation.

Ḥāẓirī is an affirmation that the saint's body has attained liberation from the space-time continuum, which also includes the "afterlife," amply described in the literature on the "spectacle of death" by religious scholars.[18] The pious are restricted to their suspended state in *barzakh* even as they enjoy God's manifold rewards. However, as immortals, the saints are free to travel and meet persons, appear and disappear across time and space to guide and help humanity. The saints are alive to instruct and guide and, in this function, continue to light the way for travelers on their spiritual journey.

---

16 Mona Siddiqi, "Death, Resurrection, and Human Destiny: Qur'ānic and Islamic Perspectives," in *Death, Resurrection, and Human Destiny* edited by David Marshall and Lucinda Mosher (Georgetown: Georgetown University Press, 2014), 30: "its significance also lies in the inability during this period of the dead to return to the earth, to return to this life in any way until the day of resurrection.

17 Kḥidr (Urdu. *Khizr*) is described as an immortal in the Qur'ān, and is a popular figure in Sufi folklore. However, the figure of Jesus (*'Īsā*) who has the power to resurrect the dead is equally important. The Sufi saints are well known for their abilities to heal, like Jesus. Much has been written on *khidr* but I think the role of *'Īsā* in Sufi metaphysics needs to be seriously studied.

18 The many types of punishments that await those who do not follow God's law is a regular feature of *khutbas* before payer congregations in mosques, a warning for the evildoers. During the Islamization of Pakistan in the 1980s such reminders (Qurānic verses or Ḥadīth) were often broadcast on the media (PTV).

Even the invocation of the names of Sufi saints works to counter hardships, illnesses, and worries of all kinds.[19] The body of the saint is an exceptionally privileged site since it offers access to the divine through direct physical experience. It is by virtue of this grace that a saint's immortality becomes the focus of a sacred center for pilgrimage (*ziyāra*). Moreover, the special living presence of the saint at a tomb brings humans into contact with the divine, and in this intermediary role the immortal body of the Sufi saint functions therefore in an analogous way to scripture. Just as recitation or other experiential engagement with the words of scripture puts the believer in the presence of God, likewise the body of the saint, in popular devotion, is a medium of divine contact and inspiration.[20] Consequently, the celebration of the death of a saint, the *'urs*, is effectively the sanctification of the body of an immortal. To visit a saint's tomb is to journey into the presence of God, which in turn means that the ritual of *ziyāra* can be seen as a form of worship (*'ibādat*). Ritual and its related devotional poetry define/create an oral/aural culture that centers upon the living presence of saints at their shrines, a phenomenon that I describe as the metaphysics of lived Sufism (see chapter 6 below).

2.3 Sufism and Textual Islam

It is through the extended interpretive process of the community of scholars that sacred knowledge has been held to be produced in authoritative "discourse." [21] The popular ritual of *ḥāẓirī* at the *dargāh* of saints offers a view of sacred power, as well as an epistemology of sacred knowledge, which give an entirely different

---

19 One example is "*yā bābā Farīd nā ho dukh nā ho ranj*" (Oh Bābā Farīd, keep us safe from grief and misery) is a mantra for those who find themselves stuck in a thunderstorm, works to pacify the storm. In a sense the names of Sufi saints become extensions of divine names/attributes.

20 This touches upon the well-known distinction between *ilhām* and *waḥy*, the latter reserved for the infallibility of the Qur'ān. *Ilhām* of course lacks similar certainty but when it comes to popular devotion to saints such theoretical distinctions become considerably blurred.

21 E.g., "since no form of powerful knowledge can exist in isolation from material forms of power, the economic strength of Sufism was crucial to the overall profile it acquired and so ultimately the discursive, miraculous and economic power of the Sufis were each interdependent," Nile Green, *Sufism: A Global History* (Chichester: Wiley-Blackwell, 2012), 7; Also, "By treating Sufism as an abstract mystical philosophy, these scholars entirely ignored the social context of Sufism as expressed in the Sufi orders, the institutions formed around saints' tombs, and the role of Sufis in politics," *The Shambhala Guide to Sufism* (Boston: Shambhala, 1997) 16.

reading insofar as the shrines in Indus Sufism are concerned. A closer look at the ritual of *ḥāẓirī* will reveal the limits of textual Islam.

Three studies that directly concern rituals associated with the veneration of Sufi saints are those by Clifford Geertz,[22] Mark R. Woodward,[23] and Pnina Werbner.[24] The latter two scholars significantly revised Geertz's ethnographic findings with the result that their arguments can be seen as working to undermine the methodological usefulness of the interpretive anthropology for which Geertz is famous. A fourth important study that examines the relationship between power and Sufism as a ritual exchange is one authored by Omid Safi.[25] Woodward, Werbner, and Safi are all committed to explaining Sufism as an ideology of power, and rely on literary narratives to make their arguments. Woodward points out a serious lapse on the part of Geertz in his identification of an Islamic ritual (discussed below) as distinctively Javanese, and Werbner radically restructures Geertz's model to question his emphasis on local diversity.[26] Both of these studies (along with Geertz's) deal with issues of tension between the local and the global, and their distinction between textual (scriptural) Islam and popular practice (folk Islam) arose primarily out of attempts to understand the relationship between formal rituals of "orthodox" Islam and the diversity observed in the veneration of saints at Sufi shrines. In the studies identified above, the question of "ritual" is directly connected to an understanding of the boundaries of Islam.

Werbner, however, does steer away from texts, but she resorts to a structuralist reading of ritual practices to arrive at the underlying ideological will to power in the ascetic body of the saint.[27] For her, "the charisma of a saint is embodied in ascetic practices and technologies of the body, and constructed ideo-

---

22 Clifford Geertz, *The Religion of Java* (Chicago: The University of Chicago Press, 1976).

23 Mark R. Woodward, "'The Slametan': Textual Knowledge and Ritual Performance in Central Javanese Islam," *History of Religions* 28.1 (1988), 54–89.

24 Pnina Werbner, *Pilgrims of Love: The Anthropology of a Global Sufi Cult* (Bloomington: Indian University Press, 2003).

25 *The Politics of Knowledge. in Pre-Modern Islam: Negotiating Ideology and Religious Inquiry* (Chapel Hill: University of North Carolina Press, 2006).

26 In view of Geertz's emphasis on the local, Pnina Werbner explains her aim in the following manner: "I have followed Geertz in the first instance by shifting focus to locally told narratives and away from published scared texts. Unlike Geertz, however, my aim has been also to explore the underlying structures—the moral fables—animating such narratives" (*Pilgrims of Love*, 289).

27 *Ibid.*, 12: "following Levi-Strauss (1966) I take mythical thought to be a science of signs, systematically interrelated through a series of homologies and analogies. The producer of myths is likened by Levi-Strauss to a *bricoleur* who creates logical systems out of the bric-a-brac of daily life."

logically through a corpus of myths about his miracles."[28] Through their revision of Geertz's anthropology, both Werbner and Woodward interpret Islamic mysticism as implicated and involved in the political arena, and even see it as almost wholly driven by the political myths that are activated in ritual processes. I will review each of their studies, starting with Werbner, in what follows to give an overview of how ritual has been interpreted by each as a mechanism of power, an approach that treats Sufism in terms of an ideological structure.

Pnina Werbner's Ideology of Sufism

Werbner's is an ethnographic study of a modern Sufi "cult"[29] that attempts to understand the charismatic authority of a living Sufi saint and how the saint's charisma is institutionalized after his death.[30] In this study, Werbner is concerned with understanding the globalization of one particular Naqshbandi Sufi order, as a "transnational movement," spreading from Pakistan to regions such as Britain, the Middle East, Southern Africa, and parts of Europe.[31] Moreover, in a Foucauldian twist, one of Werbner's main aims is to unmask the "technologies of knowledge deployed by the saint, through which he exerts control of his close followers."[32] Issues of human subjectivity, emotions, and mysticism are viewed through her structural analysis of Sufism as an ideological force. For Werbner, tension between the global and the local is summarized in the following way:

> [A] transplanted Sufi fable of perfection generates a powerful symbolic field whenever it encroaches. Like Capitalism or modernity, Sufism is a distinct discursive formation which shapes the very cultural habitus it invades, while authoritatively constituted by its everyday familiars.[33]

Therefore, "Sufi Islam, despite its apparent diversity and concrete localism, embodies a global religious ideology within a social movement which everywhere

---

28 *Ibid.*, 25.

29 Despite several issues with such a term, Werbner seems to be unperturbed by its usage, and neither explains its precise application in the Naqshbandi Sufi order she studied.

30 *Ibid.*. The shrine of the Sufi saint is in the valley of Ghamkol *Sharīf*, Pakistan.

31 In Werbner's description of the Naqshbandi order, Pakistani Sufism "represents a counter-globalizing trend to political Islam. It is traditional and sentimental. It stresses personal renewal, group amity and mutual support and tolerance. It is accommodation to worldly politics and encourages thrift and worldly success," *Ibid.*, 7.

32 *Ibid.*, 12.

33 *Ibid.*, 82.

fabulates the *possibility* (if not the actualization) of human perfection."[34] In the study of Islam, Werbner's stress on the ideological unity of Sufi Islam is a major departure from earlier ethnographic studies on Muslim societies. But there is a more important angle to Werbner's work, and that is her explanation of how the *faiz* ("grace/favor") of the saint is accessed through ritual at the shrine of the Sufi saint.[35] In Werbner's opinion, "the attitudes of Sufis to temporal power and authority are highly ambivalent, a mixture of moral superiority and practical accommodation, but they are never outside of politics."[36] She also says, "In regarding Sufism as discursive formation, I start from the assumption that discourses are not simply negotiated between equals, but are persuasively imposed in any given social context by the socially or culturally powerful."[37] Werbner's most notable contribution lies in literally extending Foucault's notion of the "panopticon surveillance" to explain the saint as knower of esoteric, secret, theosophical knowledge, and how narratives and fables in legitimation of his uniqueness contribute to giving him absolute authority over his followers.[38] Thus she argues that

> the powerful coerciveness of Sufi knowledge, its effect of certainty, stems from its location within a broader universal religious ideology or episteme, in Foucault's sense. The episteme promotes self-discipline through an experienced sense of globalized, panopticon surveillance.[39]

Omid Safi's Exchange of Baraka

Omid Safi, a historian of Sufism, argues that even prior to the Ayyūbids, the ideology of Saljūk Islam entailed a mutually beneficial relationship between

34 Pnina Werbner, "Sufi regional Cults in South Asia and Indonesia," George Stauth and Samuli *Schielke, Dimensions of Locality: Muslim Saints, their Place and Space* (Bielefeld: Transcript Verlaq, 2008), 30.

35 Werbner distinguishes between *baraka* and *faiz:* "In ethical terms, then, *faiz* is the light of generosity, kindness and concern emanating from the saint and communicated to his followers. Through *faiz* a *pir* creates the tie, *rabita*, binding him to his followers. *Faiz* is the embodiment of his spiritual power, *ruhaniyat*, another key term which is used to express the *pir*'s spirituality as a powerful force," *Pilgrims of Love*, 251.

36 *Ibid.*, 5.

37 *Ibid.*, 82.

38 *Ibid.*, 286.

39 *Ibid.*, 286.

the Sufis and the sultans.[40] In Safi's opinion, the Sufis dispensed their *baraka* in return for patronage and the promise of justice by the sultans. The political order of the Saljūks was consecrated by the *baraka* of Sufi saints. This argument is based on the study of hagiographical "discourse" and *tarikh* narratives that instead of seeing Sufism as an oppositional force, interprets it as an integral part of the larger socio-political order. Safi's work occupies an important place in Islamic Studies; it examines the relationship between Sufi saints and the Saljūks through a revisionist reading of saintly hagiographies, including works of history (*ta'rīkh*),[41] to identify the social process of reciprocity between the Sufis and the rulers. Safi defines it in the following way: "*baraka* is, as much an anything, about power: the spiritual power of the saint, the power of the saints to interact with mighty rulers, and the power to lend them legitimacy."[42] As described by Safi, sainthood is, if anything, a matter of public discourse: the "charisma of the saint is remembered, perpetuated, and transmitted through the very collection of these [hagiographical and "*ta'rīkh*-based"] narratives."[43] Safi then makes a point regarding the argument I seek to problematize: "many Sufis saw as their task and duty the 're-arranging' and 'improving' of the affairs of humanity in this 'visible universe'."[44] He maintains that issues of ritual efficacy and reciprocity are ultimately about social order and power, and in the case of an exceedingly important political phase of Sunni Islam, sealed through the patronage of the Saljūks in exchange for their blessings of saints:

> the saints themselves are often depicted as extending their *baraka* to the Saljūks in exchange for promises that those who lived under the saints' *wilāya* would be treated justly.

---

40 Omid Safi, "Bargaining with *Baraka:* Persian Sufism, "Mysticism," and Pre-modern Politics," *The Muslim World* 90 (2000).

41 For example, on hagiography, *Asrār al-tawḥid, Ḥālāt wa Sukhanān-i Abū Sa'īd-i Abū 'l-Khayr* etc., and for sources on history, *al-Kāmil fī'l-tarīkh* and *Zubdat al-nuṣra* etc.

42 Omid Safi, "Bargaining with *Baraka,*" 275.

43 *Ibid.*, 275. Safi also says in the same article, "Yet many of the accounts in the hagiography seem to be oral narratives which had been circulated for generations before being written down" (274), which, it seems, confirms for him the correspondence between the text and practice.

44 *Ibid.*, 264. In the saintly Hagiographies it is clear that the saints have a higher status then worldly rulers, and have the ultimate power, reflecting the popularity of Sufi saints. That is why rulers had a practical interest in patronizing their *khanaqahs*, but it may not be the case that Sufis necessarily, and as a convention, were actively complicit with their political projects.

> This bargaining of *baraka* for promises of justice and patronage is one of the most ubiquitous features of Saljūk Islam.[45]

Safi's emphasis on notions of discourse, visibility, and power in understanding the role of Sufism is also in line with the Foucaudian paradigm, and in the field of religious studies, it is best represented in the work of Talal Asad (discussed below in this chapter).[46] For his part, Omid Safi's work has done much to establish a "discourse-centered" understanding of the relationship between politics, or power, and mysticism in Islam.[47]

Woodward's Theory of Divine Kingship

Mark Woodward's response to Geertz in his article emphasizes the textual tradition and how, as social practice, the connection between scholars, mystics, and rulers was consolidated in the Javanese theory of kingship. He argues that this ideology was enacted daily and on special occasions through the ritual of *slametan* to inculcate values of submission to the authority and hierarchy of Javanese society in the ritual of *slametan*. In terms of the fracture between the study of Islam as an interpretive anthropology, on the one hand, and approaching it as a textual tradition, on the other, Woodward's critique of Geertz's work supports for Islamicists the absolute necessity of starting with texts, or at least paying homage to the authority of the Islamic textual tradition. Since Woodward's critique occupies an important position in its impact on the study of Islam as a religious practice, it is worthwhile to give a brief review of his relevant article.[48] It is also directly relevant to this study because the ritual of *slametan*, in its theological or legal dimension, can be seen as analogous in some ways to the ritual

---

45 *Ibid.*," 265. He argues that the Sufis challenged the power of the sultans, which from his perspective, was a form of "spiritual blackmail," a departure from our "warm and fuzzy notion" of "saintly behavior" (266).

46 Talal Asad, *Genealogies of Religion: Discipline and Reasons of Power in Christianity and Islam* (Baltimore: John Hopkins University Press, 2009). Talal Asad's Foucaldian emphasis on epistemic power, perpetuates a model of historical and cultural determinism, in understanding the role of the body and the self which seems to leave little space for individual agency in the transcendence of the political, a model that I find built entirely through reliance on "discourse." In his view, discourse produces reality, a perspective this project stands diametrically opposite. I do not subscribe to the view that body, as a social construction, is a matter of "discourse."

47 E. g., in "Bargaining with Baraka," 134: "*Naẓar* has long been a Sufi trope, a subtle and amorous glance exchanged between Sufi master and a disciple, a lover and a beloved."

48 "The '*Slametan*'" (above, n. 22).

of *fātiḥa khwānī* recited for the transfer of pious merit to the dead as a gift (see above, Ch. 1). Since *slametan* is a ritual that involves also the *Fātiḥa*, a review of Woodward's critique helps give a clearer idea of not only the theoretical significance of the ritual but also the sociological issues at stake in the study of Islam.

Woodward's target is Clifford Geertz's analysis of *slametan*, which presents this communal-feast ritual as an example of religious heterodoxy, a syncretic practice, removed from the "scripturalist" orthodoxy of the religious scholars (*santri*). *Slametan*, for Geertz, is not and could not be part of textual Islam, but Woodward identifies its ritual legitimacy in the context of the textual tradition and its social utility in preserving the political order of Javanese kingship, putting political power and mysticism in a tight embrace. Geertz was the first scholar to make an in-depth study of the *slametan* ritual meal and attendant recital of the Fātiḥa. However, at the time Geertz was unaware that "Slametan is an example of a ritual complex that links blessing (*barakah*) and food and extends from Arabia to Southeast Asia."[49] In *The Religion of Java*, Geertz identifies the ritual of *slametan* (Javanese: *slamet*, from Arabic: *salamat*) as indigenous to Javanese Islam.[50] *Slametan* is the hosting of a communal feast, during which prayers (Qur'ānic prayers, notably the surah of *al-Fātiḥa*) are recited over food, which is then consumed by all present and distributed to others. [51] In Geertz's analysis, the *slametan* feasts are seen as a common syncretic practice of social life in the villages rooted in animism, and therefore far removed from Islam. Geertz interprets the communal feasts as the core of the local, or *abangan* version of Islam as distinct from traditional textual, or *santri* Islam. The former is perceived by Geertz to be little concerned with legal obligations of the *Sharīʿah*, while the latter is believed to be wholly preoccupied with them.[52] In support of his argument

---

49 Woodward, "The '*Slametan*'," 55: It also follows that "(3) it is not primarily a village ritual; (4) its religious and social goals are defined in terms of Islamic mystical theory; (5) most of the modes of ritual action it employs are rooted in essentialist texts including the Qur'an and Hadith; [and finally] (6) elements of *slametan* derived from pre-Islamic traditions are interpreted in Islamic terms."

50 Despite shortcomings of Geertz's analysis, it was a unique study of its kind, as none had previously examined the ritual of *slametan* in Java with such anthropological detail and richness.

51 Woodward, "The '*Slametan*', 65: the prevalence of ritual feasts were observed by early orientalists such as R. Burton and Snouck-Hurgonje during their travels to Mecca and Medina, and in fact "Snouck-Hurgonje offers these parallel ritual systems as support for his theory concerning the South Indian origin of Southeast Asian Islam."

52 Clifford Geertz, *The Religion of Java*, 6: "The *santri* religious tradition, consisting not only of a careful and regular execution of the basic rituals of Islam—prayers, the Fast, the Pilgrimage—but also of a whole complex of social, charitable, and political Islamic organizations."

that the *slametan* ritual is actually embedded in textual Islam, Woodward emphasizes the importance of Ḥadīth in the legitimization of *slametan* as an Islamic mode of practice.[53] Moreover, and apart from the Ḥadīth, "the textual notion of charity, and particularly the distribution of food, informs ritual practice in many Muslim societies and is associated with the veneration of saints and Prophet Muhammad."[54] It is precisely this dimension of *fātiḥa* for the Sufi saints as charity that the present inquiry aims to explore. If people believe that the Sufi saints and their tombs possess the power to solve their own worldly problems and are able to bestow spiritual stations (*maqāms*) upon the living in their mystical quests, then why are saints in need of human charity?

In response to Geertz's interpretation, Woodward's article on the *slametan*, in mapping the relationship between textual Islam and popular Muslim piety, offers a significant revision of Geertz's description of *slametan* as an animistic rite. For Woodward, the *slametan* is not an animistic practice, although it draws upon local elements; rather, it is understood by all Muslims to be an Islamic practice. He argues that instead of evaluating the "Muslimness of a community," how it does or does not conform to preconceived notions of *sharī'ah*-minded piety—as Geertz's model does—scholars must tend to the "interpretation of Muslim texts and tradition."[55] The two most important textual sources in discovering God's law are the Qur'ān and Ḥadīth (*Sunna*). Woodward especially discusses the role of Ḥadīth in providing legitimacy to *slametan*, thus undermining Geertz's critical differentiation between *abangan* and *santri* versions of Islam. Woodward claims that the hold of textual Islam is so engrained in the everyday lived experience of the Javanese that for the *Kejawan* majority "most are unaware that in hosting a *slametan* they are recontextualizing Hadith."[56]

Woodward's seeks to put forward a strong case for revision of Geertz's approach in arguing that "contemporary Javanese fields of religion must be understood in light of fields of meaning established by larger Muslim tradition."[57] In his interpretation, a compelling reason for this methodological turn can found in the way Ḥadīth tradition functions to provide legitimacy to the ritual of *slametan*. This means paying close attention to the role of the *santri 'ulamā'*, whom Geertz sidelines. The *santri* owe their training to theological schools, known as *pesantren* in Javanese—or *madrassa* (Ar. madrasa) in South Asia— where they

53 Woodward, "The '*Slametan*'," 56: "particularly on the collection known as *Riyādh as-Ṣāliḥin* (Gardens of the Righteous)."
54 Woodward, "The '*Slametan*'," 64.
55 Woodward, "The '*Slametan*'," 62.
56 *Ibid.*, 78.
57 *Ibid.*, 55.

specialize in the religious sciences. "Many *santri* can trace their intellectual lineage to the ninth-century collector of Hadith, Bukhari," which establishes a sense of personal connectedness between the *santri* and the Prophet. [58] Indeed, as pronounced by an Ḥadīth itself, "in truth an expert in Hadith can be termed a member of the family of the prophet, even though in fact he has never met him."[59] So the student/teacher connection embodied in the Islamic learning tradition is a source of blessing, *realized* through the person-to-person transmission of Ḥadīth going back to the Prophet.[60] Thus the *sharī'a*, interpreted by the *ulamā* in "the ritual system it establishes, provides the framework for the religious lives of the "orthodox" community."[61]

The second pivotal facet of Woodward's revisionist argument as to how the textual tradition of Islam plays a central role in Muslim popular piety is his interpretation of Sufism. In his analysis, the "*kyai* are simultaneously Sufis and *'ulamā*, while the *pesantren* is both a theological school and center of mystical learning and practice."[62] Woodward does not differentiate between the *'ulamā* and the Sufi saints, as is typically done in the case of Middle Eastern Islam, and he defines Sufism as part of orthodox Islam in Java prior to the twentieth century because "mystical interpretations of Islam have served as paradigms for devotionalism, social order, and social life."[63] The all-encompassing influence of Sufism is such that even ordinary people are conversant in the metaphysical knowledge pertaining to matters of mystical union, soul, and divine essence. This also means, in his account, that Sufism is not restricted to the cult of saints and miracles. [64] Hence, Woodward sees Sufism as encompassing "esoteric mystical systems and popular saint (A. sing., *Walī*) cults."[65] This interpretation challenges the approach to Sufism that defines its popular dimensions as having solely to do with hagiography, miracles, and devotion to saints. More notable is how he sees the esoteric doctrines of mystical Islam as also providing religious

---

58 *Ibid.*, 57.

59 *Ibid.*, 58, citing the Indonesian translation of the Hadith collection, *Riyadh as-Ṣāliḥīn*..

60 For a thorough exposition of Islamic traditionalism and the importance of connection to a chain of authorities in Islam, that Woodward is describing, see William A. Graham, "Traditionalism in Islam: An Essay in Interpretation," *The Journal of Interdisciplinary History*, 23.3 (1993): 495–522. It remains until now the most comprehensive study on the topic.

61 Woodward, "The "*Slametan*," 56.

62 *Ibid.*, 60, here Woodward briefly also refers to the institutional role of mystical brotherhoods.

63 *Ibid.*, 55.

64 *Ibid.*, 59.

65 *Ibid.*, 58.

legitimacy for the "Javanese theory of kingship in which the Sultan is described as the "perfect man"."[66]

In Woodward's opinion, the entire political edifice of kingship in Javanese Islam is given over to the mystical teachings of Ibn al-'Arabī, the twelfth-century Sufi known for his metaphysics of the unity of being (*waḥdat al-wujūd*). The radical strains of esoteric or theosophical mysticism are eclipsed by what Woodward terms the "moderate Sufi view," in which all mystics must submit to *sharī'ah.* In his view, even Ibn al-'Arabī, whose writings contain a thread of antinomianism based on his theosophical views, "did not actively denounce Muslim law."[67] One influence from Ibn al-'Arabī in Java, Woodward argues, has been in the Javanese state cult, where the sultan is believed to be the perfect man (*al-insān al-kāmil*)[68] who, although he is himself exempt from religious obligations, must fulfill his religious duty to preserve the *sharī'ah.* Thus, the mystical notion of union of the soul with the divine essence, with the sultan as the perfect man, characterizes the hierarchical nature of the state as well, and is "often used to describe the relationships between the sultan and his subjects."[69]

Woodward goes on to argue that the state ideology of the sultan as the perfect man, as developed and espoused in mystical Islam, is expressed through the ritual of *slametan.* As it is, "most royal rituals are elaborate *slametan. Slametan* are held in conjunction with royal life-crisis ceremonies, or *rites de passage* (birth, weddings, circumcision, funerals, etc.). State ceremonies are *slametan* performed on a massive, imperial scale."[70] The political importance of *slametan* is such that myths governing its origin link it to the consolidation of Islam as the official religion of Java through the collective effort of saints and Muslim kings.[71] For Woodward, the mystical doctrine of Islam is the foundation for the political structure of Javanese Islam:

> if the history of the *slametan* is understood as that of a state ceremony that spread to the general population rather than an animistic response to a foreign religion, the process of Islamization can be viewed as one in which Islamic states strove to redefine an existing court-village axis in Islamic terms. The links among Islamic texts, royal, and popular *slametan* suggest that non-Javanese Islam texts and modes of ritual action served as a

66 *Ibid.*, 59.
67 *Ibid.*, 60.
68 *Ibid.*, 59.
69 *Ibid.*, 83.
70 *Ibid.*, 66.
71 *Ibid.*, 83.

> model for the formulation of a Sufi imperial cult in Central Java and this served, in turn, as a model for popular piety and ritual performance.[72]

Nevertheless, we might ask how state ideology, mystical Islam, and local tradition all come together to define the form and function of the *slametan* ritual itself? That is, how does the ritual communicate the religious values and ethical mores in the ritual, alongside its mystical dimensions, all of which are said to be involved in the performance of the *slametan?* Below I will consider each aspect of Woodward's argument and discuss how he explains the multivalent nature of the *slametan* ritual.

The most salient feature of Woodward's argument is that the sultan in Javanese Islam is also a saint, and this has a special kind of effect on the social order below. In his estimation, the "religious and social goals [of the *slametan*] are defined in terms of Islamic mystical theory."[73] But what allows the state to take on the moral and religious authority, that is, to be able to absorb fully the mystical interpretation of Islam as its political ideology?[74] This question is particularly important—although Woodward does not comment on it—because throughout much of Islamic history, the Sunni religious scholars have been extremely wary of the sultan claiming complete authority on legal matters, let alone claiming divine status as *insān-i kamīl.* What needs to be factored in is that the *'ulamā* have been most interested in claiming the mantle of sainthood. The precedent for this pattern, in a way, had to do with the theological and legal disputes of the Abbasid period, in the aftermath of which the Sunni *'ulamā'* retained their authority in religious matters and, in theory at least, even over the Caliph—this pattern too is an integral part of the larger Muslim tradition, and in the way, religious authority came to define the limits of political authority. (I deal with the competition over Sufi symbolism in chapter three in my consideration of shrine theopoetics.) Nevertheless, Woodward believes that "concern with humility extends to the highest levels of the royal and *santri* hierarchies."[75] It is in fact more than just a concern, rather a mode of socio-cultural practice that permeates all aspects of Javanese life—so much so, that "even in secular contexts, Javanese social discourse is punctuated by expressions of humility and modesty."[76] The regard for humility, again, is rooted firmly in the Qur'ān and Ḥadīth, and it is

---

72 *Ibid.*, 85.
73 *Ibid.*, 55.
74 Nowhere in his article does Woodward use the word "ideology" to describe the political role of the state.
75 *Ibid.*, 77.
76 *Ibid.*

also "among the cornerstones of Sufi mystical practice."[77] Therefore, for example, on "ritual occasions the Sultan and the *Pengulu* wear patchwork coats to simulate the rough clothing of Sufi saints."[78] As I understand it, Woodward here is convinced that the Sultan is actually perceived by the Javanese majority as the perfect man, and although he wields political power, he transcends it altogether. This kind of transcendence over worldly power proves that both external (*zaḥirī*) and esoteric (*batinī*) kingship truly reside with none other than the Sultan himself.[79]

As the perfect man, the sultan constitutes the channel for blessing, because the function of the saint, as "friend of Allah," means that the saint becomes a vehicle for divine favor, a source of blessing for the people. This form of mediation between creator and the world is explained through the concept of *tawassul* in the Islamic tradition (Even though Woodward does not explicitly mention its relevance). So the Sultan is the living presence of the saint/Sultan, whose mediation with God on behalf of the people is a source of blessing, tranquility and peace for the entire community. Rather than discussing the concept of *tawassul*, Woodward places his emphasis on the notion of mystical union between the servant (A. '*abd*) and the Lord (A., *rabb*) as the justification of the theory of kingship. He sees it also replicated in structuring the relationship between the subjects and the sultan. Apart from the concern for humility among the Javanese, "submission to religious superiors facilitates the attainment of higher mystical states leading to union with Allah."[80]

Woodward examines two myths specifically that he thinks help in understanding the origins of the *slametan*. These tell of the significance of saints and the responsibilities of Javanese kings to ensure order and prosperity for the entire community. In both myths, the ritual of *slametan* brings much needed relief at the time of bad harvest, and in one of the two examples, establishes social tranquility among different segments of the community, as well as serving as a conversion ritual. In the hosting of Javanese state ceremonies "the *santri* hier-

---

77 *Ibid.*

78 *Ibid.*

79 In the case of Mughal Empire, Akbar is a notable case of a ruler attempting to claim divine kingship in the manner of saints. The *'ulamā* for their part also lay claim to being authentic representatives of Sufism. More then any of the Sufi orders, the *Chishtī* Sufi order in India has always been wary of the court, even if there has been a history of interaction between the two sides, but such issues are difficult to reconcile only through a study of textual material.

80 *Ibid.*, 82.

archy played a vital role in the formulation and dissemination of *Kejawen* Islam."[81]

The *slametan* as a royal cult and the Qur'ānic cosmology are explained in the following manner in Woodward's words:

> Muslims pray toward the *Ka'bah* in Mecca, which is located on a cosmic axis directly beneath the throne of Allah. In this sense, prayers moves toward the center and upward while blessing moves downward and toward the periphery. The *slametan* replicates this pattern on a local level in that prayers move upward and blessing descends into the rice cone and from there to the local community. *Slametan* food, therefore defines a local model of the Islamic cosmos and is among the means through which blessing is attained and distributed.[82]

Finally, what is the form of *slametan* and how does it work as a ritual? To begin with, Woodward defines the *slametan* as a "ritual meal at which Arabic prayers are recited and food is offered to the Prophet Muhammad, saints, and ancestors, who are implored to shower blessings on the community."[83] This description is extremely vague, and theologically incorrect from the *santri* point of view. It is precisely the idea that the dead can directly "shower blessing" on the living which the religious scholars are bent on rectifying. The belief that the prophets, saints, and ancestors can directly return the favors of the living is tantamount to *shirk* among the *'ulamā'/santri* because no one is to be implored or prayed to other than God. Hence Woodward's opening lines explaining the *slametan* convey an entirely misleading interpretation of the ritual. Only in describing the role of *slametan* in funeral ceremonies does Woodward discuss the informant's view that it helps to ease the "transition between life and death and to ensure the tranquility of the dead person."[84] Woodward explains this as one of the features of *slametan*, failing fully to grasp its centrality as transfer of merit to the dead.

Secondly, Woodward does not specifically explain the mechanism by which the prayers that are recited and the food that is offered reach the dead. He only mentions in passing the "*Kejawen* notion that saints and ancestors consume the *halus* (subtle) essence of *slametan* food."[85] Instead, Woodward discusses the Islamic form of salutation, an exchange between two Muslims in return for "*al-salām 'alaykum*" (peace be upon you) the other person responds by saying "*wa-'alaykum al-salām wa-rahmat allah wa barakatuhu*" (on you be peace, bless-

81 *Ibid.*, 85.
82 *Ibid.*, 77.
83 *Ibid.*, 54.
84 *Ibid.*, 68.
85 *Ibid.*, 65.

ings and the mercy of God). The response conforms to the Qur'ānic injunction to return a salutation with a better one, which then becomes a source of blessing for both persons.[86] Woodward also explains the social and psychological aspects of *slametan* by discussing the polyvalent meanings expressed in the Arabic, and Qur'ānic, word *salām*, "used as a salutation for humans, spiritual beings, saints, and angels in many ritual contexts."[87] Woodward also explains the ritual of celebrating the *mawlid* as having to do with how angels and God offer *salāms* to the prophet. Likewise, the community also offers *salāms* to the prophet. In any case, "Muhammad is also the only 'spirit' who is always invited to even the simplest *slametan*."[88] Furthermore, in the mystical sense, Woodward links *salām* as a state of tranquility and peace to the notion of *nafs muṭma'inna*, which means the "tranquil soul."[89] In his view of Javanese Sufism, "*mutmainah* [sic] is associated with intuitive knowledge of Allah and is the condition to which those who have attained mystical union return."[90] In this manner, and through the textual importance of *salām*, Woodward continues to associate its different, but related, shades of meanings with the ritual of *slametan*.

The problem with Woodward's discussion is that he aims to explain every dimension of religious life, and its interplay with political order, as a seamless ritual affair. *Slametan* is built upon the ethos of humility, submission to authority, notions of communal tranquility and solidarity; all in harmony with mystical interpretations of union, which also serves as the foundation for political and social order in Javanese Islam. How each of these features fit together in Woodward's account has been discussed above. It is important to note that for Woodward, each of these dimensions and social practices are enshrined in the textual sources of Islam, such as the Qur'ān and Ḥadīth, and for this reason the "traditional *santri* and the *Kejawen* Muslims consider the *slametan* to be an Islamic ritual."[91] Still, Woodward mentions in a footnote, that, "while reformist[s] denounced such practices [as the *slametan*] as "worshipping rocks and trees," an attitude of tolerance prevailed."[92] The problem is if all the sections of the community find the *slametan* to be an Islamic ritual, then, why do the reformist

---

86 Woodward, *Ibid.*, 67. Here Woodward references an early account of travels to Egypt, E. Lane, *Manners and Customs of the Modern Egyptians* (London, 1908), 203.
87 *Ibid.*
88 *Ibid.*, 69.
89 *Ibid.*, 70.
90 *Ibid.*
91 *Ibid.*, 62.
92 *Ibid.*, 62, n. 39.

santri need to warn people against “worshipping trees and rocks”? Who among the Javanese, according to the *santri*, are believed to be worshipping rocks and trees, and exposing themselves to the gravest sin in Islam, that of idolatry, *shirk?*

Woodward’s analysis fails to distinguish between the ordinary dead and the special dead, and to consider the possibility of how this distinction might matter in the performance of *slametan* ritual. That is, Woodward’s interpretation involves the undifferentiated application of charity or *ṣadaqa* as carried out through the *slametan* to prophets, saints, and ancestors. Although Woodward does mention a few details surrounding the *slametan* that have the potential to instigate further inquiry into some of the contradictions and tensions in the observance of the ritual, his analysis falls well short of investigating such tensions. For example, there is a reason why the reformist *santri* “also refer to it [*slametan*] as *seḍekah* (charity; Arabic *ṣadaqa*) and emphasize the distribution of alms.”[93] It also explains why the “*santri* often state that prayers are the most important part of the ritual while *Kejawen* Muslims stress food and blessing.”[94] It is not because “spontaneous piety,” as Woodward believes, has more religious merit than mandatory giving of *zakāt*, as interpreted in *sharī‘ah*—even though obligatory acts of worship are undoubtedly encouraged in Islam—but because of what the *slametan* for saints symbolizes, and how its food possesses spiritual efficacy not permitted the *slametan* for the ordinary dead (including ancestors—unless the ancestor is regarded as a saint). Hence, in Woodward’s own words, the “*Kejawen* Muslims feel that that the food distributed in the *slametan* is of greater importance than the *zakat.*”[95] It also goes to explain Woodward’s statement: “it is often stated that the *zakat* is a legal duty, not a pure gift. Conversely, *slametan* food is referred to as a “gift of the heart.”[96] Woodward mentions a few other differences between *santri* and the *Kejawen* Muslims without examining those differences in any significant manner. Another example is his observation that “while traditional *santri* consider the *slametan* to be important, they understand it as a supplementary source of blessing. For *Kejawen* Muslims the reverse is true.”[97]

---

93 *Ibid.*, 66.
94 *Ibid.*
95 *Ibid.*, 64.
96 *Ibid.*
97 *Ibid.*, 82. Much of these differences have been discussed in chapter 1, see the section on the work of the Barelwī scholar, ‘Allama Ṭaḥir al-Qadirī.

Summary Remarks on the Focus on Discourse

As discussed above, Werbner, Safi, and Woodward have each called for examining Sufism through the prism of the political sphere. In this, the study of texts, discourses, and narratives, along the lines of Foucault's method of inquiry, has dominated their study of ritual and its salience for the way in which the charisma of the saint is circulated, appropriated and perpetuated in the community. In Werbner's study, the creativity of the saint and the emotions of the followers are all inscribed in a cultural grid, by which the uniqueness of the saint is recognized. For Safi, the saint is an active force in the "visible" world through which his charisma becomes a mode of exchange. (For Woodward, the role of the saint and the sultan are intertwined.) Sufism in each of these studies is a political force that exerts itself in the social realm.[98] In these three studies on Sufism, in one way or another the notion of sacred exchange is visibly entrenched in the political structure. Each study emphasizes the power of discourse in understanding the dynamics of sacred exchange. The problem, I argue, is that none of these approaches pays any attention to the issues of incongruity or ambiguity that may concealed from the "visible" world of discourse? I want to conclude the present chapter by arguing that the example of ritual incongruity in Pakistan offers an alternative view to those of the aforementioned studies. I want to show how devotional practice breaks away from the textual narrative of *ziyāra* in Muslim scholarly, especially legal, thinking. This divergence has much to do with the cosmology of shrines and the relationship between scriptural textuality in the service of ideology and the embodied, experiential Sufism of the Indus region.

2.4 The Textual Biases of Recent Scholarship on Sufism

To pose the question once more, why has scholarship on Sufism treated one of the most important features of *ziyāra* as epiphenomenal to its studies? By not paying attention to a metaphysics of the presence of Sufi saints at shrines as it is celebrated by devotees, scholarship on Sufism has proven unable to account for it as *lived* religion. The fault is in the methodology employed to frame Sufism, and especially in how the role of discourse as power in the study of religion more

---

98 In other words, lived experience is shaped by the discourse and its modes of reasoning by those who subscribe to its authority. Discourse is ultimately about power and knowledge transmitted through the technology of ritual to regulate the body. Its power lies in the way individuals take it upon themselves to then discipline their bodies to achieve its intended, transcendental, aims.

generally has also dominated in the study of Sufism specifically. As a result of focusing on Islam as a textual tradition as the standard for evaluating how Muslims themselves view their religion, historical studies have come to treat Sufism more as a "literary genre" than as a "religio-historical" phenomenon.[99] In doing so, scholars have neglected to study ritual as performance and to see it as embodied action on its own terms. For example, *ḥāẓirī* in the presence of the saint means that the devotee approaches the shrine in a gesture of submission, love, care, and attention, and this bodily action brings/results in the experience of encountering the living presence of the saint. In other words, the physical gestures are laden with meaning. Meanwhile, in the reasoning of the textual tradition of *ziyāra*, the visitor should do no more than send *ṣawāb* from the recitation of *fātiḥa* for a pious soul of the saint, on account of whose piety God mediates the exchange of supplications between the living devotee and the saint in *barzakh*. In overlooking the ritual act itself as capable of carrying meaning as an embodied action, most scholars have not been able to see the actual arena of devotion, which in this case is that surrounding the immortal body of the saint. Subsequently, most scholars, in their focus on textual Islam as the realm of meaning, have closed themselves off from serious consideration of the devotional relationship between ritual and poetry. Most importantly, what has really been missed is the encounter or engagement of the mystic with the tremendously real and powerful sacred presence that has its own creative force field.

This is where anthropology and social theory, or the "sociology of sainthood," has faltered in understanding the significance of the religious or mystical world of experience.[100] In the case of Indus Sufism, an entire devotional complex of practices is associated with the metaphysics of presence at Sufi shrines. El-Zein's article published in 1977 entitled "Beyond Ideology and Theology: The Search for the Anthropology of Islam," is an invaluable contribution through its comparative discussion of folk theology and textual Islam, as well as its emphasis on the methodological obstacles involved in many academic approaches to Islam.[101] However, as valuable as el-Zein's work is for understanding folk theology as a form of knowledge and the different epistemologies within Islam in

---

99 I am borrowing here the contrast between two terms as made in William Graham, *Beyond the Written Word*, 6.

100 Note, however, that Clifford Geertz did want to overcome the division between social sciences and theology to make room for the empirical validity of religious experience, and to avoid the materialistic reductionism of the scientific study of religion.

101 Abdul Hamid el-Zein, "Beyond Ideology and Theology: The Search for the Anthropology of Islam," *Annual Review of Anthropology* 6 (1977): 227–254.

relation to the use of scripture, I think that his analysis misses the nature of the competition over the meaning of revelation, or Qur'ān, or rather, the significance of interpreting Qur'ān as an open system of signs and symbols as opposed to a closed system delimited by its own textuality. Of course, this restriction of meaning to the text itself is the work of discursive reasoning over time by the community of scholars, which has stabilized meaning for the production of commentaries and supercommentaries. This is the zone of textual knowledge. Although el-Zein does mention the need for a new understanding in view of the several examples of ritual incongruities that defy easy classification, he does not dwell on the issue or offer a plausible way forward.

If there is any thinker in the twentieth century whose writings have undermined the idea of "meaning" in its familiar sense, it is Michel Foucault, namely through his notion of "technologies of the self" as a construction of power/knowledge relations.[102] His foremost concern is with the historical conditions that allow for the formation of the self, or for that matter, any mode of existence *as* existence. While we cannot go into his complex ideas here, a scholar whose contribution to the study of religion was inspired by Foucault's work and whose work is directly relevant to our discussion of ritual, meaning, and ideology is Talal Asad. Asad, like Foucault, is interested in the "historical structures in terms of which the limits and possibilities of people's lives are realized."[103] Asad's critique of anthropology in his "The Idea of an Anthropology of Islam," strikes hard at the academic enterprise for failing to undertake an analytical study of Islam, and he comes instead to the conclusion that " 'Islam' as an analytical category dissolves as well," in view of its diversity.[104] Taking McIntyre's concept of tradition to counter el-Zein and Gellner, for whom there are different "islams," Asad points towards orthodoxy as the core organizing force: "wherever Muslims have the power to regulate, uphold, require, or adjust *correct* practices, and to condemn, exclude, undermine, or replace *incorrect* ones, there is the domain of orthodoxy."[105] This is a major scholar's statement of the central

---

102 Among many notable publications of Michel Foucault, see *The Archaeology of Knowledge* (New York: Routledge Classics, 2002; first published 1969). Cf. Sara Mills, *Discourse* (Routledge, 1997), [48]: "One of the important assertions that Michel Foucault made in *The Archaeology of Knowledge* (1972) was that discourses are not simply groupings or utterances, grouped around a theme or an issue, nor are they simple sets of utterances which emanate from a particular institutional setting, but that discourses are highly regulated groupings of utterances or statements with internal rules which are specific to discourse itself."

103 Talal Asad, *The Idea of an Anthropology of Islam* (Washington D. C.: Georgetown University, 1986), 10.

104 El-Zein, "Beyond Theology and Ideology: The Search for the Ideology of Islam," 252.

105 *The Idea of an Anthropology of Islam*, 15.

importance of the learned, textual tradition of "orthodoxy" in Islam and its centrality to any understanding of Islam, a view that the present work challenges in the case of Indus Sufism and saint veneration.

2.5 Ineffability and Authority: The Unsayable

To recapitulate briefly, a consequence of approaching Sufism as a discourse and trying to develop a theoretical understanding of the nature of religious experience through the lens of authority and power has been to overlook the significance of the Sufi shrine as a source of religious experience that is beyond the authority of Islam as a textual tradition. My argument is that the immortal body of the saint that is celebrated in the popular religious culture in the ritual of *'urs* is a mystical body that moves beyond or outside the discourse of death in orthodox Islamic eschatology. The textual tradition of Islam regarding *ziyāra* is insufficient to explain or interpret actual popular practices at work. Through its special transcendence of death, the Sufi saint's body escapes the hegemony of religious authority enshrined in the textual tradition of Islam. By all traditional standards, it is a type of immortality that is not sanctioned in orthodox Islam. The death of the Sufi saint is a resurrection of the saint's eternal body, which in accordance with orthodox belief is only granted on the day of resurrection.[106] The saint's body thus transcends eschatological time, as its resurrection is instantly granted upon death, and a primordial union is enacted in the here and now. This is a revolt against "orthodox," institutionalized history, one in which past and future dissolve in the present of festival time, which is sacred time. In its associated conviction of the sacred potency of saintly bodies, the Sufi shrine constitutes one of the most centrally important sites for mystical experience in Indus Sufism.

In discourse analysis, power is the concomitant of authoritative knowledge, by which language itself becomes an instrument of power. Just about everything, including any aspect of social life, can be studied as a form of discourse analysis, as a text. But, from the perspective of Sufism, mystical piety, by its own admission, in theory, and practice, seeks *direct* experience (Ar. *dhawq*, 'taste') beyond all texts, all discourse, as a way of experiential "verification" of divine truth.[107]

---

106 "Qiyama", *Encyclopedia of Islam.*

107 For a recent study on epistemology and other related issues, See Khalid El-Rouayheb, *Islamic Intellectual History in Seventeenth Century: Scholarly Currents in the Ottoman Empire and the Maghreb*, also an earlier article describes three different types of mysticism in Islam, see

Even in the case of ritual devotion at Sufi shrines, the visitor's relationship with his/her Sufi saint involves an aspect of what William James famously termed *ineffability.*[108] In recent years, the appeal to the "ineffable" has been severely criticized because of the way it is seen to impose a foreign category from a "Protestant and western-centered model of mysticism" on the Islamic data in various studies on Sufism. Omid Safi dismisses its descriptive value in that the category of mysticism fails to account for the political significance of Sufism in the public sphere. The assumption is that Sufism "can be recognized only when it is acknowledged socially."[109] Secondly, this approach is held to have been invalidated because of the way it privileges the private, or the personal, over the social in contradiction to the socio-historical experience of Islam.

Devotees visit shrines for *ḥāẓirī* at the *dargāhs* of saints both *to be seen* by the saints and to receive the grace of this being seen, their *naẓar-i karam.* God's friends are the ones who are truly alive, but this belief in their special immortality is not part of the legal writings on *ziyāra. Ḥāẓirī* establishes a direct relationship between the visitor and the saint for whose presence the journey is made in the first place. The absence of this integral devotional practice in textual Islam raises the question as to what this exclusion means for approaches that emphasize its centrality in the study of Sufism. The saint's living presence as the locus of sacred authority challenges the logo-centric view of Sufism. This is the imaginative leap from the authority of scripture as text to the quasi-scriptural authority of the immortal body of the saint—a leap that studies of Sufism have failed to make. The next chapter argues for understanding the reformist project as a conscious agenda for normalizing or regularizing in a more "orthodox" fashion the understanding of the entombed bodies of God's friends. Devotion to the sacred body as immortal has ever been the target of reformist polemics over *shirk.* This was and remains the main objective in the regulation of popular devotion by the ulama, and it remains the same under the present Ministry of *Awqāf* in Pakistan. The present argument for "re-centering" of the Sufi shrine is a challenge to previous scholarship on Sufism whenever it has relegated the shrine's devotional significance to the periphery.

The overall aim in the present work is to examine the ritual space at the intersection of Islamic law and popular practice in order to understand better

---

Majid Fakhry, "Three Varieties of Mysticism in Islam," *International Journal of Philosophy* vol. 2 4 (1971), 193–207.

108 William James, *The Varieties of Religious Experience: A Study of Human Nature* [esp. 342–343]

109 Omid Safi, "Bargaining with *Baraka:* Persian Sufism, "Mysticism," and Premodern Politics," *The Muslim World* 90 (2000), 264.

ziyāra to Sufi shrines (and ultimately popular Sufism as well). At one end of the spectrum, we have a formalized interpretation of *ziyāra*, seen in legal discourse over the performance of *du'ā*, but at the other end, as a lived practice, the *dargāh* is a sacred space for *ḥazirī* in the living presence of the Sufi saint. However, only the former interpretation prevails, as a "*say-able*" and therefore meaningful action in the sense of possessing a rational explanation. The latter, even when it is used in in everyday speech, lacks the equivalent discourse, or language, and is therefore better characterized as the *unsayable*. It is rendered unsayable because of the charge of *shirk* for anyone directly communicating with Sufi saints at their tombs. The devotional poetics of *ḥazirī* are embodied in the way visitors attune their bodies and enter into a communicative space with the living presence of a Sufi saint. One only has to imagine for a moment how such a belief and mode of practice could be discussed in the light of the dominant narratives of science, let alone the Muslim reformist views of Islam today.

In order to understand the issue of the *unsayable* or the *unspeakable*, which is vital for seeing the kind of politics involved in attempts at regulation of experience, and for also in order to make room for agency from below, it has been necessary in the present chapter to argue that various approaches have failed adequately to observe, and to discuss, the central most significant ritual mechanism at work in *ziyāra* to Sufi shrines in modern times. The logic of *du'ā* (in the eyes of the Sunni *'ulamā*) defines the linguistic limits of the ritual practices, and consequently the narrative employed to define *ziyāra*. As odd as it may seem, the invisibility of the poetics of *ḥazirī* is directly connected with the rationalization of the ritual of *ziyāra* to Sufi shrines.

Islamicists emphasize the textual tradition as a means through which to understand the nature of religious experience, and they are suspicious of interpretations that do not account for rituals in the context of their textual significance. Anthropologists have also given in to the textual tradition as an explanation of practices on the ground that mirrors and/or reinforces the interpretations of orthodox Islam. Vis-à-vis Islamicists who focus on texts, anthropologists are more interested in understanding how things are in practice and often breaking with established paradigms of authority—in short, how things are amidst the lived dynamics of life. In this the ethnographer listens to the narratives and stories of the people, but without much concern for how the underlying constructions of authority might be shaping things from within.[110] Nonetheless, in Muslim societies,

110 Although a large amount of literature in anthropology now deals with the issue of researcher's relationship with her or his informants, the issue of establishing trust still presents us with many challenges.

the historical and intellectual significance of textual authority, as derived from its sacred sources, does play a pivotal role, and in the end, anthropologists have too often been led to draw from theories that are external to the actors, often inevitably rehearsing earlier fallacies that marred the work of Arabists and Islamicists of the 19th century. Meanwhile, the Islamicists, in their attention to texts, and analysis of their discursive features and modes of logic, have too often overlooked how practices might be escaping or subverting their intended purposes. In our case, there is the logic of *du'ā* at Sufi shrines, which is extensively treated by Muslim scholars, and in the field of practice, the poetics of *ḥaz-irī*, which in sense represent a different theology of intercession than the one espoused by the *fuqahā'* and are thus not reflected in the textual tradition of Islam according to the *fuqahā'*, or conform to the law of Islam, as *fiqh* would have it.

The next chapter argues for going back to the human body to understand better the relationship between Sufism and Islamic reformism. In it, I will show how the issue of reform has revolved around the regulation of the physical body in trying to discipline and erase the mystical body, especially the immortal body of the saints, which has been held to be at odds with the sacred texts. Textual authority is a way to discipline the body of saints, and through it, the devotional body of devotees for whom the body of saints serves as a locus of mystical experience. In contradistinction to numerous scholarly works arguing for congruity between Sufism and Islamic law, I have presented a case for ritual incongruity. That is, I have tried to make a case here for dissonance between the textual categorization of the ritual of *ziyāra* to Sufi shrines, on one hand, and its actual devotional practice, on the other. This approach tries to show how text-centered approaches are blind to what may be taking place outside of the text, invisible to the ideology of the text. The question of silences, the unspeakable or unsayable, is also part of the everyday which our reliance on narratives may not fully reveal until we take lived practice properly into account.

Chapter Three
Ritual and the Politics of Sufism: Governing the Tombs

In mainstream Sunni view, as we discussed in Chapter 1, the ritual function of *fātiḥa* at graves is a legitimate way to transfer pious merit through the performance of *du'ā* (supplication) and, in Barelwī interpretation, to do so in return for intercessory supplication by the saint, whose friendship (*wilaya/walaya*) with God gives them the power of intercession (*wasīla* or Ar. *tawassul*).[1] In mainstream legal theory, the *du'ā* as a petition for personal and collective welfare is directed to God, not to the saint, since praying *to* a saint, in strict-constructionist Islamic legal view, would amount to *shirk* (idolatry).[2] However, seeing it only from the perspective of technical textual debate on the issue is to miss the emotive dimensions of *ziyāra*. Hundreds and thousands visit shrines in Pakistan every day because, in each instance, they believe there is a spiritual and material benefit to be gained from an embodied, proximate relationship with shrine saints. Thus, there is, on the one hand, the authority of textual Islam sanctioning the recitation of the Qur'ānic text in the ritual of visiting a shrine or tomb as orthodox ritual practice and, on the other hand, the authority of praxis, namely in the lived metaphysics of people who frequent the *dargāh* in veneration of the saint. Indus Sufism is inseparable from the metaphysics of *presence* at the center of its religious devotion and poetry. This metaphysics is based on oral folklore in which the sacred body of the Sufi saint, immortalized in the shrine, is the means of connecting human beings to God.

---

1 The Deobandīs agree with the Barelwīs over the legitimacy of the ritual in principle but disagree with its popular use, giving it a ritual status that is not in line with their interpretations. The other deeper rift between the Deobandī and Barelwī is over the ontological status of the Prophet Muhammad. This controversy erupted after Shāh Ismā'īl's tract emphasized the transcendence of God to the extent of positing the possibility (*imkān*) of God creating another person like the Prophet Muḥammad if he so desired. This was not well received by Indian Muslims. Despite their shared legal affiliation with the Ḥanafī *madhhab*, the Barelwīs detest what is seen as the latters' improper attitude toward and belief in the Prophet. See Usha Sanyal, *Devotional Islam and Politics in British India: Ahmed Riza Khan Barelwi and His Movement, 1870–1920* (Delhi: Oxford University Press, 1996).

2 An example of how the polemics of *shirk* in modern Islam are related to issues of political authority can be found in Mawdūdī's antagonism to the idea of democracy. In his emphasis on the sovereignty of God as the only abiding principle in Islam, he rejects as *shirk* any form of political system that gives precedence to the will of the people. See Fazlur Rahman, "Implementation of the Islamic Concept of State in the Pakistani Milieu 6.3 (1967): 208.

https://doi.org/10.1515/9783110781496-005

In Indus Sufi cosmology, the shrines of the saints (*'awliyā Allāh*) are emblems of divine kingship, a privilege bestowed by divine grace (on God's friends), and not accessible to worldly rulers and kings.[3] In modernity, scripturalist projects, religious reformism, and the "Islamic" state of Pakistan have each sought, by means that include strategies of selective appropriation, to suppress the lived cosmology of Sufi shrines and their "royal" authority in an attempt to preserve its respective authority.

The present chapter will trace the "orthodox" religious concerns about Sufi shrines in the 20th-century emergence of Pakistan, where we will trace how the Ministry of Awqāf has systematically served as an instrument for the suppression of shrine-based Indus Sufism. The Pakistani state took over custodianship of Sufi shrines early on to bring their ritual space under its bureaucratic control. Much has been written about the opposition of scripturalist reformers, especially the Deobandīs and the Aḥl-i-Ḥadīs, to popular Sufism, but the ritual mechanisms and the legal arguments these groups have employed (and have had endorsed by the government) for the regulation and suppression of the body have been far less well understood and dealt with specifically in scholarship.

The "Islamic" state of Pakistan, in its search for the scripturalist purity (*pāk*) of Islam and in defining its body politic as possessing a homogenous religious identity, has tried systematically to destroy the prevailing cosmology of Indus Sufi shrines and the theopoetics of the *insān-i kāmil* as the locus of divinity or God's presence.[4]

---

3 As argued earlier in chapter 2, it is common for scholarly studies to view this difference between saints and kings as only a feature of discourse in Sufism, and to reject such a distinction in reality. While the language of Sufism did become a common discursive mode in medieval Islam, most previous studies have failed to consider adequately the nature of the divide, and how despite imperial overtures, and patronage, and even attempts to be seen as saints, the kings never succeeded making their graves sites of pilgrimage. In emphasizing the distinction between the spiritual realm of the saints, and the political power of rulers in view of the ritual of *ḥāẓirī*, I differ with Nile Green's argument (in *Making Space: Sufis and Settlers in Early Modern India*); Green reads the shrine-architecture *only* as an instrument of Islamization ("space-making") by the Mughals through their extensive patronage, in continuation of the precedent set by the Sultanates. Shrines served, in Green's account, to construct and reinforce perpetually the power of the rulers through its ritual space.

4 The reassertion of scriptural authority after the loss of Muslim political power in the 19th century is also one of the causes of what is commonly described as the "fragmentation of religious authority." Muhammad Qasim Zaman has written much on this issue in the case of India and Pakistan, especially in the role of the traditional Deobandī scholars in keeping up with modernity. However, the phrase, "fragmentation of authority" has been most effectively presented by Dale F. Eickelman and James Piscatori, *Muslim Politics* (Princeton: Princeton University Press, 1996). A new intellectual class of scholars challenged the interpretive authority of the *'ulamā*,

In chapter 2, I argued that the ritual significance of *ḥāẓirī* at the *dargāh* of saints has been passed over in the theoretical approach in current scholarship that seeks to study religious experience through texts. Scholarship has largely, if not wholly, avoided any serious engagement with questions of embodiment, which has led to a lack of recognition of the epistemological significance of the Sufi shrine. *Ḥāẓirī* at the *dargāh*, and its connection to the broader theopoetics of ritual devotion, remains both properly to be described and adequately to be studied. Simply put, this omission has prevented current scholarship from grasping the sacral function and ontology of the saint's body. In other words, the special immortality accorded to the body of the *insān-i kāmil* (The Perfect(ed) Human) has been ignored. Therefore, the encounter of worshipper and saint as a site of sacred power has been overlooked in studies on Sufism. In gauging the importance of Sufi shrines in Islamic history, scholarship has been remarkably oblivious to their religious and even socio-political logic as centers for a major mode of widespread *popular* devotion ("popular" being a word that has been just about banished from the more recent study of Sufism).[5] Ritual devotion directed toward Sufi shrines has been crucial in the communication of the values of sainthood. The wisdom embodied in devotional practices (i.e., *ḥāẓirī* and

blaming them for obscuring the "authentic" meaning of the Qur'ān, for their political complicity in supporting illegitimate, monarchical regimes of power. One of the founders of so-called "Islamism", Mawlānā Mawdūdī (d. 1979), believed that only the caliphate of the "Four-Rightly-Guided" caliphs was legitimate, after which the monarchy prevailed in much of Islamic history, and with a few, although notable, exceptions, like Sirhindī and Walīullāh, the *'ulamā* as a franchise kept the Muslims in a state of submissive stupor (*Khilāfat o Mulūkiyyat* [Karachi: 1966]). For this reason, the Islamists, not unlike the modernists, have sought to recover the golden age of Islam and have wholeheartedly believed in the office of the "caliphate" as part of a divinely ordained political order of Islam, a perfect scripturalist state. Many scholars have by now pointed out the fallacy of this perspective, in that the Qur'ān proposes no political system, but the notion that there is a prefabricated model of Islamic governance has been hard to put down. In the case of Pakistan, there has been an intense power struggle between various groups over the meaning of "Islam," the question of political legitimacy and formal or workable agreement over the institutional structure of the state has never been fully resolved. It is the search for a Qur'ānic system of governance that has preoccupied the Islamists and the liberalists alike. At the same time, the traditional Sunni *'ulamā* religious authority rests in their *isnād*-based, textual knowledge of *ḥadīth* sciences, the Sunna of the Prophet.

5 Since scholars have refused to pay attention to shrine ritual, specifically in terms of *ḥāẓirī*, they have failed to conceptualize the importance of popular religion. Instead, they have concentrated on patronage as a nexus of knowledge and power to explain why shrines became popular. Thus studies such as Nile Green's *Making Space* [cited in n. 3 above], and Nathan Hofer, *The Popularization of Sufism in Ayyubid and Mamluk Egypt, 1173–1325* (Edinburgh: Edinburgh University Press, 2015) continues to explain the social function of Sufism in terms of its relationship with political power.

samāʿ) continues still to shape the religious culture of *ziyāra*. What has been missed is how *ḥāẓirī*, as a ritual devotion, gives ordinary folk direct access to the divine presence; the performative culture of Indus Sufism has not been taken into consideration.[6]

This chapter will examine specifically the attempts on the part of an influential strand of religious reformism, the 19th century *Ṭariqa-i Muḥammadiyya*, to regulate popular Sufism.[7] Ideologically, this movement is connected to the reformist tradition of Shāh Walīullāh of Delhi (1703–1762) and his son, the *muḥaddith* Shāh ʿAbdul ʿAzīz (1746–1824).[8] Their denunciation of Sufi rituals and prac-

---

6 In many ways, the entire debate in Islam over Sufism is about the regulation and control of the body. In fact, Islamic law is overwhelmingly concerned with regulating the body through ritual. The present chapter attempts to explain how this regulation takes place.

7 All the parties concerned with reform of Islam agree that the totality of *religious* knowledge must be derived from scripture. There is no natural law to discover in Ashʿarite theology, for everything in the universe, including morality and ethics, must be understood through divine revelation. In traditional Sunni orthodoxy, revelation confirms the veracity of what does or does not conform to God's law. In this the Islamists and modernists both accept that divine revelation constitutes the final word, but they also assume that this means that whatever science and rationality has confirmed must necessarily be present in the Qur'ān as well. The Islamists and modernists, who reject traditional exegesis and *fiqh* methodology, attribute the success of the "west," at least in science, to the proper exercise of reason. Their critique is that the pre-modern *ʿulamā* preferred unthinking obedience to tradition and memorization and imitation of their predecessors [*taqlīd*] to creative and systematic study of God's text. However, traditional *ʿulamā* criticize the modernists because to them they are making the Qur'ān, divine revelation, subservient to human reasoning, especially individual reasoning (*raʿy*). These modern disagreements in some ways are reminiscent of the disagreements between Asharitie and Muʿtazilite theologians in the Abbasid era. See Richard C. Martin, Mark R. Woodward, and Dwi S. Atmaja, *Defenders of Reason: Muʿtazilism from Medieval School to Modern Symbol* (London: Oneworld Publications, 1997).

8 Shāh Walīullāh is seen as the precursor of all shades of Islamic reformism in the modern period, for modernists and Islamists his call for *ijtihād*, as interpreted by them, has been crucial in arguing for their interpretive authority. The Deobandī *ʿulamā* claim genealogical decent from him, and even the Barelwī *ʿulamā* who oppose the Deobandī, accept his son Shāh ʿAbdul ʿAzīz (d. 1824) as part of their intellectual genealogy. One of the reasons for the latter's widespread fame is his attempt to synthesize Islamic intellectual traditions at the time of Mughal decline, writing on *ṭasawwuf*, philosophy, politics, ḥadīth, and *usūl al-fiqh* and producing even a Persian translation of the Qur'ān (*Fatḥ al-Raḥmān bi tarjamat al-Qur'ān*). See J.M. Baljon, *Religion and Thought of Shāh Walī Allāh Dihlawī 1703–1762* (Leiden: 1986). Walīullāh belonged to an elite family of scholars, his father, Shāh ʿAbd al-Raḥīm (d. 1718) was involved with *Fatāwā-yi ʿĀlamgīrī* on the order of Mughal emperor Awrangzīb (1658–1707). He is also known for his attempts at the reconciliation of two opposing theosophical views: *waḥdat al-wujūd* and *waḥdat al-shuhūd*. The former is associated with the name of great mystic, Ibn ʿArabī (d. 1240), and

tices as *shirk* was an effort to appropriate and reorient the popular image of Sufi saints and its metaphysics for propagating their ideology. It was a program that combined violence, scripture, and Naqshbandī mysticism, even though Sayyid Aḥmad Barelvī saw Sufism as subsuming all the traditions within his *Ṭarīqa-i Muḥammadīyya* and aiming at the political revival of Islam.[9]

3.1 The Politics of Scripture: The *Ṭarīqa-i Muḥammadīyya* in India

The *Ṭarīqa-i Muḥammadīyya*, launched in the early phase of 19th-century reform in India, mounted an outright attack on the veneration of saints at Sufi shrines. The popularity of Sufi *dargāhs* of saints was posited as the reason for the political decline because Muslims had diverged, in Shāh Ismāʿīl's view, from the pure creed of Islamic monotheism, the Oneness of God (*tawhīd*). The *Ṭarīqa-i Muḥammadīyya*, in line with Shāh Walīullāh's reformist teachings, was an active effort to prevent Muslims from *shirk* involving what these reformers believed was an illegitimate veneration of saints at their shrines.[10] Its admirers see the movement as the implementation of Shāh Walīullāh's program of reform. But in his role as a reformer, Shāh Walīullāh did not organize a political campaign; instead, he sought to offer a grand synthesis of Islamic intellectual thought, and he indeed wrote on every aspect of the religious sciences.[11] In any case, according to the *Muḥammadīyya*, this veneration is an outright violation of *tawḥīd*, and the tract *Taqwīyyat al-īmān*, authored by its chief ideologue, Shāh Ismāʿīl, is a rejection of the *dargāh* or *darbār* of saints, as in any sense representative of or pos-

the latter with Shaykh Aḥmad Sirhindī (d. 1624), known as *mujaddid alf-thānī*. In addition, scholars underscore his importance for the revival of *ḥadīth* in India.

9 Shāh Walīullāh also claimed multiple initiations so as to claim authority in all of them. This was a trend that predates the 18th century, but little has been written in any great detail on its significance. As the orders multiplied there was competition among them. For example, it is common to hear the Naqshbandī followers claim that their *ṭarīqa* begins where the journey of others ends, and the Qādirīs insist that the foot of their founder, ʿAbdul Qādir Jīlānī (1077 or 1078–1166), is on the neck of every saint.

10 In the introductory note by the translator, Shāh Ismāʿīl, *Taqwīyyat al-īmān*, trans. by Mir Shahamat Ali, *The Journal of the Royal Asiatic Society of Great Britain and Ireland*, 13 (1852): 311.

11 On his views on the shrine see J.M.S. Baljon, "Shah Wali Allah and the Dargah" in *Muslim Shrines in India: Their Character, History and Significance* (Oxford: Oxford University Press, 2003), 189–197; and *Religion and Thought of Shāh Walī Allāh Dihlawī: 1703–1762* (Leiden: E. J. Brill, 1986).

sessing the authority of divine kingship.[12] As the polemical content of *Taqwīyyat al-īmān* illustrates, the subject of *shirk* in the writings of this reformist movement involves specifically the issue of divine kingship in Islam.[13] In particular, what rattled these reformers was, from their perspective, a case of misplaced identification on the part of ordinary people of the source that truly represents God's authority on earth.[14] The most authoritative source for Shāh Ismā'īl could be none other than the divine word of God, the Qur'ān. Shāh Ismā'īl argued that the entire purpose of sending down the Qur'ān was to convey God's commands in a language and a style that is easy to understand even for ordinary, rank-and-file Muslims.[15]

---

12 *Taqwīyyat al-īmān* was translated/written in 1824; it is the first part of his Arabic work entitled *Radd al-Ishrak*, and its subject is *shirk*. The second part deals with *bida'* and became the basis of *Tadhkīr al-Ikhwān*, a commentary on the work in Urdu written in 1834–1835 by Shāh Ismā'īl's followers. While the other standard text identified with the movement, *Sirāt al-Mustaqīm*, contains the conversations of Sayyid Aḥmad, the ideas contained in *Taqwīyyat al-īmān* belong to Shāh Ismā'īl. On the basis of Nūr al-Ḥasan Rashīd Kāndhalawi's findings ["*Taqwiyat al-īmān aur Shāh Ismā'īl Shahīd ke Khilāf bar-pā Shūrish Tārīkh aur Ḥaqiqat ke Ā'īna meñ*," *al-Furqān* in installments from Ixi, no. 7 (1991) to Ixi, no. 10–12 (1994): 170–196], Marc Gaborieau dates *Radd al-Ishrāk* to 1798–99, long before Shāh Ismā'īl began to associate with Sayyid Aḥmad. The movement of *Ṭarīqa-i Muḥammadīyya* is believed to be initiated by Sayyid Aḥmad but this new dating of its main ideological texts, including the pushing back to 1818 of the translation known as *Taqwīyyat al-īmān* means that the "reformist doctrine of the movement was codified in the *Radd al-Ishrāk* and *Taqwīyat al-īmān*," 457: Marc Gaborieau, "Criticizing the Sufis: The Debate in Early-Nineteenth-Century India," in *Islamic Mysticism Contested: Thirteen Centuries of Controversies and Polemics*, ed. Frederick De Jong and Bernd Radtke (Leiden: Brill, 1999), 452–467.

13 Shāh Ismā'īl identifies three categories of *shirk:* i.e. *'Ishrāk fi'l 'ilm, 'Ishrāk fi'l tasurruf, 'Ishrāk fi'l 'ibadat.*

14 Believing that the Muslims had gone en masse off the righteous path is a contradiction of an important concept of the *'ijmā* of the community in Sunni Islam as a means for safeguarding Muslims from error. Of all the reform movements in India and Pakistan, the Aḥl-Ḥadīs alone reject the value of *'ijmā* as a viable principle of authority.

15 It was sly of Shāh Ismā'īl to argue that the Qur'ān was an open book that offered itself readily to the ordinary because the *'ulamā* mediated this meaning. A layperson was advised to follow a school of law (*taqlīd*), and the meanings of the Qur'ānic verses or *surahs* were transmitted by popular and the more distinguished preachers during their *khutbahs* in mosques. The preacher's voice from the pulpit was the primary means of transmitting the meaning of the Qur'ān; this served to identify the Qur'ān with the voice of the preachers. The *Ṭariqa-i Muḥammadiyya* used print to spread their teachings, but it was much later in the 19th century that print technology gave a new class of intellectuals the means to challenge the authority of the *'ulamā*. Modern education and the spread of literacy are directly responsible for new interpretations and commentaries on the Qur'ān. See J. M. S. Baljon, *Modern Koran Interpretation* (Leiden: Brill, 1961).

The aim of *Taqwīyyat al-īmān*, a violently polemical tract printed in 1818, was the education of the ordinary folk who visited shrines in veneration and worship of the living presence of Sufi saints. Specifically, the anthropomorphic notion of kingship, the power of a sovereign and his royal court with viziers and ministers ruling in place of an absent God, needed to be rectified. In this polemical tract, Shāh Ismā'īl reminds the Muslims of India of the centrality of scripture for understanding the meaning of God's kingship.[16] In discussing, therefore, what he takes to be a pervasive problem of idolatrous practices (*shirk*), he attempts to convince the Muslims of the "correct" nature of the relationship between human beings and God:

> A servant is but a servant. No one can put his foot out of the limits of servitude, nor can he rise beyond the rank of a servant. As he is happy, from a notion of His boundless compassion, so he is equally trembling, night and day, before His awful majesty.[17]

It is the misrecognition over time of this relationship between God, and his creatures that Shāh Ismā'īl believes has misled the Muslims into observing polytheistic rituals and customs in the manner of the Hindus. The God of which Shāh Ismā'īl wants the Muslims to be mindful is a mighty deity whose power instills fear and demands obedience.

Taqwīyyat al-īmān is based on the ontological relationship between God and his creation. Yet surprisingly, this is an aspect of the tract that has not received any consideration in existing studies on this important polemical work against Sufism.[18] Once we re-center the significance of the shrine, we begin to realize the purpose and logic of his tract's focus on the notion of kingship in detail, to prove that the anthropomorphism implied in such a designation does not apply to God. *Taqwīyat al-īmān* was written to refute the popular idea that the *dargāhs* of Sufi saints carry the authority of divine kingship, something that is evident also in the way the populace invokes the saints' names in times of both grief and joy. Shāh Ismā'īl explicitly names this invocation of saint's

---

16 The tract is replete with references to Qur'ānic *surahs* to support his viewpoints.
17 Shāh Ismā'īl, *Taqwīyyat al-īmān*, trans. by Mir Shahamat Ali, 340.
18 None of the summaries discuss at any length Shāh Ismā'īl's views on *alastū rabbikum* (in the Qur'ān, referred to as the primordial covenant, or *rūz-i alast*) or the matter of prostration to ask why that would be important in a treatise written for the common Muslim. After all, *Taqwīyyat al-īmān* was translated into Urdu by the author for wider circulation. Harlan O. Pearson [*Islamic Reform and Revival in Nineteenth-century India: The Tarīqa-i Muhammadīyah* (New Delhi: Yoda Press, 2008)] is the only author to mention it, but he does not weigh in on the theological or cosmological significance of the primordial covenant between God and the entire progeny of Adam.

names as *shirk*. As mentioned in chapter 1 above, the iconography of the shrine serves as a *nishanī* of divine kingship, and *'urs* is the ritual celebration symbolizing the ontological union or ultimate realization of *tawḥīd*. The invocation of names at the *dargāh* is the key ritual act symbolic of the metaphysics of presence (in the person of the *insān-i kāmil*) in Sufi folklore. It means that the *essence* of the relationship between God and the human being in the saint's body is between the lover and the beloved. Thus, if any ritual can be said to reflect the practical manifestation of *wujūdī* metaphysics, of ontological monism, it is the celebration of *'urs*.

Since *Taqwīyyat al-īmān* is directed at the problem of *shirk*, it was clearly essential to Shāh Ismā'īl to attack what he considered to be the locus or symbol of non-scriptural authority.[19] In popular devotion, the *dargāhs* of Sufi saints were identified as representing Divine Kingship. The transcendence of death by Sufi saints as celebrated in *'urs* bestows on them an honor that supersedes the royal status of earthly kings and even that of the religious scholars, an honor not even bestowed on martyrs who die in battle.[20] This was a problem for *Ṭarīqa-i Muḥammadīyya*, a movement that was seeking to glorify violence in their doctrine of *jihad* against the Sikhs as the ultimate spiritual path that best harmonizes the *batīn* and the *zahir*. Since the Mughal Empire had lost political vigor, the topic of kingship or the relevance of its usage by Shāh Ismā'īl only makes sense in the context of the authority of Sufi shrines. For this reason, Shāh Ismā'īl's tract attempts to negate the cosmology that allows for the existence of the potential in the human form or body to realize an intimate bond of love with God.[21]

In Shāh Ismā'īl's interpretation, the Qur'ānic account of God addressing the progeny of Adam with *alastū bi-rabbīkum* ("Am I not your Lord?" [7:172]) firmly establishes the servitude of humans to God.[22] This verse is interpreted in Sufi poetry as a primordial expression of love, but Shāh Ismā'īl only sees it as a com-

---

19 Shāh Ismā'īl, *Taqwīyyat al-īmān*, trans. by Mir Shahamat Ali, 346: "Neither should they walk round the tombs, or respect the jungle (forest) around, by abstaining from hunting, or cutting trees or grass. By performing such ceremonies, and expecting therefrom worldly or religious profit, one becomes guilty of the crime of Shirk".

20 See David Cook, *Martyrdom in Islam* (Cambridge: Cambridge University Press, 2007)

21 Joseph Norment Bell, *Love Theory in Later Ḥanbalite Islam* (Albany: State University of New York Press, 1979; Binyamin Abrahamov, *Divine Love in Islamic Mysticism: The Teachings of al-Ghazālī and al-Dabbāgh* (New York: Routledge Curzon, 2003).]

22 As will be discussed in chapter 5, Bulleh Shāh treats this primordial event in his verses as God's passionate love for his creation, which also stirs in the depths of each person's soul for God. It is on the basis of this mutual longing for each other that has been the subject of religious devotion and folklore for centuries.

mand from God to recognize His Lordship (see Chapters 4 and 5). A rather interesting component of his discussion includes the narration of the prostration of the angels before Adam, at God's command, which Shāh Ismā'īl argues is analogous to the permissibility of incest (marriage with sister) during Adam's time:

> If any say that in former days they prostrated themselves before some of the creatures, as the angels did before Adam, and Jacob before Joseph, and that, hence, if they may accord the same usage towards any of the Aulīya, or holy men, there is no matter; it is an error. Because, in the time of Adam, they married also with their sisters; on the same principle, then, such reasoners ought to marry their sisters also,—God; whatever He has ordered, His servants must acknowledge with heart and soul, without raising any objection, such as making references to old days, for, by bringing forward such reasons, one becomes an infidel.[23]

Much of *Taqwīyat al-Īmān* discusses the inappropriateness of depicting God as a king with a royal court with ministers and viziers.[24] Shāh Ismā'īl spends much time on the refutation of such an earthly representation of God and on the issue of intercession. His tract is interspersed with references to Qur'ānic verses and *surahs* that he feels deny the permissibility of such a mode of devotion. The entire *Taqwīyat al-Īmān* relies on convincing the audience that the literal word of God contradicts prevalent practices among the Muslims. Shāh Ismā'īl was trained by the traditionalist Shāh 'Abd al-Azīz and saw himself as continuing the work of Shāh Walī Allāh.[25] The reassertion of scripture by Shāh Ismā'īl needs to be understood in view of the collapse of the political authority of the Mughals. It must be borne in mind that the *'ulamā* were a class of elite families who over time had become part of the socio-political structure of the Mughal empire. Their identity depended on their administrative and juridical function as experts in Islamic law. With no power to enforce *shari'ah* under the changed political conditions, the direct appeal to scripture in recovering their authority became essential for the traditional scholars.

In India, for ordinary folk, the Sufi shrine represented the true locus of divine sovereignty. Even if the *sajjada nishīn*s were informally connected to the

---

23 Shāh Ismā'īl, *Taqwīyyat al-īmān*, trans. by Mir Shahamat Ali, 345.

24 *Ibid.*, 343: "We should not suppose that God is like earthly kings, who attend to important affairs themselves, and leave the rest to be done by their ministers, whereby people are obliged to have recourse to the latter. But the management of God's affairs is quite different. He is self-sufficient, and can attend himself to crores of insignificant things. No has any interference in his kingdoms. The smallest things, even, should, be asked of Him alone; because others can neither give small or great".

25 Shāh Ismā'īl's *Abaqat* is a complex discussion on Shāh Walī Allāh's work on Sufi metaphysics in seeking to reconcile *waḥdat al-wujūd* with Sirhindī's *waḥdat al-Shuhūd*.

previous imperial order through state patronage, the shrines were venerated because of the living presence of the saints. The ordinary folk, in a form of devotion that continues to this day, sought directly to encounter and to venerate the saints at their tombs. The ritual and performative space of the shrines as a locus of lived experience of the saints' living presence is the basis for a collective consciousness that is not encompassed by textual Islam. It sustains an oral and connected to the metaphysics of presence by which it takes on virtually scriptural status in terms of its authority. The immortal body of the saint as a focus of religious devotion was directly in conflict with the textual literalism regarding the Qur'ān in the work of Shāh Ismā'īl, and this work was a continuation of attempts at a traditional synthesis for Islamic revival since the time of Shāh Walī Allāh.[26] The rituals of lighting a lamp, tying a thread, or distributing particular kinds of sweets or food are meant to appease the saint. These ritual acts involved in devotion to the saints are performed because they are believed to listen to the ordinary worshipper. They are the wellspring of love and desire for God and manifold forms of folk art and poetry.

Shāh Ismā'īl's attempt to criticize the notion of God as the king had everything to do with the popularity of Sufi shrines as the representation of Divine sovereignty:[27]

> Although God is the King of kings, yet He is not proud like an earthly king, who from arrogance does not attend to the prayer of a distressed subject, so that the people were obliged to have recourse to his minister, and seek his patronage to obtain their points. Nay, he is very benevolent and merciful; there is no need of mediation with Him. He attends to everyone who remembers Him, whether there is any to intercede in his behalf or not. He is pure and holy-natured, and higher than all. His court is not like those of worldly rulers, to which none of the subjects can have access, so that, their only recourse being to the Amīrs and Wazīrs, they are necessarily very submissive to the latter, and hand solely on them. But God is very close to His servants, if the humblest of them be heartily devoted to Him, he will always find Him present, whether in the time of need, or otherwise; there is no curtain between Him and a servant, beyond the latter's own negligence. God is present with all.[28]

26 Shāh Walī Allāh aimed to achieve a grand synthesis, to bring about a holistic reconciliation in Islamic law and between theosophical views on Sufism. Ḥadīth and *tasawwuf* were a way to find common ground that cut across legal schools and maintain the orthodox status of Sufism. The emergence of the different schools of reformist thought in the 19th century could not preserve this synthesis once the traditional scholars lost their elite standing with the collapse of the Mughal empire.

27 See also Shāh Ismā'īl, *Taqwīyyat al-īmān*, trans. by Mir Shahamat Ali, 363.

28 *Ibid.*, 342.

Shāh Ismāʿīl's point is that Sufi shrines or holy persons as representative of a hierarchy within the Divine order is an invalid concept from a proper Islamic perspective. Since God is directly accessible to everyone, the very concept of intercession in Islam is nonsensical. In framing his point-by-point refutation of what the author of *Taqwīyyat al-īmān* believes are vast, complex, interconnected practices of *shirk*, the Qur'ānic verses serve in his tract as proof of God's transcendence of the ways people have come to think of his majestic power.[29] Shāh Ismāʿīl's motive is to show how scripture contradicts the prevalence of those practices, and that those rituals and rites cannot be taken as a means of knowledge about, let alone knowing, God.[30] People have mistaken a servant for the master, according to Shāh Ismāʿīl, which is tantamount to taking a "sweeper" for a king.[31]

---

29 *Ibid.*, 321: "Because the term Shirk does not only signify to set up equals with God, but it also embraces the performance of rites of devotion to others, and the ascription to them of certain attributes which belong to God alone, such as omnipresence, and power of controlling the universe, prostration, vows, sacrifice, and invocation in the time of need. Those who ascribe any of these attributes, or perform any of these rites to any of God's creatures, through regarding them as below the Almighty, and considering them His servants, will be verily guilty of Shirk".

30 *Ibid.*, 326:"The following instance is well illustrative of the fact. Suppose a servant or subject of a King to be guilty of theft, or of sleeping on watch, or of irregularity in attending the Darbār, or of running away from the field of battle, or of failure in the payment of government dues, or of similar crimes; he is certainly accusable before his sovereign, but it is optional with the latter either to forgive or to chastise him. But there are some crimes of a rebellious or treasonable nature: as, for instance, to address an Amīr or Vizīr, or Chaudrī, or Kānūngo, or a sweeper, or a chamār, with such titles as are peculiar to the king,–to prepare a crown or throne for him,—to call him the Vicegerent of God (Zill-i Subhāni),—to bow to him in such a manner as is observed towards his majesty only, or to fix a day of Jashn (great rejoicing), and offer him nazars; these are heinous crimes, and the king must punish the perpetrator: if he neglects to do so, he is not a wise monarch, and such kings are called shameless by the wise men. Hence, we ought to be much more fearful of the Lord of Kingdoms, the Most High-minded Monarch".

31 *Ibid.*, 348: "Idolatry is of two sorts; one is to make an image after the name of any creature, and then to worship it. In the Arabic language, it called "Sanam," or image. The other is a "Thān," or monument; *i.e.*, after the name of any of God's creatures. In the Arabic idiom, it is called "Wasan," or a monument. It includes tombs, temples, sepulchres, catacombs (Lāhid), as well as Chāri (standards), and Ta'zias, besides the 'Alam and Shaddās (standards) of Imām Kāsim, Mahdi of Pīr Dastgīr, Chabutrā, or platform of the Imām, or sitting-spots of the teachers and pīrs. All of these things are worshipped by many people, where they go and make vows and offerings. So are the monuments of martyrs, as well as colours and guns, to which they sacrifice goats, and also take oaths."

The Incarnation of the Prophet Muḥammad: Sayyid Aḥmad Barelwī.

The *Ṭarīqa-i Muḥammadīyya* sought to affect the practical implementation of a scriptural focus as a means to restore the sovereignty of God. However, there is more to how the movement implemented its political program, which arguably was central to the movement's ability to attract a sizable following among the people. In particular, the valorization of the hyper-masculine body of Sayyid Aḥmad as a mystic/prophet/warrior was put forward in the public sphere as the ideal prototype of the true believer. Ultimately, the most powerful catalyst in attracting people to the *Ṭariqa-i Muḥammadīyya* cause was the propagation of the notion that Sayyid Ahmad was in direct communication with the living spirit of the Prophet in a way that allowed him to be just like him—as I argue below. The nature or metaphysics of this "living" connection, in fact, have him as the second coming of the Prophet. It was more than just a matter of his perfection in *imitatio Muhammadi* by following the Prophet's *sunna*; Sayyid Aḥmad's body was claimed to be identical with the sacred body of the Prophet.[32] The paradox is that the model of Sayyid Aḥmad as a semi-divine figure contradicts the scriptural purity Shāh Ismā'īl advocates in *Taqwīyyat al-Īmān.* [33] Claiming to be, or being seen, as a spiritual person in communion with the Prophet was not unique to Sayyid Aḥmad during his time. It was part of the larger Naqshabandī-led cycle of reform that was sweeping across the Muslim world.[34] As

32 Yet, as pointed out above, the polemical thrust, as far as texts and sermons are concerned, of the movement, was on the superiority of adhering to the word of God, the Qur'ān. From the point of view of traditional Islam, in order to follow the Qur'ān, the *'ulamā* find the practice of the Prophet's *sunna* vital to understanding the meaning of the Qur'ān—see below the discussion on traditionalism. According to Barbara D. Metcalf, *Islamic Revivalism in British India*, 60) the well-known Muḥammad Ikrām's *Mauj-i Kauṣar* (Lahore, 1968 reprint) posits the following: "In the life of Sayyid Ahmad they saw parallels to the life of the Prophet, of whose family he was. Both were uneducated; both were subject to trances and dreams; both awoke people from forgetfulness of God's law. Sayyid Ahmad's two faithful lieutenants were likened to the Prophet's own companions: 'Abdu'l-Hayy to the quiet and dignified Abu Bakr, and Muhammad Isma'il to the brave 'Umar."

33 Harlan O'Pearson captures the paradox well, 86: "Sayyid Ahmad claimed a special sanctity received from a higher authority through his personal spiritual experiences, and at the same time he preached that God's commands expressed in scriptures were the only pure and divine source. The dilemma was that he could not claim to receive divine authority without being worshipped as divine himself." *Islamic Reform and Revival in Nineteenth-century India: Tarīqah-i-Muhammadīyah* (New Delhi: Yoda Press, 2008).

34 John O. Voll, "Hadith Scholars and Tariqahs: An Ulama Group in the 18th Century Haramayn and their Impact in the Islamic World," *Journal of Asian and African Studies* 15.3–4 (1980), pp. 270: During the eighteenth century the new style of Sufism saw litanies as directly linked

was the case in other regions, in India the *Ṭarīqa-i Muḥammadīyya* was launched as a Sufi order grounded upon direct communicationfrom the Prophet to Sayyid Aḥmad – so much so, that Sayyid Aḥmad was his very incarnation. In addition to Sayyid Aḥmad's initiation by Shāh 'Abdul Azīz (1746–8124), the son of Shāh Walīullāh, he "claimed that his real initiation was that of an *'uwaysī*, one directly initiated by the Prophet, Fāṭima, or 'Alī, and finally God Himself."[35] Here he is putting forward his own body as the perfect exemplar, indeed having the same body as that of the Prophet. Thus he is the trans-historical leader of the *Ṭarīqa-i Muḥammadiyya.*[36] Far from being just an "ordinary" believer, it was as if, as far as the cult of his following was concerned, Sayyid Aḥmad was no "ordinary" person by any standard. Far from it: he was the incarnation of Muḥammad—what can be described as a "resurrected" body of the Prophet present in a new age to help Muslims revive and renew (*tajdīd*) their religion, and also, regain once again political glory in an historical time of crisis.

According to *Muḥammadī* reformers, this crisis was created because Muslims had fallen prey to idolatry (*shirk*), developed mistaken ideas in their experience of finding God through the sacred bodies of saints and visiting their shrines for

with the Prophet, usually being defined for the founder in a vision of Muhammad. In this way, the founder could claim that his was a *tariqah muhammadiyyah.* This was reinforced by hadith studies by hadith studies, since the leader also argued that the practice of the tariqah conformed strictly and correctly to the practice of the Prophet. A tariqah muhammadiyyah also claimed a distinctive silsilah which was direct from the founder to Muhammad with no intervening links. This muhammadiyyah silsilah was usually supplemented by more traditional one and the founder claimed that the regular tariqahs with which he was affiliated were united in the muhammadiyyah spirit of his own order or movement."

35 Marc Gaborieu, "Criticizing the Sufis: The Debate in Early-Nineteenth-Century India," in *Islamic Mysticism Contested: Thirteen Centuries of Controversies and Polemics*, ed. by Frederick De Jong and Bernd Radtke (Leiden: Brill, 1999), 454. This account is written only in *Sirāt al-Mustaqīm,* a Persian text containing sayings of Sayyid Aḥmad, by Shāh Ismā'īl and 'Abd al-Ḥayy, which represents him as a saint. As Harlan Pearson (*Islamic Reform and Revival*, 106) has written: "Sirāt al-Mustaqīm represents the most complete statement of the comprehensive ideology of the movement. Written in Persian in 1818 A. D. by Muhammad Ismā'īl and 'Abd al-Hayy and based primarily on the sayings of Sayyid Ahmad Brēlwī, Sirāt al-Mustaqīm was the first known printed tract to be issued by the movement in Calcutta in 1823 while Sayyid Ahmad was performing Hajj."

36 As important it was for the movement to invoke Prophetic authority for proving the superiority of its way for gaining popular support, it is also necessary to consider the controversy generated by Shāh Ismā'īl in his declaration that God could make another Muḥammad if he wanted to, in proving the majesty and independence of God. In a religious culture that held the Prophet as the reason for the creation of this world, such a view necessarily was taken as a grave offence. The later conflict between the equally Ḥanafī Deobandī and the Barelwī scholars was over the exact ontological, status of Prophet Muḥammad in relation to God.

pilgrimage. According Shāh Ismāʿīl, how could the "dead" help the living, or have the kind of powers attributed to them?[37] Since the aim of the movement was to counter the living presence of "resurrected" saints at their tombs, nothing short of a still more powerful "resurrection" of the Prophet was needed. It may appear odd, but the logic here is that if Sayyid Aḥmad is Muḥammad *himself* in the flesh, having appeared again to the Muslims, then the claims of other Sufi saints are automatically nullified. Thus, it was equally important for the movement, if it were to succeed, to convince ordinary folk of the supernatural powers and ritually to communicate to them the charisma of Sayyid Aḥmad.[38]

As I will argue in the rest of this section, the strategy to appropriate the "metaphysics of presence" was provided much earlier by the leading Naqshbandī scholar of India in the 17th century, Aḥmad al-Fārūqī Sirhindī (1564–1624). Sirhindī attacked the *wujudī* metaphysics of Ibn ʿArabī, arguing based on his own mystical experience (*kashf*). *Kashf* was, in his view, a convincingly higher ontological state of witnessing (that of "servitude") that showed him the fallacy of believing in the experience of the "Oneness" of Existence under the influence of *waḥdat al-wujūd*.[39] His theory, accordingly, is called *waḥdat al-shuhūd*. It confirmed for him the necessity of worshipping God as a servant, as stipulated in Islamic law. Moreover, to reform Sufism, Sirhindī offered his theory of *tajdīd* and the special role assigned in it for the *mujaddid*, the "renewer" whose coming is prophesied in ḥadīth. Sirhindī's revisionist metaphysics was meant for the intellectual class, the elite scholars, and his theory (see below) provided the intellectual justification for the type of mystical experience he claimed to have, which

37 Metcalf, *Islamic Revivalism*, 58, makes this clear: "Underlying the reformist discussion was the insistence that all ceremonies with dead be understood to have the purpose of *isal-i sawab*, the 'transfer of merit' for the performance of a commendable deed to the deceased. The purpose was not to be propitiation of the dead or expectation of favors. Anything else, wrote Muhammad Isma'il—in a simile characteristically Indian—was 'as putting the crown of a king on the head of [an untouchable] Chamar.'"

38 Public spectacles of him giving *bay'a* to the people, renewing their souls and energizing their bodies, was one of the ways he heightened the effect of his presence among the community.

39 Cf. Burhan Faruqi (*The Mujaddid's Conception*, 106): "Mujaddid [Sirhindī] points out that Ibn 'Arabī is speaking at the stage of *Tajallī-i-dhātī* or vision of Being, *i.e.*, the stage at which the mystic feels that he is directly apprehending the being of God. But the mystic discovers the error involved there-in only when he outgrows that stage...Ibn 'Arabī took the world as identical with God because he did not pass beyond this stage. It was the highest stage of his mystic progress. Had he advanced further, he would have realized that God is beyond all *kasfh-o-shuhūd* or intuition and experience."

merited his being called the *mujjaddid-i Alf-i thānī.*[40] The point is that his theorization of mysticism, his system of theosophical mysticism, served to justify the grandiose nature of his claims among the scholars and literati, claims that could compete, and essentially be seen to trump, the "heretical" *wujudī* metaphysics of Sufi shrines. However, Sayyid Aḥmad was certainly not a scholar like Sirhindī; but this was used to show his uncanny similarity with the Prophet (who too is seen as having been illiterate, *ummī*). There is plenty of evidence to show that Sayyid Aḥmad claimed to be an incarnation of Prophet Muḥammad, his second coming in flesh, and was seen as such by his followers. His charisma brought him followers, who pledged allegiance to him, and followed him in into battle to achieve martyrdom; but his claims are also a source of confusion as to his exact nature or spiritual identity.[41] For example, after he died in 1831, the Patna branch of *Ṭarīqa-i Muḥammadī* described him as the Mahdī in its ideological literature.[42] The movement was committed to *jihad* to establish a pure form

40 Annemarie Schimmel, *Islam in the Indian Subcontinent*, 92: "Ahmad Sirhindi had been called *imām-i rabbānī* and *mujaddid-i alf-i thānī,* because he seemed to have arrived in India as the renovator who was to expected at the beginning of the second millennium after Muhammad." According to Burhan Ahmad Faruqi's *The Mujaddid's Conception of Tawhid* (India, Lahore: Sh. Muhammad Ashraf, 1940), p. 3: "It was Mullā 'Abdul Ḥakīm of Sialkot (d. 1067 A. H.), the most illustrious scholar of the day and the Shaikh-al-Islām of India, who was the first to apply to Shaikh Aḥmad the epithet of Mujaddid-i -Alf-i-Thānī. Indeed all the divines and mystics of eminence have acknowledged him as such. For example, Shāh Walī-Ullah and his son Shāh 'Abdul 'Azīz, among a host of others, always speak of him as Mujaddid-i-Alf-Thānī."

41 S. A. A. Rizvi, 74: "The Breakdown in Traditional Society," P. M. Holt, Ann K. S. Lambton, and Bernard Lewis (editors), *The Cambridge History of Islam* 67–96. (Cambridge: Cambridge University Press, 1977), p. 74: "On 11 January 1827 he assumed the title of *imām* and a sultan, and urging them to help him without any fear for their own thrones. Shāh Ismā'īl also wrote a treatise on the subject."

42 Ghulam Muhammad Jaffar, "Tariqah-i-Muhammadiyah Movement and the Propagation of its Ideas," *Islamic Quarterly* 43.2 (1999), pp. 138: "Similarly, Maulawi Wilayat Ali compiled a collection of forty *aḥadīth* called *Arabain fi al Mahdiyin.* This collection of *Aḥadīth* deals with the appearance of the *Mahdi.* Maulawi Wilayat Ali was an ardent believer in the coming of *Mahdi,* and in fact, considered Saiyid Ahmad as the *Mahdi.*" Pearson (*Islamic Reform and Revival*, 62) describes this in more vivid detail, "In Risāla-i Da'wāt, Wilāyat 'Alī invited Muslims to accept the teachings of Sayyid Ahmad Brēlwī. But he also introduced a new element in the doctrines of the Ṭarīqah-i Muḥammadīyah by asserting the continued earthly existence of their former leader. To judge from this tract, the Patna Muhammadi had begun to perpetuate the belief that Sayyid Ahmad had not been killed at Balakot in 1831. Indeed Wilāyat 'Alī attributed his death's reports to Satan's influence. Sayyid Ahmad did not die; God had ordered him to pray and fast in the mountains where anyone could see him 'like the bright sun illuminating the universe." Cf. Hafeez Malik "In another work in Arabic entitled *Arba'īn fil-Mahdī'īn,* Wilāyat 'Alī related the various traditions of the Prophet involving the advent of the messiah. The belief among

of Islamic government under the leadership of Sayyid Aḥmad.[43] In order to understand the reformist program of the Ṭariqa-i Muḥammadīyya fully, to comprehend its practical implementation and the strategies used to achieve its ambitions for purification of Islam and creation of a religio-political state, it is necessary to pay special attention to the image and identity or role of Sayyid Aḥmad as a special kind of *mujaddid*[44] who was an incarnation of the Prophet.[45] Naturally, if it could be established or appear to the masses that Sayyid Aḥmad

---

the Patna Muhammadi that Sayyid Ahmad Brēlwī had not been killed at Balakot developed into a conviction that he would reappear as the Imām Mahdī to lead the faithful Muhammadi to victory." In any case, Gaborieau notes that because of the political conditions, millenarian ideas were part of the religious environment; thus he points out that "Shah Rafī al-Din (1750–1818), the first translator of the Quran in Urdu, wrote the *Athar-Qiyamat*, also called *Qiyamat nama*, a pamphlet of popularization in Persian on the end of the world and the arrival of Mahdi".

43 Harlan Otto Pearson, *Islamic Reform and Revival in the Nineteenth Century India: The 'Tariqah-i Muhammadiyah'*, 51: "His name was then read as the sovereign ruler in the Friday *khutbah* (sermon) during the congregational prayers among his followers on the frontier."

44 Marc Gaborieau has rightly connected Sirhindī to Sayyid Aḥmad also in this sense: "But another more striking resemblance has not, until now, attracted attention: this conception of a return to the golden age in the middle of time is formally analogous to the conception of the second millennium *mujaddid* developed by Ahmad Sirhindi (1564–1624), see Friedman, 1971: 13–21). The main mystical affiliation of Sayyid Ahmad Barelwī was to Naqshbandiyya Mujaddidiyya founded by Sirhindī (Gaborieau 1999a): it would not be surprising that the doctrine of the local Mahdi had its origin." Furthermore, he is also correct that "studies published in India and Pakistan for sixty years carefully erase the extreme aspects of the character, especially the pretense of Mahdi. Thus the school of Sayyid Abu'l-Hasan 'Alī Nadwī (born in 1914), which nevertheless insists on the religious dimension of the character, minimizes his spectacular mystical experiences and his post-hum career of Mahdi (M. Ahmad, 1975: 40–41 and 285–286)."

45 In his comparison between Wahhabism and Mahdism, John O. Voll ("Wahhabism and Mahdism: Alternative Styles of Islamic Renewals," *Arab Studies Quarterly* 4.1/2 (1982): 115) describes two styles of renewal: "Renewal can be more message-oriented or more man-oriented.... This can provide us with an analytical spectrum for examination of the process of Islamic renewal, the endpoints of which are a scriptural-message orientation that de-emphasizes the role of the human agent and a leader-oriented reform in which the human agent plays a central role."

In terms of *emphasis*, the "scriptural-message" orientation applies to Wahhabism while the "man-oriented" orientation fits the case of Mahdism:

Wahhabism and Mahdism provide us with almost paradigmatic cases of the two styles of tajdid. Wahhabism is a clear example of the message-oriented style of movement which places little emphasis on the character of the leader. Mahdism, in contrast, stresses to a much greater extent the charismatic nature of the *tajdid* leadership and the special role of the leader in the process of religious revival" (119).

However, in the case of *Ṭariqa-i Muḥammadiyya*, both styles of renewal work in tandem and are complementary to each other.

was more than just someone in communion with the Prophet, namely the *mujaddid* and even a second coming of the Prophet (especially at a moment of crisis when Muslims had lost all political power to the British), then it would be compulsory for Muslims to follow such a divinely guided leader. [46] A spiritual authority that could be shown to be superior to that of the "resurrected" Sufi saints of the *dargāh* was needed to appeal to ordinary Muslims and wean them away from their belief in the living presence of saints at their tombs.[47]

This meant the implementation of strategies, in the valorization of his body, to show ordinary folk that Sayyid Aḥmad was indeed divinely guided and directly ordained by the Prophet in his program of religious revivalism;[48] hence the movement's name: *Ṭarīqa-i Muḥammadī.*[49] To "reform" Islam, by calling itself the "order or way of the Prophet", the movement was establishing its superiority over other Sufis who were seen simply as "friends of God" (*awliyā' Allāh*).[50] So, in

---

46 Shāh Ismā'īl, *Taqwīyyat al-īmān,* trans. by Mir Shahamat Ali, 311: Description of an account of the mystical experience of both Shāh Ismā'īl and 'Abd al-Ḥay upon praying for the first time behind Sayyid Aḥmad: "They were told to offer the first portion after him, and to complete the remainder separately, when they were both so deeply immersed in contemplating God, that it was not over till the dawn of the morn appeared".

47 It must not be forgotten that the main target of the *Ṭarīqah-i Muḥammadī* was the ordinary people, for whom their saints were alive in their tombs and able to communicate and help them. At the most basic level of translation, the reformers were trying to present the people with the following choice: would they rather follow their "saints" or follow the Prophet? In effect, framing the issue in this manner (the Prophet or the saint?) in front of the masses can yield only one answer. It is further important to note that "it is significant that Sayyid Ahmad's movement did not denounce Sufism, as did the movements of the contemporary Wahhabis in Arabia and the Fara'izi in Bengal. Indeed, his position among his followers was primarily that of Sufi *sheikh* to his disciples. To emphasize, however, the centrality of the Law, he gave initiation not only in traditional orders but in the *tariqat-i muhammadiyyah*, the way of the Prophet, which was characterized not by internal devotions but by external obedience to the Law." 57.

48 Muin-un-Din Ahmad Khan, "Ṭarīqah-i Muḥammadīyah Movement: An Analytical Study." *Islamic Studies* 6.4 (1967), 385: "In a letter, Shah 'Abd al-'Azīz (d. 1823) commended Sayyid Aḥmad to Munshī Na'im Khān as God-sent and as one of the blessed and chosen *mediums* through whom regeneration of mankind is effected by the Providence."

49 Metcalf, *Islamic Revivalism,* 57: "It is significant that Sayyid Ahmad's movement did not denounce Sufism, as did the movements of the contemporary Wahhabis in Arabia and the Fara'izi in Bengal. Indeed, his position among his followers was primarily that of Sufi *sheikh* to his disciples. To emphasize, however, the centrality of the Law, he gave initiation not only in traditional orders but in the *tariqat-i muhammadiyyah,* the way of the Prophet, which was characterized not by internal devotions but by external obedience to the Law."

50 Muin-un-Din Ahmad Khan, "Ṭarīqah-i Muḥammadīyah Movement: An Analytical Study." *Islamic Studies* 6.4 (1967): 375: "the propriety of this term was questioned as early as A. D. 1829 by an opponent, Mīr Muḥammad 'Alī of Madras, who asserted that Sayyid Aḥmad had no right to designate his own *Ṭariqah* as "the Ṭariqah of Muḥammad"." In discussing Arthur F. Buehler's

the case of *Ṭarīqa-i Muḥammadīyya*, it was not just the case of Sayyid Ahmad's receiving instructions directly from Prophet Muḥammad, but his being a new incarnation of the Prophet himself—making his *tariqa*, therefore, the most "authentic" of all existing Sufi orders, most of which had become corrupt over time through wrong beliefs and ritual practices.[51]

The aim of Shāh Ismā'īl's *Ṭaqwīyyat al-īmān* (see above) was to penetrate the consciousness of the ordinary believer to argue in favor of the supreme authority of scripture over all other forms of authority. It also sought to project Sayyid Aḥmad as a Prophet-*mujaddid* figure, a more fundamental and practical means of attracting the masses.[52] The main aim of the movement was to undercut the authority of Sufi shrines as places of pilgrimage (*ḥāẓirī*) and the rituals at the shrines by which ordinary folk took their petitions into the living presence of saints.[53] In the eyes of these reformers, the Muslim masses are mistaken in such adoration, having fallen into error due to their social interaction with Hindus over the centuries, which has sapped their faith, or *imān*. As a consequence of this perceived poor spiritual health of the Muslims, Shāh Ismā'īl's work was,

---

work, Butrus Abu-Manneh, "Transformations of the Naqshbandiyya 17th-20th Century: Introduction," *Die Welt des Islams* 43.3 (2003): 304: "For Sirhindī, the *sharī'a* is the organizing principle for all of Islamic life, and both *ṭarīqa* and *ḥaqīqa* are subservient to it. Sirhindi emphasizes the primacy of *nubuwwa* (prophethood) over *wilāya* (sainthood or closeness to God). Anybody who thinks differently has a "distorted vision"."

51 As it happened to be the case, "Sayyid Ahmad was born into a *sayyid* family recognized as descendants of the Prophet, and was respected for his family's origin and its reputation for piety and learning. As a boy he himself showed little inclination toward the family tradition of scholarly activity, but distinguished himself for his physical strength, his generosity, and a religious bent of mind," Metcalf, *Islamic Revivalism*, 53.

52 Devin DeWeese, "Intercessory Claims of Ṣūfī Communities During the 14th and 15th Centuries: 'Messianic' Legitimizing Strategies on the Spectrum of Normativity" 217–218: "The competition that encouraged such increasingly expansive claims was itself the product both of the increased 'public' participation in Ṣūfī communities, with widening social circles involved in the devotional and communal aspects of ṣūfī life, and of the increased opportunities, reflecting the interests of ruling elites needing legitimation and seeking an avenue for channeling the sentiments and support of constituencies tied to ṣūfī leaders."

53 Cf. Metcalf, *Islamic Revivalism*, 58: "He condemned the discipline of repeating the name of a saint as well as that of *tasawwur-i shaikh*. He criticized prostration at shrines and pilgrimages that entailed special dress, circumbulation, sacrificing animals, burning lights, and sanctifying water. He held that the common Fatihah ceremony, consisting of Qur'anic readings and distribution of food at shrines or graves, be performed simply, without lights and without restrictions as to the date or kind of food or condition of the participants. He condemned the consecration of ritual dishes offered as propitiation to the daughter of the Prophet and to saints, and the dedication of sacred animals to them."

for a good reason, entitled *Ṭaqwīyyat al-īmān* ("Strengthening of Faith") and written specifically to address this malaise that had befallen the Muslims.

The idea was to show that the *Ṭariqa-i Muḥammadīyya*, through the mystical "medium" or body of Sayyid Aḥmad, was in present times the *only* legitimate Sufi order that deserved to be followed, if Muslims desired salvation. It is important to note that it was not a reformist call for a "new" interpretation of Islam in the face of a changing world, but a renewed emphasis on following the *sharī'a* by observing all the rituals and practices mandated by Islamic law.[54] In other words, the supposedly mystical presence of the Prophet in the figure of Sayyid Aḥmad—his second coming—is not the result of his training or expertise in the traditional or textual sciences, yet his vision reaffirmed the authoritative status of the traditional Islam of the text-oriented *'ulamā*.[55]

What is significant in all of this is that Sayyid Aḥmad was not an expert in the ḥadīth sciences nor trained as an*'alim* like Shāh Ismā'īl. Shāh Ismā'īl had taken *bay'a* to Sayyid Aḥmad and hence was under his spiritual tutelage.[56] So the latter's knowledge, or rather connection with the Prophet was not "ḥadīth-based" as such, but his words in his *Sirāt al-Mustaqīm* were treated as equivalent to Prophetic speech (*ḥadīth*).[57] In traditional Islamic sciences, the function of the

54 In this construction, the *Tarīqa-i Muḥammadī* can be seen as a continuation of the earlier strand of Naqshbandī sufi thought promoted by figures like Sirhindī and Shāh Walī Allāh, in which the synthesis of ḥadīth and *tasawwuf* is seen as correct form of Sufism.
55 Muin-un-Din Ahmad Khan, "Ṭarīqah-i Muḥammadīyah Movement: An Analytical Study." *Islamic Studies* 6.4 (1967), 375: "This *Ṭariqah*, in fact, was the logical culmination of the spirit of reform introduced into Indo-Muslim mysticism two centuries earlier by Shaykh Aḥmad Sarhindī. Shaykh Aḥmad's *Ṭariqah-i-Mujaddidīyah* aimed at guiding the steps of mysticism back to *Sharī'-ah* and [it] stood for nothing less than taming the *leviathan* of Sufism with the loop of Sharī'ah."
56 Ghulam Muhammad Jaffar, "Social background of the Tariqah-i Muhammadiyah Movement," *Journal of the Pakistan Historical Society* XLIII.4 (1995): 365: "The first support which Sayyid Aḥmad gained was from the close relatives of Shāh 'Abd al-Azīz. Mawlawī 'Abd al-Ḥay'ī, the son-in-law of Shāh 'Abd al-Azīz, Mawlawī Muḥammad Yūsuf of Phulat, the grandson of Shāh Walī Allah's elder brother Shāh Ahl Allah, Mawlawī Wajīh al-Dīn, Ḥakīm Mughīth al-Dīn and Ḥafīz Mu'īn al-Dīn, were the first people to make *bay'at* (oath of allegiance) at the hands of Sayyid Aḥmad. The *bay'at* of these learned men was a signal for the religious leaders and the students of Shāh Walī Allah's family to accept the leadership of Sayyid Aḥmad. The members of Shāh Walī Allah's family and many religious leaders in India, who were closely connected with each other by virtue of their teacher-student, *pīr-murīd*, and matrimonial relationships, accepted the leadership of Sayyid Aḥmad."
57 Cf. Harlan Otto Pearson (*Islamic Reform and Revival*, 108–109): "Clearly the two authors of Sirāt al-Mustaqīm composed this work to establish the unchallenged authority of Sayyid Ahmad Brēlwī as sole leader of the Tarīqah-i Muhammadīyah. The work explicitly claims a prophetic and saintly nature for Sayyid Ahmad Brēlwī. In addition, much of the book is devoted to a description of Sufi beliefs and practices. For this reason it stands in contrast with the popular Urdu

ḥadīth sciences (*'ilm-i rijāl*) is seen as the most reliable chain of transmission for preserving the knowledge of the Prophet and his praxis (*sunna*)—see below. In Sunni Islam, ḥadīth-science is one of the fundamental sources for *sharī'a*, and it is essential for understanding the Qur'ān. *Ḥadīth* preserves the knowledge of the Prophet in the form of written reports, a science which forms the basis of the authority of the *'ulamā* and is central to the Sunni identity as being the *ahl-Sunna wa Jamā'a* (the people who follow the practice of the Prophet and the community [of Muslims]). It is how Sunni orthodoxy sees itself in the mirror.

On the other hand, *tasawwuf* calls for direct experiencing of the divine reality. Sufi saints, such as the 18th century Bulleh Shāh, have mocked the scholars for claiming to understand the nature and efficacy of mystical experience (of God) through their reliance on texts. The paradox in the case of Sayyid Aḥmad's claims is that while he confirmed the authoritative status of Sunni orthodoxy, in which the *'ulamā* are the "heirs of prophets," to convince the people of his lofty spiritual station he not only drew on popular models of sainthood and the "metaphysics of presence," but tried to surpass all spiritual hierarchy in claiming to be a second appearance of the Prophet Muḥammad.[58]

The Prophet Muḥammad and the Two Mims

In terms of textual sciences, or as far as metaphysical theory is concerned, the rationale for how Sayyid Aḥmad could be identified with the Prophet had

---

literature of the movement which condemns nearly all practices commonly associated with Sufism. The pure Islamic doctrine in the popular literature was based entirely on the authority of God from the Qur'ān and the authority of the Prophet from hadīth. These sources are not neglected in *Sirāt al-Mustaqīm*, but a third source of pure doctrine, Muslim saints, is the chief concern of this work. Inspired by God, the Prophet, and the wisdom of the founders of the major Sufi orders, Sayyid Ahmad Brēlwī himself proceeded to establish his own order, the Tarīqah-[i] Muhammadīyah. When Sayyid Ahmad Brēlwī was questioned by Hakīm 'Atā'ullāh of Rampur on the need for creating a new Sufi order, Sayyid Ahmad explained that the other orders were the inner spiritual nature of Islam (bātinī) while the Tarīqa-i Muhammadīyah represented the outer (zāhirī) relation to the Prophet. Just as the active Prophet was superior to the reclusive saints, Sayyid Ahmad's tarīqah reached to common Muslims by stressing outer personal behavior in accordance with belief in the words of God and example of the Prophet. At the same time, in order to reach the spiritual and intellectual elite of Indian Islam, the tarīqah encompassed and transcended the highest levels of Sufism."

58 The *Taqwīyyat al-Īmān* was directed to the ordinary people on the dangers of *shirk* and the importance of *tawḥīd. The Ṣirāṭ al-Mustaqīm* was a compilation of his discourses, *malfuzāt* style, on his theory of Sufism.

been formulated much earlier by Sirhindi in his theory of the *mujaddid.*[59] Sirhindī's prophetology had made it possible to show how the *mujaddid* in the present day and age represented the "order of the Prophet." Sirhindī's presentation of the mystical nature of the Prophet Muḥammad allowed him reconcile the *ḥadīth*-based "presence" of Muḥammad with the mystical idea of being directly in communion with the spirit of the Prophet. The Prophet was *present* (alive) to the believers, primarily through the figure of the *mujaddid*, who serves as a "Prophetic link or medium" between God and the rest of humanity.[60] For Sirhindī, as well, the Sufi shrines could not possibly serve such a function—and he strove to counter the popular belief in the "resurrected" bodies of saints and their transcendence of death upon their *wiṣāl.* Since in Sunni orthodoxy, it is held that revelation (*waḥy*) received by the Prophet is the only source of drawing close to God, and in turn the knowledge about the Prophet is contained in the ḥadīth, the claim of the mystics, and the faith of the people that the "friends of God" (*'awliyā Allāh*) receive their knowledge from God (by undergoing the same experience as the Prophet) had to be proven false.[61] This is the task that Sirhindī undertook in offering his theory of Sufism to bring it in harmony with Islamic law. [62]

---

59 Sayyid Aḥmad's mysticism was a continuation of Sirhindī's version of Sufi metaphysics described as *waḥdat al-shuhūd*. It was on the basis of his Sirhindī's metaphysics that Sayyid Aḥmad also could claim exalted mystical experiences, as long as those experiences confirmed the sanctity and inviolability of Islamic law, which meant being fully convinced of the necessity of the ritual worship of God in the capacity of a servant. It meant that there could be no possibility of union, because of nature of God's existence is entirely, and incomprehensibly, different from finite nature of human existence [i.e., between the *aṣl* and *zill*, see M. Reza Pirhbai, *Reconsidering Islam in South Asian Context* (Leiden: Brill, 2009) 59.] For our case, the most interesting feature to take note of is his redefinition of Muhammadan Reality:

60 Sirhindī's *mujaddid* served the same function as that of a living Shi'a Imām, since the Imām of the Shi'a is in "Greater Occultation," and the image of Sirhindī's *mujaddid* is quite similar to that of a living (*ḥaẓir*) Ismā'īlī Imām in the present to guide the community.

61 Cf. Burhan Ahmad Faruqi, *The Mujaddid's Conception of Tawhid* (Lahore: Sh. Muhammad Ashraf, 1940), 79: "Now it must be borne in mind that on the principles of Islamic mysticism the reliability of Kashf is to be measured by the criterion of the spiritual experience of the Prophet of God; for that was the highest and truest experience. This gives us a standard, so to say—an internal evidence of the truth or otherwise of the findings of a mystic."

62 Cf. Annemarie Schimmel, *Islam in the Indian Subcontinent* (Leiden: E. J. Brill, 1980), 93: "Ahmed Sirhindi's teachings develop logically out of his aversion to Akbar's policy of reconciliation, which is, in his eyes, anti-Islamic. This explains the frequent discussions about the 'perfections of prophethood' and the 'perfections of sainthood', and the emphasis on *qurb al-farā'iḍ*, proximity to God as obtained through legally prescribed works (which is the way of the Prophets) against *qurb an-nawāfil*, proximity reached through supererogatory works (the way of the saints). That again is connected with the problem of 'intoxication', the station of saints, and 'sec-

Sayyid Aḥmad's lack of scholarly credentials over against Sirhindī's deep traditional learning actually helped in propagating his identity as a being more like the Prophet, one whose knowledge of divine mysteries is not indebted to "books."[63] It is accurate to say, considering the geopolitics of South Asia in the 17 and 18th centuries, that Sayyid Aḥmad's "Sufism" saw itself to be in competition with mystics like Bulleh Shāh (see chapter 4 for the latter's poetry criticizing orthodox rituals of Islam). Despite the change in political circumstances with the collapse of Mughal power, Sirhindī's concerns over what he saw as false ritual practices are precisely what Sayyid Aḥmad saw himself as trying to fight and overcome. Indeed, the rituals of devotion performed in veneration of saints at Sufi tombs were the main target of both these "renewers" (*mujaddids*) of Islam.

How did Sirhindi's metaphysics combine *ḥadīth* and *tasawwuf?* Annemarie Schimmel has summarized his interpretation of Muḥammad's name and the Prophet's manifestation (second coming) as "Aḥmad":

> Sirhindi claimed that Muhammad possessed during his lifetime two individuations, the bodily-human and the spiritual-angelic, symbolized in the two *m* of his name. By the end of one millennium the bodily manifestation had completely disappeared and the first *m* was replaced by an *alif*, in Islamic letter mysticism the symbol of *ulūhiyyat*, Divinity. This *Muhammad* became *Aḥmad*. The 'perfections of Prophet-hood', which have been gradually disappearing since Muhammad's time, will reappear in persons who deserve such a blessing because they are the Prophet's heirs and followers... One can, however, be quite

---

ond sobriety', the station of the prophets. These distinctions then form the basis of Ahmad's theory of the superiority of *wahdat ash-shuhūd* over Ibn 'Arabi's *waḥdat al-wujūd:* the mystic, intoxicated, experiences unitive ecstasy and sees nothing but God; in spiritual intoxication which does not change after his 'return', he feels that 'Everything is He', *hama ūst*, if he is a poet, or tries to build up and elaborate the doctrine of Unity of Being if he is a philosopher. As for the wayfarer who follows the Prophetic example, he will reach the same station but will finally return, purified and transformed, into the world (*sair ilā'l-ashyā*); then he knows that the unitive experience is subjective, not objective: not 'Everything is He' but rather "Everything is from Him (*hama az ūzt*)—that is his credo."

63 Annemarie Schimmel, "The Books of Life-Metaphors Connected with the Book in Islamic Literatures." *The Book in the Islamic World* 85: "The Mystics had always pretended to be averse to bookishness, and many of them pretended, also, to follow the example of the "unlettered Prophet," or else know only one letter of the alphabet, that is, *alif.* But we cannot deny that they probably wrote more books than those whom they criticized for their literary and scholarly activities, so that the output of mystical writings in Arabic, Persian, Turkish, Urdu, as well as the regional languages of Indo-Pakistan, Southeast Asia, and Africa, seems to have no limits. Yet a book was only a medium which should be studied under the supervision of a master who would know what to teach the disciple and how to explain the difficulties, the inner meaning, according to time-honored and often experiential methods. That is why one finds numerous remarks, especially among Sufis, against the use of books."

> sure that his word *Muḥammad Aḥmad shud*, 'Muhammad has become Ahmad' points to his own name, Ahmad. He also regarded himself as the *qayyūm* upon the world rests, a rank that is meant by *amāna* (Sura 33/72), the trust which heaven and earth did not accept. As *qayyūm*, he is superior to the *quṭb*, the highest member of the generally accepted mystical hierarchy.[64]

The person named "Aḥmad" was thus the embodiment of the Prophet Muḥammad and the one who upheld the law of the Prophet without introducing a new *sharī'a*. As for the *mujaddid*, ḥadīth confirms the role of such a figure is to play in post-Prophetic history.[65] Tradition posits that the entry of the *mujaddid* guarantees that Islam will not stagnate.[66]

---

64 Annemarie Schimmel, *Islam in the Indian Subcontinent* (Leiden: E. J. Brill, 1980), 92.
65 Burhan Ahmad Faruqi's *The Mujadddid's Conception of Tawhid* (Lahore: Sh. Muhammad Ashraf, 1940), 1: footnote on "Mujaddid-i-Alf-i-Thanī," cf. the "Preliminary" section of the book: "The word Mujaddid can be translated as the Reformer, the Regenerator, or the Renewer. I prefer Renewer. The idea of Mujaddid has its origin in the ḥadīṯẖ: *An allāh yab'asa fī hazah al-ummat 'ala rā'is kul mā'it sunnat min yajadid lahā amar dīnahā*. "God will, on the eve of eve of every century, raise a person in this nation (Islām) who would renew the religion": Abū Dāūd (202–275 A. H.). It is maintained that many persons have accordingly been the Mujaddids of their centuries, *e.g.*, 'Umar b. 'Abdul 'Azīz (d. 101 A. H.)—First Century; Imām Shāfī Muḥammad b. Idrīs (d. 101 A. H.)—Second Century; Ibn Suraij (d. 306 A. H.)—Third Century; Imām Bāqillanī Muḥammad b. Ṭayyab (d. 403 A. H.) or Imām Bāqillānī Muḥammad b. Ṭayyab (d. 403 A. H.) or Imām Asfrāyyinī Aḥmad b. Muḥammad (d. 406 A. H.)—Fourth Century; Imām Ghazzālī (d. 505 A. H.)—Fifth Century; Imām Fakẖruddīn Rāzī (d. 606 A. H.)—Sixth Century Ibn Daqīq Al'īd Muḥammad b. 'Alī (d. 702 A.H.)—Seventh Century; Imām Bulqīnī Sirājuddīn (d. 905 A.H.)-Eighth Century; Jalāluddīn al-Suyūṭi (d. 911 A. H.)—Ninth Century, and others if the subsequent centuries. (*Cf.* AM, Vol. IV, p. 181) ['Awn-ul-Ma'būd fi Sharḥ Abū Dāūd (Ar.), by Mawlwī Shamsul Ḥaq of Patna, Anṣārī Press, Delhi, 1323 A. H.]. However, it is noteworthy that only Shaikh Aḥmad has claimed the dignity of the Mujaddid-i Alf-i-Thani for himself. Khwāja Kamāluddin Aḥsān has quoted two ḥadīthes [*aḥādīth*] in *rauḍat al-qayūmīya:* "A man will arise at the beginning of the 11th century, who will be a great light and whose name will be same as mine; (he will arise) amidst tyrant kings; thousands of men will enter Paradise through his intercession." [and] "There will be a man in my nation who will be called a 'conjoiner,' through whose intercession there will enter Paradise so-and-so." It is believed that these predictions were made about Shaikh Aḥmad. (See RQ., Part I, pp. 37–38.)."
66 Ella Landau-Tasseron, "The "Cyclical Reform": A Study of the mujaddid Tradition," *Studia Islamica* 70 (1989): 82: "Ever since the Prophet died a gap was created in the Muslim community, which could only partially and inadequately be filled. This is expressed by statement such as: "the scholars are successors to the prophets", or: "had it been possible that God would send another prophet it would have been so-and-so" (this was said about 'Umar, al-Ghazāli, al-Juwaynī, Ibn Ḥanbal). The cessation of revelation with the death of the Prophet had to be compensated for in various (deficient) ways; the *mujaddid* could be another solution to the same problem. The very terms used in the *ḥadīth: inna allāh yab'aith* "God will send", appear to point to a connec-

Sirhindī's unique conceptualization of Muḥammad's Prophethood (*nabuwwa*) worked to forge a connection between *ḥadīth* and *tasawwuf* in a way that enabled him to appropriate for himself, through the notion of the *mujaddid*, all the spiritual functions and powers that had come to be commonly associated with the Sufi saints. In particular, his primary target was the living presence of the shrines of Sufi saints, who were the objects of people's adoration and ritual devotion. The model of the *mujaddid* worked for Sirhindī, and later scholars following in his path argued for the superiority of the *Ṭariqa-i Muḥammadīyya*. In terms of traditional Islam, Sayyid Aḥmad is an example of a Naqhsbandī religious leader who, amidst the turmoil of the 18th century, was able to claim the kind of *messianic* sanctity that he did for himself in the context of Sunni orthodoxy.[67] His notion of the *mujaddid* established the superiority and privilege of the body of an*'alim*, as well as that of the "*sayyid* (descendant of the Prophet)," since the *mujaddid* belongs only, as he must according to the *ḥadīth*, to the class of the *'ulamā*.

3.2 Nationalization of Sufi Shrines in Pakistan

In the late 1950s, the state of Pakistan—then under the rule of General Ayūb Khān— passed an ordinance following the suggestion of the late Justice Javed Iqbāl—the son of philosopher-poet Muḥammad 'Allamā Iqbāl. Javed Iqbāl had earlier written a book entitled the *Ideology of Pakistan*, and it was he who oversaw the passing of *The West Pakistan Waqf Properties Ordinance* of 1959 (*All Pakistan Legal Decisions 1959*).[68] This Ordinance handed over the control and man-

---

tion with prophecy. Also by analogy one may reach the same conclusion because usually the *mujaddidūn* are supposed to come from the circles of religious scholars; since "the scholars are successors to the Prophets", the *mujaddid* too is a successor."

67 The *kashf* of people like Sirhindī, even if it raised objections by other scholars because of what they saw as being excessively grandiose, even as outright heresy, allow the orthodox scholars, particularly of the Naqshbandī variety, to claim the mantle of messianic-sainthood in their circles. Ironically, it must be kept in mind that Sirhindi's rejection of Ibn 'Arabī's metaphysics is carried out on the basis of *his kashf*, and ironically, is without *any textual basis* of the sort he claims all mystics should abide by!

68 Ewing, *Arguing Sainthood*, 70: "The act was superseded by West Pakistan Waqf Properties Ordinance of 1961 (Government of West Pakistan 1961–62) and, under Bhutto's regime, by the Auqāf (Federal Control) Act of 1976 (*All Pakistan Legal Decisions* 1976), each of which further extended the authority of the Auqāf Department."

agement of *waqf* properties, such as shrines and mosques, to the state.[69] In his book, Justice Javed Iqbāl argued that Pakistan should be an Islamic state in order to achieve the fullest realization of its creative potential as a state guided by a divinely ordained religion. The main target of this ordinance was to gain full administrative control of the Sufi shrines in Pakistan. Through this intervention, the state sought ultimately to reform the cosmology of Sufi shrines in support of its own political ideology. The main aim was the unification of Sufism and the state by establishing a relationship of *reciprocity* between the state and the Sufi shrines. As a consequence, the principal shrines of the Indus Sufi poets, Bābā Bulleh Shāh of Qasūr (1680 – 1752), Khawājā Ghulām Farīd (1845 – 1901), Sachal Sarmast (1739 – 1827), Bābā Farīd of Pakpattan (*Ganj-i Shakar*), Shāh Husaīn or Mādhū Lāl of Lahore (1539 – 1593), Sultān Bahū, and Shāh 'Abdul Laṭīf (1689 – 1752) are all currently controlled/employed/constrained under the state bureaucracy, managed by the Ministry of Awqāf.[70]

The state of Pakistan thus sought to insert the hand of its official bureaucracy between the tombs of the saint and the devotees visiting the shrine.[71] In contrast to earlier patterns of patronage in colonial and pre-modern Islamic history, Ayūb Khān's government aimed to remove the hereditary custodians of the shrines (*gaddī nishīns* or *sajjada nishīns*) as *baraka*-laden intermediaries between

69 Using *waqf* as an instrument of state policy has an older, pre-modern history in the Islamic world. Nile Green considers its use by the state as an effective vehicle to stabilize Muslim settlements in rural areas: "The main apparatus for this settlement policy was the endowment or *waqf*, by which either a sultan or imperial notable granted a Sufi shaykh sufficient property to not only support himself but also to attract a larger community of followers," 133. Nile Green's explanatory model for the diffusion and popularity of Sufism is driven by his reliance on material factors, so in his perspective, the policies and strategies of elite (rulers) are largely responsible for shaping the religious culture of Sufi shrines.

70 The Ministry of Awqāf continues to control the shrines, and has over the years undertaken extensive architectural innovations in what amounts to a destruction of its sacred landscape in order to "modernize" these spaces. Recently, two relevant studies have appeared in Urdu, one with an English translation as well. Both are by a single government officer, Ghafer Shahzad: *From Sufi Thinking to Sufi Shrine* (Lahore: Sang-e Meel, 2014), *Dātā Darbār kampleks: ta'mīr se takmīl tak* (Lāhaur: Idrāk). The studies are not official publications of the *Awqāf* but they do portray the reconstruction of shrines as a commendable accomplishment on part of the organization, even if both publications are meant to present a documentary history by an independent author.

71 For a history of the *waqf* Acts passed, and on the structure of the *Awqāf* Department in Pakistan, see Jamal Malik, "*Waqf* in Pakistan: Change in Traditional Institutions," *Die Welt des Islams* XXX (1990): 63 – 97.

the people and the shrine.[72] The spiritual or religious charisma of the *pīrs* translated into political influence over the people, a source of power that Ayūb Khān wanted to neutralize. He wanted further to re-channel the financial resources and spiritual power of the shrine in the service of the state.[73] In line with 'Allamā Iqbāl's proposal, modernists in Pakistan wanted use the legal doctrine of *ijmā'* as the basis for establishing the *religious* foundations of the state. As an interpreter of his father's ideas, Justice Javed Iqbāl argued, "Islam cannot tolerate the presence of medieval mysticism. Such mysticism robbed the Muslims of their healthy instincts and gave them only obscure thinking in return."[74] The only way forward in the modern age was to restore the sovereignty of God in the institution of the state in line with the democratic will of the people. *Ijmā'* has, in history, been the foremost tool of Sunni Islam claimed as going back to the early formation of the Sunni caliphate and the subsequent consolidation

---

72 As Katherine Ewing discusses in her book, General Ayūb wanted to modernize Islam, and in shrine Sufism the image of the *pīr* stood in the way of progress as understood by the modernists. Sufi Islam was seen as an Islam laden with superstition, fatalism, irrationalism, and even magic, all of which promoted social evils that kept Muslims in medieval darkness. The traditional *pīrī-murīdī* (master-disciple) relationship was seen by the modernists as a thoroughly corrupt institution. Muḥammad Iqbāl, however, was deeply drawn to Sufism, to Jīlī's theory of *insān al-kamīl*, and he took on the great Jalāl al-Dīn Rumī as his spiritual guide (*murshid*) in his spiritual quest. The body of Iqbal's poetry is the result of his communion with the immortal saint, Rumī (e.g., in his *Jāvīdnāme*). That notwithstanding, there was also much contradiction or ambiguity in his relationship with Sufism, as his poetry often shows. For example, he could criticize both Ḥallāj and Ibn 'Arabī on occasion. See Elizabeth Sirriyeh, "Strengthening the Soul of the Nation," in *Sufis and Anti-Sufis:The Defence, Rethinking and Rejection of Sufism in the Modern World* (Curzon,): 112–139.

73 Despite the attempts to neutralize the influence of the *gaddi nishīns*, the wealthy, landed elite linked with the shrine-complexes, in places like Multan and *Ucch-Sharīf*, continue to serve as *pīrs* to the ordinary folks on account of their hereditary claims on the saints. Moreover, many of these families of these regions are also entrenched in the state's political structure, which, in addition to their status as *pīrs*, is a source of tremendous political clout in Pakistan. For example, Syed Gilanī is a *gaddī nishīn* with considerable wealth and power at his disposal and was the prime minister during the government of the Pakistan Peoples Party (the Supreme Court later sacked him). He had also served as a cabinet member under Zia al-Ḥaq's dictatorship in the Islamization period of the 1980s. Some numerous such clans and families are connected to illustrious saints and possess both political power and wealth. For an account of the Chishtī relationship with political rulers and those among the Chistīs who avoided political contacts, see Richard Eaton, *Sufis of Bijapur, 1300–1700: Social Roles of Sufis in Medieval India* (Princeton: Princeton University Press, 1978).

74 Javed Iqbal, *Ideology of Pakistan*, 28

of the juridical authority of the *'ulamā* in the four schools of law during the 9th/ 10th centuries.[75]

Turning to consensus (*ijmā'*) as a political doctrine for transferring religious authority to the state also required educating the people in the "true" principles and rational ethos of Islam. Just as the modernists wanted to give the state a monopoly on the interpretation of scripture, in order to wrest authority from the *'ulamā,* they also sought to uproot the functional, widely accepted cosmology of Indus Sufi shrines.[76] This cosmology is represented in the *wujūdī* metaphysics of the doctrine of *insān-i kāmil,* which the religious scholars themselves had been preoccupied with regulating since the 19th century—indeed, the theoretical debates and reconciliatory efforts over the *shuhudī* and *wujudī* positions had long been underway. Studied mainly as theosophical debates among the learned, little effort has been made to reflect upon the wider, anthropological implications of these debates. Nevertheless, since Sufi shrines have been at the center of *popular* devotion, the regulation of their sacred space has been paramount in the scripturalist politics of Islamic reformism.[77] Thus the question of

75 Its legitimacy as a political concept has been disputed initially by the Shi'a, for whom divine guidance requires the presence of an infallible *Imām* at all times. Still, according to what was to emerge as the Sunni interpretation, *ijmā'* sanctified the entire community to ensure the infallibility principle. With the occultation of the *Imām,* the Shi'a (*ithna ashari*) have resorted to the consensus of the scholars, which was given further elaboration by the revolutionary leader of modern-day Iran, namely, Ayatullah Khomeini in his theory of *valiyat al-faqīh.* In short, the process of "routinization of Prophetic charisma" after his demise led to political rifts, unable to find a satisfactory way to reconcile religious authority and political power held by the Prophet. It is critical to point out that the first basis of political authority is not based on a Qur'ānic verse but predicated on *ijmā'.* As the gap between political power and religious authority widened due to historical developments, this was eventually reconciled by the jurist's doctrine of necessity. Moreover, and the most significant of all, the textual domination of the scholars has been the main vehicle in the institutionalization of Prophetic authority, which establishes not only their authority but also provides the state with legal and bureaucratic means to rule—by the time of the Ottoman empire the scholars had been fully formalized within the state structure. In any case, the overlap between textual Islam, backed by the coercive power of the state, a close connection was maintained; what is known as the *siyāsa-sharī'ah* model of governance.
76 The last time such a dispute over scripture regarding the authority of the state vis-à-vis the religious scholars took place was in the 9th century, during Caliph Ma'mūn's inquisition (*mihna*). He wanted to impose the Mu'tazila's doctrine of the created-ness of the Qur'ān as orthodoxy, since it was held that admitting to its being the uncreated word of God would make it co-eternal with God, and thus a violation of God's unity.
77 Current scholarship is highly skeptical about describing Sufism as a "popular" mode of religious devotion on account of the participation of many learned scholars in the tradition. This has meant that scholars have hardly examined the role of Sufi shrines and their ritual significance. The failure to understand the materiality of Sufi shrines as loci of sacred power has con-

divine sovereignty in the modern period has been centrally concerned with the cosmology of the human body in its ontological relationship with God. In other words, the issue of "authenticity" in Islamic reformism is, if anything, essentially about the reform of the body.

The *dargāh* of the Sufi saint offers an understanding of Divine sovereignty, especially in its connection to human agency, that differs from the interpretative constructions of the *madhab*-oriented *'ulamā* and also from those of the *ghair-muqallid* (anti-traditionalist) modernists and the Islamists. If we take into account the role of ritual as an embodied performance, the devotionalism at Sufi shrines expands the impact and range of scripture far beyond its literal form as a physical "text." Furthermore, the shrine as a sacred space affirms Ewing's work on the politics of Sufism remains one of the best accounts of how the state has sought to co-opt the shrine.[78] Ewing recounts three successive phases in the history of Pakistan: Ayub Khān's modernist phase (1958 to 1969), Zulfiqār 'Alī Bhutto's socialist phase (1971–1977), and General Zia al-Ḥaq's Islamization phase. In Ewing's view, the first two of these three leaders tried to co-opt sufism for their own purposes. Ayūb Khān, lacking democratic legitimacy, wanted to reach the people through the shrine to educate the masses. Modernization theories, economic and political, had treated the traditional sectors of society with great suspicion on account of their perceived irrationality. Bhutto, however, projected himself as the friend of the ordinary person and identified closely with shrine devotionalism. Under both Ayūb Khān and Bhutto, government leaders visibly participated in the *ulamā'*-sanctioned ritual ceremonies to gain popularity among the masses. Ewing argues that all the governments drew selectively on some aspect of Sufi doctrine to justify their policies in support of their respective sociopolitical visions. In a shared interest of sorts, the "secular governments of Ayub Khan and Bhutto chose to identify themselves with the doctrines of Sufism in order to create a link with religious authority."[79]

All three governmental regimes issued pamphlets giving historical information and biographies of the Indus saints that presented them as enlightened reformers, ordinary mortals who performed great tasks for the welfare of their communities. Ewing's conclusion is that the aim of these governmental affirmations of the Sufi doctrine of *wahdat al-wujūd* was to enact a merger: "the government

cealed the embodied iconography of shrines. The argument for "re-centering" the Sufi shrine is that it draws attention to its symbolic meaning and how it brings ordinary Muslims to experience the divine presence.

78 Katherine P. Ewing, "The Politics of Sufism," *The Journal of Asian Studies* 42, no. 2 (1983).
79 *Ibid.*, 253.

is 'immanent' in its citizens."[80] The saints were described as law-abiding, conscientious persons who through their piety converted the masses to Islam. The function of the Awqāf in all three regimes was to supervise devotional activities and to use the revenue collected from the saints' shrines (*nazaranas*) for welfare activities. The government built guesthouses, toilets, libraries and expanded enclosures for managing the rush of pilgrims (although none of these facilities have been adequately maintained). The better guesthouses in places like Shehwan were and are usually reserved for government officials and politicians. All these actions were meant to show that the state is not against the shrine, even if the *gaddi-nishīns* had been displaced. Since Ayūb and Bhutto were against the Islam of the *'ulamā*, their support or promotion of Sufism was used as a counterfoil against the latter's influence. But their interpretations of Sufism were guided by their modern views of it, which admitted admiration of Sufi poetry and the Sufi inclination to universal humanism, but were unwilling or unable to appreciate the devotional practices and cosmological views of Sufi shrine worship on its own terms.

Ayūb Khān wanted to modernize Islam, and he created institutions to support that socio-political vision.[81] The purpose of these institutions was to present a view of Islam compatible with modern, rational ideals such as those espoused by many Islamic modernists.[82] "Traditional" Islam in the eyes of Aligarh modernists like Sir Sayyid Ahmed Khān did not have the right answers for modern Islam. Such modernists argued that there was an urgent need for Muslims to undertake *ijtihād* in order to retrieve the original ethos of Islam.[83] In their view, traditional Islam had failed the Muslims because of its inflexible adherence to *taqlīd* (slavish imitation) which, the modernists claimed, prevented Muslims from modern, scientific progress. The two scholars during Ayūb Khān's rule invited to oversee this task of modernizing Islam were Ghulām Aḥmad Parvez and Fazlūr Raḥman, both of whom turned directly to scripture as the main source for their interpretations, which were critical of traditional Qur'ānic exegesis.[84] In place of traditional commentaries, the modernist scholars aimed at pro-

80 *Ibid.*, 261.

81 For a complete list and discussion on the function and aims of these institutes can be found in Abdur Rauf, "Specialized Cultural Institutions," in *Renaissance of Islamic Culture and Civilization* (Lahore: Sh. Muhammad Ashraf, 1964).

82 See Ahmed Aziz, *Islamic Modernism in India and Pakistan, 1857–1964* (Oxford: Oxford University Press, 1970).

83 J. M. S. Baljon, *The Religious Thought of Sir Sayyid Ahmed Khan*. Leiden: E.J. Brill, 1949.

84 Ghulām Aḥmad Parvez's rejection of Ḥadīth as a viable source for interpretation had earned him the wrath of the *'ulamā*, campaigning extensively against his participation in the public de-

ducing a systematic and rationalist exegesis that would provide the foundation for a social, political, and economic reform needed to reinvigorate Islam as a modern ethical force. Sufism and shrine culture were seen as a degradation of Islam, the antithesis of its rationalist ethos, and removed from scripture in spirit and practice.

The resulting creation of new institutions was a means to wrest authority away from the traditional and Islamist scholars.[85] A large number of books and periodicals were published to argue the case in the public sphere. The Central Institute of Islamic Research (CII) was set up in 1960 to carry out advanced and comprehensive research in Islamic intellectual history and religious thought.[86] It drafted syllabi and an educational program for the study of Islamic history, theology, law, mysticism, and this program was seen as one of *tajdid* (reconstruction) in line with a liberal and rational approach to all these areas of thought.[87] As for the Institute of Islamic Culture, founded earlier in Lahore in 1950, "one of [its] major objectives [was] to resolve the conflict between the scientific and the religious attitudes."[88] "Shah Wali Ullah Academy" was created in 1963 by the West Pakistan Awqāf Department with funds provided by a "lady of Sujawal."[89] All of these were state institutions that played an active role in the dissemination of Islamic publications, books and journals in Urdu and English, and all were part of a phase of literary activity intended to bridge the ideology of Pakistan and Islamic religion. These institutions continued to serve under sub-

bate on Islam. The Aḥl-i Qur'ān published their ideas in *Tulu-i Islām* published from Lahore. Fazlur Rahman, after heading the Central Institute of Islamic Research (established in 1960) at the behest of Ayub Khan and publishing his thoughts in a series of articles arguing for the reinterpretation of Islamic sacred sources, was compelled to resign and to leave the country to return to academic life in America.

85 The traditional *'ulamā* and the Islamists following Mawdūdī also published their own religious literature to counter what each group saw as a threat to their vision of Islam. Essay after essay was published in rebuttal of their opponents' arguments and to argue the authenticity of their own interpretations.

86 Abdur Rauf, "Specialized Cultural Institutions," in *Renaissance of Islamic Culture and Civilization in Pakistan* (Lahore: Sh. Muhammad Ashraf, 1964), 211. "Clause (1) of Article 197 of the first Constitution of Pakistan, which was promulgated in 1956 had provided that "the President shall set up an organization for Islamic research and instruction in advanced studies to assist in the reconstruction of Muslim society on a truly Islamic basis."

87 See Abdur Rauf, *Renaissance of Islamic Culture and Civilization in Pakistan* (Lahore: Sh. Muhammad Ashraf, 1964, for a list of the works to be included in the study of the religious sciences, 214.

88 Abdur Rauf, *Renaissance of Islamic Culture*, 226.

89 *Ibid.*, 230.

sequent regimes, each time supporting the view of the government in power, as discussed by Katherine Ewing.

Islamization under Zia al-Ḥaq took a different turn from that under his two predecessors in its relationship to Sufi shrines, even while the Ministry of Awqāf consistently retained its power over shrines under the new regime. Zia al-Haq's vision of Islam was congenial to Islamist hermeneutics of scripture and more in line with Mawdūdī's political theology of the state than with that of the traditionalists. The Deobandī *'ulamā* now became prominent with their direct involvement in providing human resources for the Afghan war against the Soviet Union. The doctrine of *jihad* entered the *madrassas* as part of their teaching curriculum, at least in *madrassas* directly receiving Deobandi funding. During the Islamization decade (1979 – 1989), the reform of Sufi shrines consisted of presenting the Indus Sufi saints as religious scholars, *'alims* trained in Islamic law, *shari'āh*-minded and unambiguously Sunni. Furthermore, unlike the previous governments, Zia al-Ḥaq's regime was not interested in giving visibility to the Sufi shrines in the state media.

The religious groups, the traditional *'ulamā*, and the Islamists complemented/ strengthened the various governmental interpretive efforts to define a socio-political vision of Islam by forming the Pakistan National Alliance.[90] General Zia al-Ḥaq, after removing Bhutto, set in motion a plan to reverse the earlier programs of liberalization and to assert the supremacy of Islamism as the state religion.[91] The agenda of Zia al-Ḥaq aimed at an implementation of *shari'āh*, which was a political shift that greatly empowered the religious groups, even if they did not agree with each other. In any case, Zia al-Ḥaq's government emphasized the idea that the original saints were themselves *'ulama*, trained religious scholars, which was compatible with the ideology of both the anti-shrine Deobandī and the pro-shrine Barelwī *'ulamā*.[92] The Sufi saints of the Indus were projected now as scholars, modeled in the image of Shāh Walī Allāh, who has remained an inspiration for all the reformist groups that first formed in the 19th century.

In succession, modernist Islam, Islamic socialism, and Islamism as political phases in the history of Pakistan have used Sufi shrines in advancing their agendas. The state through the instrument of *waqf* (pl. *awqāf*) has continued to oversee the Indus shrines. The Ministry of Awqāf has in recent years undertaken architectural renovations that have destroyed the earlier spatial intimacy of the

---

90 see on the politics of Nizām-i Mustafa, Tahir Kamran "The Pre-History of Religious Exclusionism in Contemporary Pakistan: Khatam-e-Nubuwwat 1889 – 1953," *Modern Asian Studies* 49.6 (2015), 1840 – 1874.

91 The states that identified themselves as Islamic organized to revive the idea of Pan-Islamism

92 Ewing, "Politics of Sufism," 264.

graves of saints. The sacred ecology of these shrines was hardly considered in the destruction of their physical sites in the name of modernization. In various ways, all three governmental ideologies have used Sufism in the service of their politics in tandem with their rejection of its grassroots religious culture and rituals. The embodied devotion to saints is anathema to the "scripturalism" of the modernists and the Islamists as well as the *'ulamā* (see also the discussion of the Barelwī case in Chapter 1 above).

In the forging of a relationship between Islam and the state after Partition in 1947, the issue of religious authority took on an added dimension. After the political collapse of the Mughal Empire, scriptural sources had been insufficient to provide a unity of vision that the religious scholars wanted to propagate. In doctrinal and legal disputes, as noted above, the sacred authority of the shrine emerged as a major problem for the religious scholars. Their reassertion of scripturalism came into direct conflict with the devotional culture of Sufi shrines with their veneration of the immortal bodies of saints.

3.3 Pakistan (Nation-State) and Islam

In Pakistan, ritual practices at the *dargāh* offer an insight into everyday religious devotion that needs to be considered apart from the various textual constructions of Islam. The crisis of religious authority in the modern period revolves around lack of agreement on a shared language of hermeneutics in interpreting the scriptural sources. In this effort, each intellectual position claims to be the most authentic version of Islam, the one most suitable for the modern age, and the one most in tune with the "foundational" truths of a divinely revealed religion. The *dargāh* as a sacred space has been at the center of reformist concerns because its rituals are seen a threat to the religio-political identity of Muslims in India. Even though the Barelwī *'ulamā* defend *ziyāra* in legal terms, that is, as a ritual of *fātiḥa*, yet, as I argue throughout this dissertation, the ritual of *ḥāẓirī* goes beyond this interpretation. There is a consensus among scholars on the legitimacy of reciting or offering *fātiḥa* at Sufi shrines as, in principle, a prayer for the dead. However, as I have argued above, despite a central concern with ritual and the role of intercession (*wasīla*), the legal polemics are very different from the theopoetics of lived Indus Sufism. Moreover, this distinction between textual Islam, including textual Sufism, and the theopoetics of the *dargāh* is one between two different modes of knowledge. In the metaphysics of presence, the *dargāh* connects the ordinary person with the divinity through the sacred and immortal body of the saint. The oral and aural culture of Indus Sufism involves what I have termed a metaphysics of presence actualized in the ritual practice of devotees.

After the Partition of India in 1947, and the transfer of power from the last British Viceroy, Lord Mountbatten, to the first Governor-General of Pakistan: Muḥammad 'Alī Jinnāḥ (known as *Qā'īd-i Aẓām*, "the Great Leader"), Pakistan remained part of the British Dominion for its first decade until the ratification of its first constitution in 1956. The father of the nation, Jinnāḥ, died in 1948, restricting his tangible legacy to the colossal effort of the founding of Pakistan. The practical task of forging the ideological and administrative structure if the state took place after his death. Consequently, nationalist historiography has never been able quite satisfactorily to reconcile the Islamic basis of the state with Jinnāḥ's own westernized identity.[93] Jinnāḥ's body language, manner of dress, accent, and gestures presented him, at least in public perception, as entirely comfortable in his westernized persona—only late did he begin to wear a *sherwanī* (a knee-length buttoning coat to the neck).[94] As a result, Jinnāḥ's body has been an ideological battleground—although traditional Muslim scholars have never seen in him an "Islamic" role model.[95] The arch Islamist Mawlānā Mawdudī (d. 1979) refused even to accept the religious legitimacy of Pakistan until much later. Lib-

93 Jinnāḥ, a westernized Indian lawyer, had little knowledge of Islamic law, or for that matter, of Muslim intellectual thought and history. In this he was very different from Muḥammad Iqbāl, was far more knowledgeable when it came to Islam as well as comparative philosophy and theology. Jinnāḥ's statements on Islam were based on what he saw as "universal" principles: justice, equality, freedom as an individual, and collective rights, all of which he took to be completely in accord with the social ideals of Islam. Jinnāḥ was on the same page with the modernists who argued that Islam was inherently "modern," "democratic," "scientific," "rationalist," and so forth, based on assumptions of the innate perfection of "Islam". So Jinnāḥ's positive view of Islam had to do with his belief in its intrinsic moral rectitude, which made it well-suited for present times. But it was his Western lifestyle (he was, for example, fully at home among the wealthy Bombay socialites, where he met his wife, Rutie Jinnāḥ) and political savvy that stood him in good stead in negotiating with the British and perhaps served him well in outflanking seasoned politicians like Nehru and Gandhī.

94 Hermann Kulke and Dietmar Rothermund, *History of India*, 3rd edition (New York: Routledge, 1998), 284: "At the session of the Muslim League in October 1937 Jinnah underwent a fateful metamorphosis which he also expressed outwardly by abandoning his fashionable Western suit and donning, for the [first] time, a North Indian *shewani* (long coat) complemented by the typical fur cap which soon became known as the 'Jinnah cap'. In this attire he concluded a pact with the powerful chief minister of the Panjab."

95 Since the 19th century, the split between the westernized class, educated in modern sciences with knowledge of the English language from their training in schools set up by the British for educating Indians, separated them from the *madrassa*-trained students. See Mushirul Hasan, ed., *Knowledge, Power & Politics: Educational Institutions in India* (New Delhi: Lotus Collection, 1998).

erals emphasize the speeches and statements that show him unequivocally in favor of a democratic, "secular" state, while religious groups cite his references to Islam.[96] Since their inception, the religious groups were distrustful of the western educated class, who in turn despised the traditional scholars. The difference between the Islamists and the modernists is not so much in their methodology regarding scripture as it is in their different visions of Islam. [97] Now that the Muslims had a homeland of their own, the state, as an ideological construct with coercive power, presented a new set of challenges. The main issue was how to conceptualize the relationship between religious authority, as the realm of *sharīʻah*, and the political jurisdiction of the state.

In principle, the state represents the collective will of its people as the "nation" (*qawm*), an identity inscribed on the basis of territory, a common language, and a shared culture and history.[98] The political demand for Pakistan was made in the name of Islam because it was held that Muslims are not just a religious community distinct from the Hindus but constitute a different nation.[99] Yet the Muslims could hardly agree among themselves as to what this religious identity exactly means. Religious authority had become fragmented in the 19th century as the traditional *ʻulamā* were blamed for the "backwardness" of Islamic intellectual thought. Since scriptural sources, specifically the Qur'ān, were held to be the literal word of God, reformers advocated going back to its "authentic" message. In the modern age this reassertion of scriptural authority came to undermine the textual authority of the traditional scholars in novel ways. The emergence of a new class of scholars and thinkers was unacceptable to those trained in the *isnad*-paradigm of Ḥanafī *fiqh*. Furthermore, the *ʻulamā* were divided among themselves over both epistemological and theological issues into groups such as the Barelwīs, the Deobandīs, and the Aḥl-i Ḥadīs. Intermediary groups like

96 The case for a democratic, secular state with equal rights for all its citizens, irrespective of their religious identities, is certainly much stronger by all historical accounts.

97 Muhammad Qasim Zaman, "Religious Discourse and the Public Sphere in Contemporary Pakistan," *Revue des mondes musalmans et de la Méditerranée* 123 (2008), 57: "The Islamists, too, usually prefer to locate authoritative religious norms not in any historically articulated Islamic discursive tradition but squarely in what they take to be the plain and original meaning of Islam's foundational texts. It is, however, not so much their approach to the foundational texts as their self-conscious refusal to rethink Islamic norms in conditions of modernity *and* their commitment to the public implementation of these norms that set them apart from the modernists."

98 The Deobandī *ʻalim*, Ḥasan Madanī, sided with composite nationalism because it did not to him such a position did not imperil religious identity of Muslims.

99 The two-nation theory is the ideological basis of Pakistan. Muslim separatism is traced back to Syed Aḥmad Khān (1817–1898), and Muḥammad Iqbāl (1877–1938).

the Nadwat al-ʻUlamā produced well-respected individuals, such as Shiblī Naumanī and Nadwī, but were institutionally unsuccessful in achieving a consensus in the face of so much factionalism. The modernists wished to do away with traditional methods of learning and to foster a climate of *ijtihād* to wrestle with the multitude of challenges thrown at Muslims in the modern age. But the *ʻulamā* continued to resist such thinking through their own engagement in the print sphere as well as their political activism. Thus the issue of scriptural interpretation, that is, the hermeneutics of the sacred text, was among the central religio-political disagreements in 19th-century India.[100]

If scripture was the supreme text for Muslims, then, what was to be done with the Constitution of Pakistan? While Muslims are exhorted to follow God's law, the *sharīʻah*, the Constitution was without question the work of human beings—of males to be more precise—but it was and is a necessary component in the organization of the state.[101] Pakistan inherited the British colonial administrative structure and a Western legal system that continues to inform the guidelines for the judiciary.[102] The civil service had no training in religious law, and as it was, the political case made for Pakistan was largely the work of a Western-educated class of thinkers.[103] The egalitarian values of Islam made the justification for a democratic Pakistan imaginable. However, democracy bequeaths power to the people, who ratify the legitimacy of the government through an electoral process. In opposition this, the religious groups insisted that God's law should be the source for deriving legal norms. Since God is believed to possess the ultimate authority, His divine word in scripture is the supreme source for authority. It is not the people's will that is deemed important, but the conformity of their will and actions to the dictates of God's revelation and the law derived from it.[104]

---

100 Modern Qur'ānic exegesis done by thinkers such as Mawdūdī (or earlier, Syed Aḥmad Khān) and Yusuf ʻAlī, is not accepted by the traditional *ʻulamā*, although Mawdūdī's *Tafhīm al-Qur'ān* is more easily available than the traditional commentaries even today. See J. M. S Baljon, *Modern Koran Interpretations* (Leiden: E. J. Brill, 1968).

101 One of the leading scholars of Islamic legal theory, Wael Hallaq, has recently argued that the formation of a modern Islamic state is a contradiction in terms. See *The Impossible State: Islam, Politics, and Modernity's Moral Predicament* (New York: Columbia University Press, 2012).

102 The sharīʻah courts exist but play a peripheral role in the overall legal structure of Pakistan.

103 Vali Nasr, *Islamic Leviathan*, 45: "Pakistan was from the outset ruled by a bureaucratic and military elite who had been in the service of the British Raj, in alliance with the intermediaries that had also supported British rule."

104 Muslim judges (*qadī*) stopped functioning within the legal system under the British raj. With a new political system in place, the role of the *Qadi* in the public sphere has become irrelevant.

In any case, the Constitution of 1956 set the paradigm for confrontation with and accommodation of Islam in the newly created body politic. Since Pakistan was won in the name of Islam, as a separate homeland for the subcontinent's Muslims, this recognition was taken seriously by the religious leadership. Islam meant different things to different groups, and there was a cultural gap between the western-trained leadership of the Muslim League and the religiously trained scholars. The state defined itself in Islamic terms but the religious groups were extremely suspicious of its having any role in interpreting Islam for the public. The *'ulamā* had always considered this to be their privilege and justified their political role by their claim to be the inheritors of Prophetic knowledge and tradition; consequently, they were unwilling to let the rulers pass Islamic decrees without their approval. The rulers often had pushed back on this matter of authority, tilting the balance at times, but the *siyasa-sharī'āh* equilibrium had prevailed for centuries. The *'ulamā* were gradually institutionalized through the formalization of the *madrassas* and their appointments as *Qadī*s (juri-consult) or in other official posts. Despite their dependence, the religious scholars liked to see themselves as an autonomous class alongside the political class of the rulers. This equilibrium was broken after the British dismantled the Mughals, in the aftermath of which the reassertion of scripture became a political symbol for the *'ulamā*. Scholars have been unable to find an epistemological middle ground to replace the one lost in the 19th century. Could the modern state provide the political unity, or at least a new equilibrium of sorts, for a community divided over the meaning of Islam? The Islamists advocated a capture of the power of the state, seeing it as an instrument of power useful for the implementation of Islam. Mawdūdī's interpretation of the Qur'ān was an attempt to show how the Prophet's mission was a step-by-step model for the creation of an Islamic state, yet even Mawdūdī was at first opposed to the idea of Pakistan as a territorial entity. Mawdūdī's interpretation was not acceptable to the traditional scholars, but it was equally critical for them to define Pakistan as an Islamic state. The older, 19th-century debates over the religious sphere were now to be waged in the context of the new state of Pakistan.

3.4 Sovereignty and Scripture in Pakistan

The tension between scripture, on one hand, and the Constitution of the state, on the other, has to with the issue of sovereignty.[105] Insofar power is concerned, the state requires no higher or metaphysical authority to exert its legal control over whatever falls within its territorial limits. That is to say, as a mechanism of power the state derives its sovereignty from itself, a political culmination of a historical development. Nevertheless, in the face of pressure on the Constituent Assembly by its religious factions, the preamble of the Objectives Resolution affirmed the following:

> In the Name of Allah.... Whereas sovereignty over the entire universe belongs to God Almighty alone and the authority which He has delegated to the State of Pakistan through its people for being exercised within the limit prescribed by Him is a sacred trust; This Constituent Assembly representing the people of Pakistan resolves to frame a constitution for the sovereign independent State of Pakistan: Wherein the State shall exercise its powers and authority through the chosen representatives of the people; Wherein the principles of democracy, freedom, equality, tolerance and social justice, as enunciated by Islam, shall be fully observed; Wherein the Muslims shall be enabled to order their lives in the individual and collective spheres in accord with the teachings and requirements of Islam as set out in the Holy Quran and the Sunna.[106]

Sovereignty belongs to God, and all the different major religious groups in Pakistan agree.[107] The practical application and implications of this are where disagreements surface, even to the point of bringing into question what the notion of sovereignty in Islam means in practice. So the real dispute over the compatibility of divine sovereignty with the power of the state is effectively over the idea of vicegerency—the earthly representation of God by a human. The core of all de-

---

105 There is already plenty written on the various problems faced in framing the Constitution in Pakistan.

106 "The Objectives Resolution," *Islamic Studies* 48.1 (2009): 91. The Objectives Resolution was presented to the Constituent Assembly in a speech by the Prime Minister, Liaquat Ali Khan, March 7, 1949.

107 Mawlāna Shabbir Aḥmad Usmānī (as translated by Manzooruddin Ahmed) defines sovereignty in the following way:

An Islamic state is not a state in its own right, with authority inherent in it. It is a state to which authority has been delegated. The real sovereignty belongs to God. Man is his vicegerent on earth and discharges his obligations along with other religious duties within the limits prescribed by God, thus exercising his authority within a more comprehensive Divine sovereignty. (In *Pakistan: The Emerging Islamic State* (Karachi: The Allies Book Corporation, 1966), 90)

The preceding text is cited from *Constituent Assembly of Pakistan Debates*, V (1949) (7th to 12th March, 5th session).

bates on the relationship between state and religion in Islam is about determining the ontology of the human subject. Modern-liberals, traditionalists, and modern-Islamists each turn to scriptural sources as a means to validate the legitimacy of their interpretations and to invalidate the positions held by their opponents. Reform of the thoughts and actions of the individual human body has been central for religious reformers since the 19th century, with a particular focus on both holding and acting on "correct" beliefs and the contractual duty to worship God. The project of sovereignty in the modern period has been fundamentally connected to the reform of the body in modern Islam.[108] The reassertion of scripture in the modern period is an attempt at building a consensus and, subsequently, a system of governance by *resituating* scripture as the source of regulation for socio-political activity.

3.5 The Body and Islamic Reform

The modern state wants its citizens to have a body that is rational, in order for it to be able to contribute to a productive economy. In the case of Pakistan and its creation as a modern state, Islam was seen as a rational religion that would choose to reject mysticism as a viable model for social action. The modernist position made revelation subservient to the dictates of reason (a similar debate had been waged in the early centuries of Islamic history) and gave scant room for bodily practices associated with saint veneration. For the *'ulamā,* it was paramount that the worshipper's body follow all ritual prescriptions according to Islamic law, and *tasawwuf* as an experiential mode of knowledge was only acceptable as long as it maintained its subservience to Islamic law. Of the traditionalist religious parties, the Jama'āt-i Islāmī and the Aḥl-Ḥadīs have been farthest from acceptance of religious devotion at saints' shrines. The only movement among these modern groups that has sought to define itself as a layperson's Islam is the Tablighī Jamā'at. The Tablighīs claim to be the most authentic representatives of Sufism, emphasizing love and inner piety in their practice of *sharī'āh.* Finally, however, the Tablighīs are no different from the Deobandīs in their views on veneration of shrines.[109] All these movements have actively sought to define the

---

108 Muhammad Qasim Zaman's essay traces the genealogy of the concept in Mawdudī's thought.

109 There are multiple other factions and sectarian outfits that have been directly linked with militant activities. The *'ulamā,* besides being organized as *wafaq* boards that oversee their network of institutions, centralizing authority of the *madrassas*, also have their political wings. See Yohanan Friedmann, "The Attitude of the *Jam'iyyat-i 'Ulamā'-i Hind* to the Indian National Move-

public sphere in *shar'ī* terms. Still, as much as these different movements have emphasized the necessity of following prescriptive commands of *sharī'āh* in everyday life, which involves also a conscious turn to scriptural sources above all, the impact of these movements has not diminished popular devotion to Sufi saints at their shrines, which has continued to have its own momentum. The popularity of devotion to the living presence of Sufi saints is a much larger reality that has eluded any containment by the various modern reform groups. This is evident in the great influence of the *sajjada-nishīns* as leaders of their Sufi orders and hereditary links to the saints. Yet this religious devotion directly connected to saints' shrines has been largely omitted or downplayed in historical accounts of Sufism.

In the public religious sphere represented by these scripturalist movements in the modern era, Islam has been univocally projected as universal and homogenous. Such an ideal of Islam as a single, worldwide, unified community is also affirmed in the ideology of Pakistan as an Islamic nation-state. Nationalist historiography has seized upon a narrative of an undivided, primordial Muslim community after the Arab invasion of Sindh in 711 C.E. If figures such as Shāh Walī Allāh and Sirhindī are treated as *mujaddids*, sent to renew Islam at the end of an historical cycle, the militant movement of Sayyid Aḥmad Bareilly's Barelwī *jihad* is equally honored for its zeal in the name of religion and in fighting *shirk* among the Muslims. The image of the ideal Sufi saint has consisted in a juxtaposition of the figure of the scholar, such as Walī Allāh, and the charismatic warrior epitomized by Sayyid Aḥmad Bareilly Barelwī. Another thinker, Sir Sayyid Aḥmad Khān, despite his rivalry with the traditionalist scholars, is credited for his modernism, and also honored for first conceiving the two-nation theory in the Subcontinent. As it is, the Aligarh leadership played a major role in the political movement for Pakistan, laying the groundwork for a modernist vision of Islam. The ensuing nationalist historiography has become part of the curriculum in schools, and is held up as a model for what constitutes "Sufism." The vernacular traditions and folk culture of the Indus region are selectively presented and used only to corroborate or affirm the saints' orthodox status as scholars of Qur'ān and Ḥadīth. In other words, the conflation of national ideology with an essentialist history of Islam has meant the institutional dominance of textualist Islam over the popular, largely orally expressed, Indus Sufism with its religious devotion to the sacred bodies of Sufi saints.

---

ment and the Establishment of Pakistan," *Asian and African Studies: The 'Ulamā' in Modern History* 7 (1971): 157–180.

3.6 Ambiguity and Ritual

In scholarship, the issue of an ambiguous duality in Sufism as a mode of religious knowledge in modernity —i.e., the tension between textual narratives and everyday religious practice—has been studied by the aforementioned Katherine Pratt Ewing in a fine book, *Arguing Sainthood: Modernity, Psychoanalysis, and Islam*. Ewing's work has brought the political dynamics of Sufism forward for scholarly consideration, but so far, it seems, other scholars have been unable (or chosen not) to continue her line of inquiry in any substantial manner. Based on her fieldwork in Pakistan, Ewing deploys psychoanalysis, while acknowledging its limitations, to study subjective consciousness in the context of hegemonic forces that are understood to have been unleashed through modern institutions (society/family/government/orthodoxy/etc.). She chooses to do this through investigation of the image of the *pīr*, seeing, as she puts it,

> focus on the pīr as a kind of nodal point where these political and personal processes come together. At this intersection, I observe closely the extent to which ordinary people are shaped or determined by a discourse of modernity and by the ideologies that arise out of and in reaction to this discourse.[110]

Ewing's anthropological focus is on the narratives offered in response to her inquiries about Sufism, healing (and in particular the role of the mediating *pīr*), science, superstition, and religion. Ewing also brings in the figure of the Qalandar to discuss antinomianism in Islam in its literary and social functions. In continuation of her inquiry, my work seeks to contribute an analysis from the perspective of religious studies, one in which the recitation of *fātiḥa* is highlighted as a key ideological signifier that Ewing overlooks, one that allows us to rethink the question of individual reflexivity in resisting hegemonic structures of ideology in modernity. Ewing uses the method of entering into social interaction in conversation or dialogue by asking questions. The narratives that emerge from the interaction allow Ewing to conduct her psychoanalysis because "through narratives the subject visibly works to cover over the gaps created by these slippages, thereby indicating its presence while evading capture." [111] "Slip-

---

110 Katherine P. Ewing, *Arguing Sainthood: Modernity, Psychoanalysis, and Islam* (Durham: Duke University Press, 1997): 5.

111 Ewing's work (*Arguing Sainthood*) has been an important inspiration for this project. Ewing maps her inquiry over or through the body of a Pakistani *pīr* to engage in matters of "hegemony, consciousness, and the subject" (1–37). Like her book, my study is also a reaction against the inherent pessimism expressed in the "presumption of hegemony." My aim is also to resist the

pages" in the narratives reveal contradictions and paradoxes when the subject attempts to use any ideological narratives that help explain their action or belief.[112]

Ewing's main concern is to retrieve the subject or self from the grip of structuralist theories that inhibit the possibility of subjective consciousness in religious practice.[113] The notion of ambiguity is central to Ewing's project for recovering the subject, which presents itself in "slippages" in the movement of the narrative, allowing a glimpse of a reflexive action in the gaps therein: "I thereby create a theoretical space for a desiring, experiencing subjectivity that stands at a nexus of discourses. It is this space, which enables a critically conscious center of the experience, that I demonstrate and explore in this book."[114] What does ambiguity caused by the "slippages" in narratives do? As mentioned above, Ewing writes: "my focus is on the movements of slippage, the points of ambiguity, where the subject cannot be identified."[115] *Arguing Sainthood* is dedicated to revealing such "lines of flight" from discursive structures of ideology and seeks to answer the question, what does "ideology do for its part to establish its hegemony?"

> Ideology constitutes an identity as sameness within a fixed structure of signs. The interpellated subject is the identity constituted ideology, the occupier of a subject position, an ego that seeks to express fixed, unchanging truths about self and other and, therefore, hides or obliterates the inevitable slippages of the signifier.[116]

---

discourse that views the process of modernization as essentially determining the "subjective experience of the postcolonial," 4.

112 *Ibid.*, 5: "Breaking with the tendency of many recent writers to assume a determined relationship between hegemonic discourse and consciousness, I return to an older usage of the concept of hegemony as a control over public discursive space, a phenomenon that must be distinguished from consciousness. Discourses constitute subject positions, but the experiencing is a nonunitary agent (perhaps better described as a bundle of agencies) who—in part through the experience of competing ideologies and alternative discourses—operates with a potential for critical distance from any one discourse or subject position, including a discourse of modernity."

113 Ewing, *op. cit.* "Foucault's subject is not an individual or even a particular aspect of the psyche; it is merely a property of a discourse. The subject is reduced to an effect of a nonsubjective process, and individuals are seen only to the extent that they assume subject positions within a specific historical formation," 19.

114 *Ibid.*, 5.

115 *Ibid.*, 129.

116 *Ibid.*, 103.

Narratives are part of the conversations between her and her study subjects. Their use constitutes the most crucial method for Ewing as it does not involve defining or imprisoning the "other" through texts. Ewing's interpretative analysis of language uses Freud's psychoanalytic practice by "following 'transferences' of desires and fear as they were experienced in significant relationships with others in the past into the immediate relationship constituted by our conversation."[117] This also includes the author's own "countertransferences" in the relationships developed since she knows how her presence changes the field. Still, Ewing uses this interaction as a source for examining the "conflicts of the subject."[118]

The overarching framework for Ewing in investigating the interplay between ideology and the self is that of a tradition-modernity dichotomy. She aims to reveal the fractures or ambiguities in everyday practice, mainly how self-representations are anchored in prevalent ideologies. Ewing provides many examples of persons struggling to present coherent narratives in conscious relation to dominant ideologies and identities, mainly to defend the authenticity of their actions. I am not quite sure if her model goes far enough to resist postmodern theories that have decentered the self to the point that such theories allow no space for agency or reflexivity, but this is its purpose.

I want to focus on one of her examples to move to the ritual issue at the heart of the present study. It is also my aim here to move the issue of ambiguity beyond the "tradition-modernity" framework, which undeniably plays an important ideological role. As far as Sufism is concerned, issues of epistemology and ontology are easily discernable in the use of language. I agree with Ewing that the issue of ambiguity in the relationship between Islamic law and Sufism has been glossed over in current scholarship because of the structuralist theory of social action and history.[119] Embodied practices as ritual performances have received far less attention in the study of religious experience. The issue of ambiguity is better seen in terms of contestations over the metaphysics of the body in the construction of knowledge. This applies to our modern constructions as well, such as the impact of Cartesian thought on science and analytical philosophy in framing the empirical relationship between language and experience. Unfortunately, studies on Sufism rarely attempt a more holistic and encompassing meth-

117 *Ibid.*, 165.

118 *Ibid.*

119 *Ibid.*, 19. "Hegemony defined in terms of an experience of arbitrary signs as a naturalized, everyday reality is a theory of perception and consciousness. It presumes that there is a direct, causal relationship between a system of signs and the consciousness or experience of the individual, and that reality can be experienced only through a grid of signs."

od of investigation that enters into dialogue with Sufism as an epistemology of experience.

The example I have chosen is Ewing's discussion of "Zahida and Data Ganj Baksh," an encounter in which Zahida visits the shrine seeking the saint's intercession. Ewing identifies in Zahida's conversation the "ideological stance [in] her linking of the signifier "*shirk*" to the practice of visiting pīrs.[120] However, Ewing also observes that "though she does not actually use the word in the above story, the signifier organizes much of what she said."[121] The ritual of visiting Dātā Sahib works for Zahida in her experience and in the experience of others who undertook similar visits. Yet Zahida also responds to accusations that people visiting shrines are involved in worship (*puja*) in the same manner as Hindus (and therefore *shirk*). According to Ewing:

> She [Zahida] maintained the position that it is wrong to pray at shrines, an attitude that she was taught early in childhood. But her experience forced her to adopt a position that it is proper to visit the shrines of certain highly respected saints, if one is careful not to pray to the saint as an independent agent who will grant one's request without God's help (what she called "doing pūjā"). In her formulation, she partially justified visiting the shrine by equating the visit with the kind of visit one would pay to the grave of one's parents or other deceased elders.[122]

In addition to recognizing the narrative of *shirk* as having to do with the modern, reformist critique of shrines, Ewing also notes perceptively that the worshipper cannot explain the efficacy of visiting shrines as an experience. What explains such a disjuncture? To Ewing, it is because "her image of the saint is of a powerfully significant other whose presence she has experienced in ways that defy explanation within the framework of a modern reformist ideology."[123] In this, I fully concur with Ewing's explanation in her identification of modern reformist ideology:

> Ideologically systematized reform movements have striven to make Islam unambiguous. They stress that there may be no mediators between God and humans, and that all appeals for help must be made directly to God through prayer. By redefining the limits of proper Islam, reformists have severely narrowed the range of acceptable practice and belief. Such an unambiguous stance is difficult for many Muslims to maintain. New cultural forms are continually being generated in an effort to integrate at least the most blatant inconsistencies. But even in the absence of new cultural forms that serve as effective synthe-

120 *Ibid.*, 123.
121 *Ibid.*
122 *Ibid.*, 124.
123 *Ibid.*, 125.

> ses, individuals have remarkable capacity to cope with radical contradictions, in part by keeping only one frame of reference in mind at any particular moment. As a result of such strategies, even those individuals whose Islam appears to be fully objectified, expressible in systematic ideology, also live other discourses —not repressed but contextually delimited.[124]

However, Ewing misses the key ideological signifier that helps bridge the gap between its (the shrine worship experience's) *unsayable* dimension and the *sayable* narrative. Zahida reports that she visits the shrine to recite *fātiḥa*, which helps her preserve her ritual as an Islamic one that keeps her from *shirk*. The question of why the subject cannot explain the ritual action in a rationally coherent narrative is indicative of an epistemological tension in Islamic life. It is a tension to which studies on Sufism, as highlighted in chapter 2, have failed to give serious attention because of their focus on textual Islam.

The issue of language here does not just concern a historically specific phase of dominance but is related to the tension that has ancient roots. I am arguing that the notion of ambiguity should be understood in terms of epistemology.[125] Discursive narratives are insufficient to show how things are being negotiated in practice, in the fluidity of social interactions that encompass language and the movement of the body, as well as rituals and gestures. Theologically, *shirk* is the worst offense a Muslim can give because it challenges the monotheistic core of Islam. But the matter is not as literal as it seems or is made to appear in political use. The success of the reformist ideology lies in its successful interpretation of *shirk* as applicable to the visiting of Sufi shrines. The normalization of the polemics of *shirk* may have to do with a new turn in political terminology in the modern period that has facilitated such a development. The reassertion of scripture as a text to anchor political thought is what Geertz described as "scripturalism." Or, to put it another way, this reassertion gave an ideological edge to existing dominant processes through the introduction of print technology and, most importantly, the political formation of the nation-state as one based on religious orthodoxy. The resulting tension has involved a quasi-theocratic state directly in competition with the influence of the metaphysics of the shrine. It is a reassertion of religious orthodoxy with a new set of political tools and conditions

124 *Ibid.*, 126.

125 One of the qualities of Katherine Ewing's engaging work is that it is an example of dialogical anthropology in scholarship, in the serious juxtaposition of her training in psychoanalytical theories in acknowledgment of Sufi knowledge about the self: "But the problem of the constitution of the subject has also occupied scholars in the Muslim world over the centuries, particularly within the intellectual tradition now most commonly glossed as "Sufism" 35.

that also undermine its authority as a religious tradition—a grave threat to one form of ancient, perennial knowledge or cosmology that centers on the anthropocentric sacrality of the human body. This has led to the creation of political theologies that are intolerant of any plurality in religious practice.

This chapter has argued that one of the major reasons for reforming the religious culture of Sufi shrines has been the concern both of traditional religious scholars and the modern state that the ritual devotion practiced at Sufi shrines poses a challenge to the concept of divine sovereignty so central to Islam. In popular devotion, the *dargāh* as the royal court of God is represented in the immortal body of the Sufi saint. The popularity of this kind of devotion, which sees the Sufi shine as a center of pilgrimage, has stood in the way of the *'ulamā*, the modernists, and the Islamists who have sought to reconstruct the myth of political authority on a scriptural basis. The paradox for the Islamists, more than any other reformist group, has lain in their attempts to transcend historical time through a reenactment of Prophetic time in the modern age, a model they want to derive from their textual interpretation of scripture. In textual Islam, the sovereignty belongs to God. It is embodied in the scriptural texts, the interpretation of which rests with religious scholars who serve as intermediaries in defining the relationship between religious law and political power.

Yet the popularity of Sufi shrines has consistently outstripped the authority of the scholars, emerging as a force or swell from below. The popularity is not just the result of the work of Sufi orders, as has been conventionally posited in studies emphasizing the textual tradition as the source for defining authority, but because the living presence of Sufi saints has functioned as a crucial form of religious experience for the ordinary Muslim. The *'ulamā* have had to emulate the popular saintly models to enhance their authority as religious scholars. For all the rhetoric of the reformist movements about the egalitarianism of Islam as a basis for attacking the intercessory power of saints at Sufi shrines, the irony is that the veneration of shrine saints has offered the ordinary believer the most direct means of experiencing the presence of God. The saint's body removes the abstraction of encountering God through the mediation of the "text" and makes it immediately accessible to ordinary folk. And along with this experience, the vernacular verse of the *kāfī* is even experienced as scripture—as a dimension of oral culture (see Chapter 6).

Once we consider the power embodied in worship at the Sufi shrine as a mode of religious experience, we can understand the nature of the competition over authority in Islam. The immortality of Sufi saints is celebrated in the iconography of their shrines, and the ritual of *'urs* is a telling basis for the competition these sacred spaces offered to the hegemony of textual definitions of religious authority (thereby frequently becoming sites that attracted the patronage of rulers). Only by re-centering the Sufi shrine in our narratives can we begin to understand the issue of the ambiguity of Sufism as a means to experiential knowledge in the larger context of the dominance of textual Islam. Indus Sufism provides a concrete example of oral and aural, as well as bodily-experienced, popular religious culture as a parallel but alternative stream of religious knowledge to that of scriptural, textual Islam.

Just as scripture became an important political symbol for the *jihad* movement of the Ṭariqa-i Muḥammadiyya in early 19th-century India, the short history of the state of Pakistan has seen the power of emphasis on the centrality of Qur'ān and Ḥadīth. Even if there was little agreement among the different schools of interpretation of Islam, the sovereignty of the Islamic state has been widely seen as ultimately resting with God. The different movements and schools of thought inevitably clashed in state-building, and Pakistani leaders failed to achieve even a semblance of consensus among the different parties. Conceptually, sovereignty, as understood in the state's ideology, has been acceptable to all the Pakistani religious groups discussed above. Meanwhile, at the mundane, everyday level, the saints' shrines have commanded (and continue to command) the devotion of the people whose visit to their sacred threshold is a source of meaning and hope entirely apart from the ideology of state or religious establishment. Therefore, the perspective from below offers a different view of sovereignty that can be juxtaposed to its ideological imposition from above by religious or political authorities.

Chapter Four
Ritual and Bulleh Shāh

The eighteenth-century mystic, Mīr 'Abdullāh Shāh Qādirī-Shattarī of Qasūr, popularly known as Bulleh Shāh.[1] is considered to be one of the most defiant (and popular) of Sufi saints in the Indus region of Pakistan because of his critique of religious ritual. Bulleh Shāh's verses have a lyrical buoyancy and vernacular economy, and, along with his satire of religious ritualism, they present an interesting foil, even a reproach, to the politics of *fātiḥa-khwānī* in Pakistan. As such, the *'ishq* of Bulleh Shāh is above any ritual, but any act can become a ritual if done in love. This creativity is evident in his verses, which are recited so soulfully by folk singers across the Indus Valley region. In the verses below, the lover seeks ways to obtain her love by seduction, trickery, and even magic:

aik ṭonā achanbā gāwān gī
meiñ ṛuṭhā yār manāwāñ gī
aieh ṭūnā meñ paṛh paṛh phūkāñ, sūraj agan jalāwāñ gī
akhaeñ kājal, kāley bādil, bhawāñ sey āndhī leyāwāñ gī
sat sumandar dil dey andar, dil sey lehr auṭhāwāñ gī
bijlī ho kar chamak ḍarāwāñ, bādil ho garjāwāñ gī
ishq angīṭhī, harmal tārey chānd sey kafan banāwāñ gī
lāmakān kī paṛī aūpar bah kar nād bajāwāñ gī
lāey sawāñ men shawah gul apney, tad meiñ nār kahāwāñ gī
aik ṭonā achanbā gāwāñ gī
meiñ ṛuṭhā yār manāwāñ gī

In the voice of a desperate woman whose lover is under influence of another, Bulleh Shāh says that she will cast a special type of magic spell, which will immediately appease her angry lover:

> I will cast the spell again, and again, in the fire taken from the sun itself, turning the *kajal* in my eyes into black clouds, stirring a hurricane with my eyelashes. I have seven seas in my heart, from my heart I will raise a wave; I will flash like lightening and fall like thunder from the clouds. I will turn the rays of stars and the moon into my conspirators and sit upon the stool of *lāmakān* and sing. When monsoon comes, I will sleep next to my lover in a tight embrace, only then should I be called a woman.[2]

---

1 Scholars are unsure of the exact dates of Bulleh Shāh's birth or death. The dates given here are generally accepted to be correct approximations. According to one early source, he died in 1767–68 C.E.

2 Translation is mine. For Christopher Shackle's translation, see *Bullhe Shah: Sufi Lyrics* (Cambridge: Harvard University Press, 2015), 145.

https://doi.org/10.1515/9783110781496-006

In fact, the mystical body of Bulleh Shāh, and the theopoetics of the songs sung by his devotees at his tomb, is the antithesis of the disciplining of the body that orthodoxy wants to effect through the scriptural instrumentalization of *fātiḥa*. Bulleh Shāh's tomb is a place of great power that places him in the illustrious ranks of those who have realized union (*wisāl*) with their Lord and thus been made truly *alive*.

As argued in Chapter 1, the popular imagination is very much beholden to the ideas of love (*'ishq*) found in the biographies of great Sufis. If any single thread may be said to run through Sufi folklore in Pakistan, it is the topos of divine love. The vernacular notion of *'ishq* in the verses of Indus Valley Sufi poets has not been adequately explored. Ideally, *ziyāra* to Sufi shrines is rooted in this relationship of love and understood as visiting a beloved Sufi who is not dead but alive in God. The mysticism of the Punjabi Sufi, Bulleh Shāh, is the antithesis of prescriptive, legal Islam and represents the strain of Sufism that reformist Islam seeks to limit and regulate. The creative possibilities in the expression of devotional poetics are seen as threats to the normative laws of Islam.

Vernacular Sufism in Punjabī, Saraikī, and Sindhī cultures of the Indus Valley is inseparable from the shrine culture of the region. The shrines of Bābā Farīd Ganj-i Shakar, Bābā Bulleh Shāh, Khwāja Ghulām Farīd, Shāh Ḥusayn, Abdul Latīf Bhittai, Sachal Sarmast, and Sultān Bahū continue to be centers of *ziyāra* for their devotees. Amongst other representations, their shrines are symbols of Divine Love, eternal union realized by the Sufi upon death. This aspect of the Sufi saint being experienced as alive or immortal is reflected in the way Sufi verses and devotional life are generally pursued and experienced in the shrine culture of Indus Valley religious life.

Significantly, the verses of the Sufis are not described as poetry (*sha'irī*), a term that is reserved for non-religious poetry, but as *kalām*—or more emphatically as *zinda kalām* (living word). The significance of such a designation seems to have been generally overlooked in the study of Sufi verse. Sufi verses are imbued with a certain corporeality that inspires ardor (*josh*) and ecstasy (*wajd*) in an attuned listener, one who hears the (*zinda*) *kalām* with not just the ear, but the entire body, in a kind of corporeal perception.[3] In popular Sufism it is not uncommon to understand Sufi *kalām* as a source of nourishment for the body, as a spiritual sustenance, and to describe it as a form of *'ibādat*—worship.[4]

---

3 See M. Merleau-Ponty, *Phenomenology of Perception*, translated by Colin Smith (New Jersey: The Humanities Press, 1962).

4 This is particularly true of the Chishtīyya order in India for whom the *samā'* (in the form of *qawwalī*) has been central to its ritual identity as a means for reaching God. On the history of the Chishtīyya *silsila*, see Carl E. Ernst and Bruce Lawrence, *Sufi Martyrs of Love: Chishti Sufism in*

Along with the popular belief in the saint's access to divine truth, the verses of the Sufi saint are also seen as a source of guidance and religious contemplation. People visit the saint to receive guidance and help in dealing with their material lives, and the saint's verses are essential to this. Engagement with the verses sung at the shrine creates a special bond that surpasses purely literary admiration of the Sufi verse to become a more intimate and existential tie to the saint. Thus, the Sufi verses may be seen as commentary (*tafsīr*) on revelation, but even more as *standing in place* of divine revelation, becoming an extension of divine revelation itself (see chapter 6).

Just as all Sufis who realize union with the eternal One upon their death have transcended death, and just as the Sufis' devotion of love (*'ishq*) brings them intimacy (*qūrbat*) with God (symbolized in the classical Sufi literary metaphor of a moth flinging itself into the candle's flame), the visitor to a shrine similarly need only bring love in the heart as an offering.[5] Sufism is essentially about taking creative liberties, which are only learned once a person is smitten by love, entering a region where legal prescriptions become irrelevant or at least secondary.

Studies that have tried to probe Bulleh Shāh's religious identity have left out any mention of his devotional and ritual significance in the devotional practices at his shrine in Qasūr. Yet these devotional poetics are essential to grasping his hagiographic identity and, in particular, both the reason why Bulleh Shāh is seen as an antinomian saint and also the *meaning* of this perception of him for understanding his mysticism and the metaphysics of presence for worshippers at his shrine.

Should Bulleh Shāh's emphasis on religious devotion or *'ishq* be taken as a rejection of orthodox rituals? Or is his critique of prescribed rituals, as Mustansir Mir argues, directed against performing them without the correct inner disposition? Is Bulleh Shāh's verses against ritual practices essentially a critique of hypocrisy that results from not performing the prescribed rituals in accordance with their true spirit? On the one hand, Bulleh Shāh can be seen as rejecting rituals in many verses of his poetry—it is hard to offer any other interpretation, but one also finds a strong emphasis in other verses on the need for cultivating authentic piety.

South Asia and Beyond (New York: Palgrave Macmillan, 2002). Also, See Uzma Rahman, "Spiritual Power and 'Threshold' Identities: The Mazārs of Sayyid Pīr Waris Shāh and Shāh Abdul Latīf Bhitāi," Clinton Bennett and Charles M. Ramsey, South Asian Sufis: *Devotion, Deviation, and Destiny* (London: Bloomsbury, 2012).

5 For an excellent discussion on the variety of metaphors used in Sufism, see Annemarie Schimmel, *Mystical Dimensions of Islam* (Chapel Hill: The University of North Carolina Press, 1975).

The argument in this chapter is that we can better understand this paradox in the context of his image among the devotees who make the pilgrimage to perform the ritual act of *ḥaẓirī* at his shrine in Qasūr. According to oral tradition, Bulleh Shāh's identity is particularly subversive of orthodox (Islamic) rituals because his devotion to Shāh 'Inayat made him dance (*thaiyā thaiyā*) like a courtesan. Scholars have not studied the meaning of this dance with regard to its ritual, performative, significance, or context. Through dance, Bulleh Shāh found his Lord (*rabb*), or to be accurate, the *finding* of his *rabb* made him dance. As such, this ritual feature is inextricable from his hagiographic identity. Bulleh Shāh was a Sayyid (descendant of the Prophet) who embraced a *kanjar* identity (as one of a lowly caste of street singers, performers, and musicians), and to this day, devotees can be seen wearing *ghungru*—a musical ornament worn by classical (Indian) dancers around the ankles to mark the rhythmic movement of the feet. How should we interpret or understand the ritual significance of this dancing? Reaching the Divine in such a manner involves sacralizing the dance (as performed by Bulleh Shāh). The danger for Islamic orthodoxy is that if dance also is seen as a way to the Lord, then such a belief undermines the exclusive sanctity of prescriptive ritual actions (such as *ṣalat* etc.) in Islam. If other routes or paths to reach God are available, then this changes the reliance on formal, "orthoprax" rituals (as the *only* way) to find God. In other words, it would validate dance as a legitimate form of worship (*'ibāda*). Bulleh Shāh's example also grants other seekers the creative license to find their unique ways to the Divine. Since Bulleh Shāh's shrine is a sacred center for pilgrimage, dance and music are forms of the visitors' expression of devotion to the saint. It is in this religious context that Bulleh Shāh's critique of prescriptive Islam (especially its formal rituals) begins to make better sense, and the reason for attempts at suppression of shrine-worship practices by the state becomes clearer.

Bulleh Shāh's emphasis on learning about the divine through *experiential* knowledge positions him over against (Sunni) religious orthodoxy. Indeed, Bulleh Shāh's critique of institutional (orthodox) forms of ritual is applicable not only to Muslim, but also Hindu or Sikh practice/worship. Ultimately, Bulleh Shāh's mysticism is an exhortation for each person to recognize the Divine in the human self and awaken to its mystery. In this awakening, love and sincerity are all needed to worship and experience closeness to the Divine.

The presence of both Islamic and Hindu (Vedanta) vocabulary in Bulleh Shāh's poetic corpus illustrates the effective *equivalence* between these two religious traditions from his perspective.[6] There is also synchronicity between the devotional culture of Sufi shrines, with their appeal to some Hindus and Sikhs, and Bulleh Shāh's poetry. As scholars have noted, Hindus, Sikhs, and Muslims visit Sufi shrines in India, where there is a steady *confluence* of different religious communities, evidenced in shared vocabulary and devotional aesthetics. Despite their differences, these communities come together because of the metaphysics of presence (*ḥāzirī* at the shrine of the saint)—communion with the Divine, and this is evident at Bulleh Shah's shrine.[7]

The reason for Bulleh Shāh's "portability" is the popular belief among Hindus, Muslims, and Sikhs alike in him as an embodiment of the Divine. It is easy to see him as a follower of either Vedantic or theosophic Sufi (*waḥdat al-wujūd*) metaphysics because of Bulleh Shāh's emphasis on the self as the locus of Divine manifestation, or theophany since both Vedanta and Sufism emphasize the presence of the Divine in the human. Furthermore, Bulleh Shāh's focus on the metaphysics of presence is finally about having *direct* experience beyond categorization, beyond rational classifications of thought that obscure true insight into the nature of reality. Based on mystical experience of the Divine, then, Bulleh Shāh is neither a Muslim, nor a Hindu or a Sikh: "*Bullehā, kīh janāñ meiñ kaon:*"

> Bullha, what do I know about who I am?
> I am not a believer in the mosques, not do I follow rites
> of unbelief. I am not among the pure or the polluted.
> I am not Moses or Pharoah. I am not in the Vedas or in the scriptures;
> I am not in drugs or in liquor. I am not among the drunken
> reprobates. I am not in waking, nor am I in sleep.
> I am not in joy or in sadness, nor am I in polluted or
> Purity. I am not of water or of earth, nor am I fire

---

6 See S. A. A. Rizvi, *A History of Sufism in India. Early Sufism and Its History in India to AD 1600, Vol. 1* (Lahore: Suhail Academy, 2004), 322–396.

7 See Ali S. Asani, "Sufi Poetry in the Folk Tradition of Indo-Pakistan," *Religion & Literature* 20.1 (1988): 81–94. For a more recent work on the significance of poetry in South Asian Islam see Nosheen Ali, "From Hallaj to Heer: Poetic Knowledge and the Muslim Tradition," *Journal of Narrative Politics* 3.1 (2016): 12–13: "Love as a poetic theme has of course been present in the Middle East and South Asia for a long time. But as a declaration of bold resistance againt power and religious orthodoxy, the tenth century of Hallajian Love signified a turning moment. Spiritual struggles centered on the reclamation of love and defiance or orthodox clerical religion had also begun to emerge in brahminical India around the same time, led by Bhakti-poet-saints in Tamil Nadu and later in 12th century Karnataka. This dissenting, devotional poetic culture continued to emphasize the primacy of love and oneness, and spread across the landscape of India encompassing radical spiritual-saints such as Kabir in the 15th century Varanasi and Bulleh Shah in 18th century Punjab." (I am thankful to Ali. S Asani for drawing my attention to this stimulating article.)

Or air. I am neither an Arab nor from Lahore, nor an Indian
from the city of Nagaur. I am not a Hindu, not a Turk
from Peshawar. Nor do I live in Nadaun.
I have not discovered the secret of religion; nor am I born
of Adam or Eve. I have not given myself a name; nor
am I found in sitting or moving about.
I know myself to be first and last, I do not recognize
anyone else. No one is wiser than I am. Bullha, who is the lord standing here? [8]

4.1 Hagiography and Ritual: "*Thaiyā Thaiyā*"

Bulleh Shāh's verses are well recognized throughout the Indus Valley region of Pakistan for their iconoclasm and evocative language. In addition to his popularity among the masses, Bulleh Shāh has a high status in the Punjabi literary canon. The educated urban class and the unlettered rural folk of the Punjab partake equally in the pleasure of listening to his verses. In comparison to that of other illustrious Punjabi Sufi poets, Bulleh Shāh's Sufism has a unique ethos that is conveyed through a particular topos in his biographical narrative. This topos is that of a man from a *Sayyid* family (lineage of the Prophet Muhammad) who accepts a person from a lower social status, such as an Arain,[9] as his *murshid* (teacher/beloved/guide). In a bold act of transgression, Bulleh Shāh renounces his noble lineage to adopt the profession of the *mirasis*,[10] professional singers and dancers, in order to win the acceptance of his *murshid*, Shāh 'Ināyat Qādirī (d. 1728). Bulleh Shāh's Sufism is known for its ideal of an individual who forsakes his masculine, hereditary status to embrace an effeminate profession, that is, to sing and dance in the streets to find his Lord (*rabb*).

Bulleh is the diminutive of his birth name 'Abdullāh, by which he is known to the people of Punjab (although other narratives of how he got his name, and its significance, also exist). When he died, religious scholars argued over his right to burial in a Muslim cemetery, the problem being that powerful religious voices of the time had issued legal rulings condemning Bulleh Shāh's transgressions against Islam. In the view of these scholars, he had become an apostate. The issue of his burial as an apostate in a Muslim burial ground became the

---

8 Here is a translation of this entire *kāfī* by Christopher Shackle (ed. and trans.), *Bullhe Shah: Sufi Lyrics* (Cambridge: Harvard University Press, 2015; Murty Classical Library of India), 105.
9 Primarily an agricultural or farmer-gardener caste, see B. B. Chaudhuri, *Peasant History of Late Pre-Colonial and Colonial India*. (New Delhi: Pearson Longman, 2008), 195.
10 John T. Platts, "*Mīrāsī*," *A Dictionary of Urdu, Classical Hindi, and English*, 1105: "A singer (of a caste which follows that profession from generation to generation)."

focus of a raging debate at his death. Nevertheless, he was finally buried as a Muslim, and almost two-hundred-and-fifty years later, his burial shrine is today the central attraction of the city of Qasūr.[11]

Bulleh Shāh was born in the town of Uch Gillania in Bahawalpur to a respectable religious teacher by the name of Sakhi Sain Muhammad Darwesh.[12] Six years after his birth, his father moved to Malakwal, and Bulleh was dispatched to Pandoke, a village near Qasūr, for the continuation of his formal education in the religious sciences. The young boy became a student of Ghulām Murtaza, imam of a mosque in Qasūr. Later in his youth, he became enamored of Shāh 'Ināyat Qādirī, who became his *murshid* and could be said to be the Shams Tabrīz of Bulleh Shāh. Shāh 'Ināyat belonged to the Qādiriyah Sufi order and by caste was himself an *Arain*. Bulleh Shah, on the other hand, belonged to the *Sayyid* lineage. The submission of Bulleh Shāh to a man who was not only not a *Sayyid* but also from a caste of low standing was anathema to the *ashrāf* (noble, *Sayyid*) class. As was to be expected, controversy ensued. In one of his poems, Bulleh Shāh describes how his family's women-folk came to knock some sense into him. The subversive theme of a *Sayyīd* bowing to an *Arain* for spiritual guidance and grace is one of the most important motifs in the hagiography of Bulleh Shāh.

Bulleh Shāh's encounter with Shāh 'Ināyat had a transformative impact on him and is a constant theme in his poetry. The meeting of the two is explained in the following anecdote. It is said that Bulleh Shāh set out to Lahore in search of a *murshid*, and once there, he visited a garden in the city. In the garden, he saw a ripe mango on one of the trees and uttered God's name with the desire in his heart to eat the ripe fruit. All of a sudden, the mango fell into his hands. Immediately, he heard the gardener's voice assailing him for having stolen the mango. Gathering his wits, Bulleh Shāh replied that he had not climbed a tree and that the wind caused the fruit to fall in his lap. The gardener explained that he was guilty of theft because Bulleh Shāh had taken the name of his Lord just before the mango fell in his hands. This was enough for Bulleh Shāh to realize that he was not dealing with any ordinary person, and from that point onwards, he became Shāh Ināyat's disciple. A story is told that during his training, he grew anxious to visit the grave of the Prophet in Medina. As this desire grew in him, Bulleh Shāh became increasingly restless. One day, he spoke of this to Shāh 'Ināyat, who inquired why he needed to undertake such a journey. Bulleh Shāh explained that his passion for the Prophet's grave made him restless. Fur-

11 Khalid Ahmed writes about this in his "introduction" to Taufiq Rafat, (ed.), *Bulleh Shah: A Selection* (Lahore: Vanguard Publications, 1982).

12 An alternative version assigns Pandoke, Lahore District, fourteen miles South-east of Qasūr, as Bulleh Shāh's birthplace. C. F. Usborne agrees with this version.

ther, had not the Prophet himself said that visiting his grave is like having the opportunity of seeing him in person? In this response, Shāh Ināyat asked him to reflect on his request for three days. On the third night, Bulleh Shāh had a dream in which the Prophet visited him and told him to call his *murshid*. Bulleh Shāh then falls at his feet. The *murshid* is called, and the Prophet asks him to sit on his right. As Bulleh Shāh stands in front of them and is told to cast his glance upwards, he cannot differentiate between his *murshid* and the Prophet. He then wakes up in a state of astonishment and fear.[13]

An alternative, less dramatic story is that Bulleh Shāh arrived in Lahore searching for a *murshid*. Since he had heard of Shāh 'Ināyat, he presented himself to him. The saint inquired as to the purpose of his visit, and Bulleh said that he wanted Shāh 'Ināyat to take him as his disciple. In response, Shāh 'Ināyat recites the following verse:

Bullhiā rabb dā kī pāuṇā,
Idharon putṇā te udhar lāuṇa

Bulleh, finding God?
Uproot from here, plant over there[14]

This simple and direct response convinces Bulleh Shāh that he has met a true Sufi. The problem however is that Bulleh Shāh is a *Sayyīd*, and he has submitted to an *Arain*, which causes his family members (womenfolk) to intercede. Bulleh Shāh tells them in verse that anyone who calls him *Sayyīd* will burn in hell, and the ones who accept him as an *Arain* will earn a high place in God's heaven.

Every female from near and afar
Turned up to brief Bulleh Shah.
Listen Bulleh, they said, it's a sin
To condemn the prophet's kin.

13 There are hundreds of stories of Bulleh Shāh in the oral form spread out all over the region of Indus Valley. These legends and anecdotes have yet to be compiled in written form. My late grandmother and my mother often narrated the stories of Bulleh Shāh to me during my childhood. I was not even ten when my mother explained to me why Bulleh Shāh once lost the favor of his *murshid* because of his carelessness. Consequently, he had to spend years singing on the streets to convince Shāh Inayat to take him back as his disciple, and during this period Bulleh Shāh underwent a thorough cleansing (*tazkiye-i nafs*) of his heart which prepared him to receive the spiritual grace of Shāh Ināyat. My late grandmother always murmured in a barely muted sigh after one of her Sufi stories: "Ah, such stations are not easily achieved". The sigh of my late grandmother remains entrenched in my memory as a symbol of the tribulations a Sufi undergoes to realize his or her love.

14 The literal translation of two verses is by the author.

> Those who acknowledge them, you tell [say],
> Will get one-way ticket to hell.
> Does sanctity deserve such scorn?[15]

As Bulleh Shāh progressed in his spiritual training, he managed to annoy his *murshid*. Some believe this was due to his provocative verses, which openly defied Islamic law, and despite Shāh Inayat's warning, Bulleh Shāh persisted, eventually forcing his *murshid* to cut off contact with him. According to one oral tradition, Bulleh Shāh earned the displeasure of Shāh Inayat because of one moment during which he forgot his humility.[16] Bulleh Shāh had risen quickly in the spiritual realm and perhaps felt self-sufficient. As a punishment, Shāh 'Ināyat removed his grace from Bulleh Shāh, leaving him in a state of extreme anguish. Bulleh Shāh immediately realized that he was nothing without Shāh 'Ināyat. It was during this period of separation (*firāq*) that Bulleh Shāh produced many verses of lament. Bulleh Shāh then joined a group of singers and dancers (*mirasis*), traveling with them as they went throughout the region. The presence of a *sayyīd* among the *mirasis* became a matter of great consternation in Qasūr. His verses became a source of great annoyance, earning him the wrath of powerful scholars of his time.[17]

After years of living among the *mirasis*, he got an opportunity to perform with them in the presence of Shāh 'Ināyat. Some claim that Shāh 'Ināyat was returning from a mosque and heard a melodious voice. He stopped to inquire if it was Bulleh Shāh who was indeed singing. Bulleh Shāh, with shame in his eyes, replied that he was not Bulleh anymore, but *bhulla*, one who is repentant. This greatly moved the *murshid*, and he immediately embraced him. In that moment of contact (*sīna ba sīna*), Bulleh Shāh received his sainthood from Shāh 'Ināyat

---

15 A *Kāfī* by Bulleh Shāh on this occasion chastises those who call him a *sayyīd*, see Taufiq Rafat's translation (*Bulleh Shah: A Selection*) of "*Bulleh nūñ samjhāwan āiyāñ*" as "Briefing Bulleh," 74–75.

16 Other versions exist to explain the cause of Shāh 'Inayat's estrangement from Bulleh Shāh. Among the possible reasons, are that the separation may have to with a misunderstanding between Bulleh Shāh and Shāh 'Inayat, but in this we cannot rule out the role of malicious talk that made it difficult for the two mystics to meet. After all, Bulleh Shāh belonged to the *Sayyid* class, and his *murshid* was merely an *Arain*. Social distinction between the *Ashraf* and *Ajlaf* gave those who belonged to the former a sense of religious superiority over what were perceived as local or indigenous castes. See Imtiaz Ahmad, "The Ashraf-Ajlaf Dichotomy in Muslim Social Structure in India," *Indian Economic and Social History Review* 3.3 (1966): 268–278.

17 This aspect of his life is dramatized to the fullest in modern times. It projects Bulleh Shāh as a liberal man, ahead of his time, exposing the hypocrisies that plague the religious minded.

Qādirī. Taufiq Rafat translates the verses believed to have been sung then by Bulleh Shāh:

> What a dance your love has led me
> Since within me it opened its shop
> And I drank from the poisoned cup.
> Come to my help, apothecary
> What a dance your love had led me!
>
> The Sun has set; some redness remains
> Unveil your face but once again
> To the simpleton who set you free.
> What a dance your love has led me!
>
> I would not be denied. Now the boat
> is in the whirlpool, barely afloat.
> I choose to stay on, foolishly.
> What a dance your love has led me!
>
> Although it is a peacock's squawk
> In the wild, this love is all I ask.
> Inflicting this, why did you flee?
> What a dance your love has led me!
>
> Love has brought me to your door
> Of Inayat, dressed like a whore.
> I click my heels, and leap in glee
> What a dance your love has led me![18]

The combination in Bulleh Shāh's verses of the subversion of religious rituals and the employment of the vernacular Punjabi has an explosive quality. The explosiveness resides especially in the audacity of the verses in a language that is not viewed as an intellectual equal to Arabic, Persian, and Urdu, yet one that is rich in metaphysical symbolism and allegory. It is above all the use of Punjabi vernacular to express feelings about mysticism that gives Punjabi Sufism its unique sensibilities.[19] The lyrical nature of these verses, especially those by Bul-

18 Taufiq Rafat (trans.), *Bulleh Shah: A Selection* (Karachi: Oxford University Press, 2014; first published in 1982 by Vanguard Publications), 80. For another version of the translation, see J. R. Puri & T. R. Shangari, *Bulleh Shah: The Love-intoxicated Iconoclast* (Dera Baba Jaimal Singh, Dist. Amritsar: Radha Soami Satsang Beas, 1986), 191–192.

19 In comparison to the higher cultural status Persian and Urdu that replaced it in the 19th century, Punjabi never quite received due literary recognition—much like other regional languages of the Indus Valley. The identification of Punjabi with the local idiom continued to undermine its position well into the 20th century. The outright neglect of oral folk traditions in academic insti-

leh Shāh, who mocked both religious rituals and ascetic practices commonly associated with the sacred quest of finding God, carries a powerful affect. In addition to the almost Dionysian quality of Bulleh Shah's *kalām*, Punjabi verse is part of the regional oral tradition, and does not boast of an extensive textual tradition. Nevertheless, Bulleh Shāh's verses were able find their way into the hearts of the people as they were sung by both folk and classical (*gharānas*) performers, *qawwāls*, and wandering minstrels. Bulleh himself joined the ranks of *mīrāsīs*, rejecting his Sayyīd status, to play the sarangi and to dance (*nāch*) in the streets. Bulleh Shāh became the patron saint of all *mīrāsīs*, *kanjars*, and *qawwāls* in the Indus-Valley region. His uniqueness is that his mysticism emanates from and speaks to the lowest social strata. His *kalām* questions the piety of those who profess the superiority of their way and actions because of their religious authority. Bulleh Shāh's mysticism is not just a matter of subjectivity but entails a particular location in society as well. It is the story of a *Sayyīd* who joins the ranks of the *mirasīs*, consciously forsaking his blood lineage as a member of the family of the Prophet.

The unique power of Bulleh Shāh's *kalām*, then, stems from the unique synthesis that joins both the metaphysical and the socio-political. Unfortunately, many writers on Bulleh Shāh have shown a tendency to focus on only one of these dimensions at the expense of the other. In a bold verse Bulleh Shāh himself says:

> *jeḥṛā sānūñ sayyed saddey dozakh milañ sazāeāñ*
> *jo koī sanūñ rāeṇ ākhey bḥisḥteñ pīnghāñ pāeāñ*[20]
>
> Whosoever calls us *Sayyīd* will be punished in hell, and anyone who calls us *Arain* will have swings made ready for her blissful pleasure in heaven

This verse is a strong declaration of Bulleh Shāh that he has abandoned his high status in favor of the lesser caste to which Shāh 'Ināyat belonged. What is more, he puts a curse on anyone who calls him by his former title and promises heavenly bliss to all who accept his new status. What could be more honorific then belonging to the Prophet's lineage in Islam? Bulleh's choice must have irked the entire *Sayyīd* clan in Qasūr. In a sense, it is a kind of lesser apostasy, al-

---

tutes made it exceedingly difficult to grasp fully the significance of the cultural context of Bulleh Shāh's vernacular verses. Only a handful of scholars in Pakistan devoted their entire attention to the revival of Punjabi literature. Unfortunately, because of the folk identification of Bulleh Shāh's verses, his mysticism has often been viewed as that of a rustic, lacking in sophistication attributed to that expressed in Persian or Urdu verses.

20 Miān Ẓafar Maqbūl, *Terey 'ishq nachāyā: Sharḥ Kalām Bulleh Shāh* (Lahore: Maktabah Daniyāl, n.d), 189.

though interestingly, one that is without precedent in legal discourse. The *fiqh* has nothing to say of someone who is recognized as *Sayyīd* and choses to forgo his noble lineage, even turning it into a personal insult. The ingenuity, and uniqueness, of this feature in Bulleh Shāh's *kalām* that aims to express his fidelity to Shāh 'Ināyat has not been adequately appreciated in literary Punjabi discourse. Furthermore, neither has the context of this mystical utterance of Bulleh Shāh been fully explored in the context of his Sufism.[21]

4.2 Religious Identity of Bulleh Shāh

The Sufi poetry of Bulleh Shāh is well known for its hard-hitting verses criticizing religious rituals and customs. Yet his poetry is profoundly mystical, laced with religious imagery and longing. Primarily as a result of these apparently contradictory elements in his poetical corpus, literary scholars and other intellectuals have long argued about Bulleh Shāh's religious identity and questioned whether it is appropriate to associate his mysticism specifically with any a single, overarching belief system. Consequently, contemporary scholars find it reasonable to label Bulleh Shāh as a syncretic figure. They see him as an inevitable product of, or characteristic of, the intermingling of diverse metaphysical systems in the premodern religious environment of South Asia. Others go so far to describe Bulleh Shāh as having transcended the category of religion entirely, and in the process they (mis-)assign him a largely secular identity. Indeed, the inherent paradoxes of the poetry and legacy of Bulleh Shāh make the approach any modern scholar chooses to adopt in the study of this major Punjabi Sufi poet of the 18th century inevitably a challenging and, often, a contested one.

The earlier approaches to the study of Bulleh Shāh took his religious identity as the most important issue and focused on him either from a materialist perspective or from a metaphysical angle. By contrast, the present study seeks to

---

21 This rejection of *sayyid* status by Bulleh Shāh is particularly provocative at a time of renewed interest among Sufi orders in describing their way as the one directly inspired by Prophet Muhammad (e.g., *tariqa-i Muḥammadī*), thus giving it a superior status (to that of orders such as the Chishtiyya or the Suhrawardiyya etc.). We see no such claim, or pretense, to be associated with the Prophet in Bulleh Shāh's mysticism. In fact, quite the opposite, his rejection indicates the lack of any ambition to build on sacred genealogy as a means to claim high spiritual station. In chapter 3, we looked at the case of the 19th-century reformist movement led by Sayyid Aḥmed Barelwī and Shāh Ismā'īl, whose writings and activism aimed at a ritual purification of Muslims, attacking local varieties of Sufism, especially rituals associated with devotion to what they saw as "dead" saints. In his critique of orthodox rituals, Bulleh Shāh's poetry speaks directly to the reformist trends of that time that are aggressively asserting their religious program publicly.

avoid the earlier pitfalls that have marred the study of this great Sufi by examining Bulleh Shāh from both perspectives. Second, and most critical, the present study seeks to bring Bulleh Shāh's tomb as a location for *ziyāra* into serious consideration as a key to understanding the *meaning* of Bulleh Shāh's religious/mystical popularity and to combine this with consideration of a hitherto unexamined narrative biography of the saint.

As is well known, the modern controversy over Bulleh Shāh's identity has much to with the persistent religio-nationalist polemics of the twentieth century in South Asia. In the politicization of the literary heritage of the Indus Valley, the proponents of state modernization have sought to deploy local and regional tradition in the service of political liberalism or secularism, as a counterweight to the religion of the *'ulamā'* and the Islamists. Second, it is important to consider the dynamics of scholarly discourse on Sufism in the West as well, which has not been taken adequately into account in the existing literature on Bulleh Shāh. Third, there has been a conflict in interpreting Bulleh Shāh's legacy between those who emphasize the impact of local, non-Islamic influences, and those who wish to draw upon Persian Sufism as a more relevant source of his mystical thought. Fourth, and perhaps much more challenging to address, is the larger issue of the treatment of religion, especially mysticism, as a category in intellectual history. This final challenge adds another layer of complexity to an already contested area of study, especially when we are engaged in the study of non-Western histories and cultures and must constantly be conscious of the use of Eurocentric categories (It goes without saying that the use of any and all terminology in cross-cultural and cross-religious study requires diligent and acute sensitivity).

The combination of these four factors makes it a complex undertaking to approach a figure like Bulleh Shāh. Nevertheless, the interdisciplinary and intellectual legacy of comparative religion continues to be a promising one for the study of Bulleh Shāh, despite over-zealous post-modernist criticism of it.[22] In using this legacy carefully, and by highlighting existing methodological bottlenecks in the treatment of Bulleh Shāh, the approach taken in this chapter seeks to offer an alternative mode of analysis that may circumvent some of the inherent difficulties in writing about this very important 18th-century Punjabi visionary, poet, and Sufi.

In discussing the contemporary state of literature on Bulleh Shāh, Robin Rinehart has done well to analyze past scholarly writings, and to try to define the

22 See Kimberley C. Patton and (editors), *A Magic Still Dwells: Comparative Religion in the Post-modern Age* (Berkeley: University of California Press, 2000).

nature of what he calls an "interpretative impasse."[23] Because of the scarcity of scholarship on Bulleh Shāh—and in recognition of Rinehart's attempt to open up the field—we shall commence by examining Rinehart's contribution in two articles on the interpretation of Bulleh Shāh. This will enable us to utilize her review and criticism of the diverse material already available on Bulleh Shāh, and then also to comment on her contributions before taking up our own approach in the remainder of this chapter.

Rinehart's primary motive is to examine, and to challenge, the normalization of Bulleh Shāh in contemporary Hindu, Sikh and Islamic narratives.[24] As Rinehart views it, the core problem is that the study of a figure like Bulleh Shāh is hampered by the lack of credible 18th century manuscripts.[25] The earliest relevant manuscripts are from one-hundred-and-fifty years after his death and therefore present a formidable obstacle to establishing any reliable historical information about him. Furthermore, the extant biographical text corpus of Bulleh Shāh materials is riddled with innumerable inconsistencies, and it has clearly expanded over a period of time to include ever more variations and interpolations (al-

23 "The Portable Bulleh Shah," Biography, Categorization, and Authorship in the Study of Punjabi Sufi Poetry," *Numen*, Vol. 46, No. 1 (1996): 55. Cf. Rinehart's more precise comment (*Ibid.*, 59): "The problem with this method is that it creates a circular process of interpretation. Many of the details that provide the raw material for Bullhe Shah's biographies have been gleaned from his poetry; these details—now instantiated "facts"—in turn are used to explain other aspects of his poetry. Thus, interpreters use Bullhe Shah's life to explain his poetry, and his poetry to explain his life. Yet seldom can either strategy be grounded on solid historical evidence."

24 Rinehart's arguments are a response to scholars like Mustansir Mir. For example, according to Mustansir Mir, "Teachings of Two Punjabi Sufi Poets," 520–21: "What we have called the critical aspect of their poetry can be appreciated by people of any religious outlook—or for that matter by people of any outlook on religion—but the constructive side of their poetry cannot be reduced to an abstract, formless doctrine, for it consists of a positive set of values that is decidedly Islamic in structure and detail. For all their distractive remarks about the mulla and the scholar, Sulṭān Bāhū and Bulleh Shāh, themselves orthodox in belief and practice, hold the basic code of the Sharī'ah—Islamic law—to be inviolate, seeking only to imbue conduct with moral insight and spiritual meaning. The apparently bold statements about religious belief and practice that, especially in the case of Bulleh Shāh, sometimes have an antinomian ring, must not, therefore, be interpreted to mean revolt against formal religion in general or formal Islam in particular. They are rather attempts to shock people into reality through the use of the favorite technique of "allopathy"—remedying something by means of its opposite."

25 "The Portable Bullhe Shah," 59: "To be sure there are multiple hagiographical accounts which provide a wealth of alleged information about significant events in his life, but they are grounded in the aims of hagiography, not critical hagiography. The interpreters never seem to take into account that the primary source of this allegedly historical material is Bullhe Shah's poetry."

though Rinehart does not delineate nor describe the various texts it contains). In the face of such a diffuse text corpus, it is "more precise to speak of a Bulleh Shah tradition, a body of poetry attributed to Bulleh Shah."[26] Rinehart focuses on three religious interpretations—Hindu, Muslim and Sikh—and in turn discounts all of these perspectives because of their hagiographical foundations that do not meet the criteria of "critical historiography."[27] Accordingly, Bulleh Shāh is shown either as a fully orthodox Muslim or as a mystic who became enamored of Hindu pantheism or Sikh teachings later in his life and decided to abandon his intolerant Islamic faith to embrace alternative pluralist doctrines. Rinehart discredits such interpretations as products of religious identities constructed in the 19th and 20th centuries:

> if Sikhism is shown to be taken from the best of Islam and Hinduism, then it transcends both; if Punjabi Sufi poetry is strongly influenced by Hindu mysticism, then it shows that the Sufis realised that Hinduism was superior, or that Sufism is simply a derivative form of Hindu mysticism.[28]

In her opinion, scholars have previously rummaged through the hagiographical accounts to reconstruct the historical Bulleh Shāh and then set about collating a consistent body of poetry out of a very fragmented corpus. Her verdict in a nutshell is that "if the information about his life is problematic, and the corpus of his poetry is suspect (at least insofar as attributing either to a historically locatable figure), then using one to establish the other is clearly a difficult method to defend."[29] Nevertheless, the article goes on to pose the thought-provoking question, "what is it, after all, about Bulleh Shāh that is attractive to members of different religious communities?"[30]

Rinehart finds it prudent to focus on the multiple Bulleh Shāhs of the existing literary corpus to see what they reveal about the "rhetorical personae" of the poet, instead of obsessing over establishing the "real" Bulleh Shāh. In other words, Rinehart seeks to dismantle the implicit assumption of Bulleh Shāh's authorship in the 'existing interpretive discourse on his work and legacy. Her aim is to problematize the attempts to construct a homogeneous image of Bulleh Shāh

26 Robin Rinehart, "Interpretations of the Poetry of Bulleh Shah," *International Journal of Punjab Studies* 3.1 (1999): 48.
27 "The Portable Bulleh Shah," 59.
28 Rinehart, "Interpretations," 46.
29 Rinehart, "The Portable Bulleh Shah," 60.
30 Rinehart, "Interpretations," 58. This is an important question for Rinehart in understanding Bulleh Shāh's wide appeal.

in the service of one or another modern religious ideology. The solution in overcoming such insular readings of Bulleh Shāh, Rinehart suggests, is that

> If we accept the concept that Bullhe Shah's readers and listeners create their own Bullhe Shahs, we can develop a new conception of Bullhe Shah as an "author" variously created and recreated within different discursive spaces (e.g. the various discussions of national, communal, and regional identity in contemporary Pakistan and India.) This, in turn, will allow us to consider a crucial question: why is Bullhe Shah's name so powerful? [31]

Rinehart offers the following explanation:

> A Vedanta might see this passage as reflecting the union of the individual self [*ātman*] with the universal reality [*brahman*]; it might evoke for a Vaiṣṇava the pain of separation from god [*viraha*] and the joy of a possible future union; a Sufi versed in ibn-'Arabī's writings might find it a good expression of *waḥdat al-wujūd*, according to which the true self is in fact a manifestation of Allah. In the world of philosophical texts and treatises, one who studied ibn-'Arabī's theory of *waḥdat al-wujūd* and its attendant technical, philosophical apparatus might not immediately see affinities with the monistic philosophy of Advaita Vedanta, or the pain of separation from God felt by the Vaiṣṇava devotee. At that level, each tradition has its own history and sense of identity, its own vocabulary, its own idiosyncrasies and particularities. Yet a reader or listener familiar with the basic ideas of any of these traditions could certainly find affinities between them and many of the basic sentiments expressed in Bullhe Shah's poetry.[32]

And again:

> If Vedantins are inspired by certain Bullhe Shah poems, then perhaps it is because they find in those poems ideas that fit their own understanding of Vedanta. Muslims who respond favorably to Bullhe Shah's poems of course can also fit many poems into a framework that suits there own visions of Islam. These responses, in turn, generate varying senses of the "real" Bullhe Shah, relative to the interpretative framework the reader or listener brings to the poems. It is no wonder that there are so many different interpretations of the poet. What is it about his poetry that makes this possible?[33]

As is clear from her own words above, Rinehart is puzzled by the ideological rifts over Bulleh Shāh—over the need for different religious communities to appropriate him – and very much interested in finding out the nature of the (cultural) dynamics at work that give Bulleh Shāh this special ability to, as it turns out, transcend categorization—what she describes as his "portability." Rinehart does

31 "The Portable Bullhe Shah," 78.
32 "The Portable Bullhe Shah," 84.
33 *Ibid.*, 82.

admit the need to understand the social context in which the production of such constructions of these "rhetorical personae," is made possible? That is, how is the notion of "rhetorical personae" grounded in history? The search is for finding the social context—or the metaphysics of his "portability"—that can explain why Bulleh Shāh can be a Muslim, Hindu, and Sikh at the same time. Rinehart's inquiry, in my assessment, seeks to ask what kind of sociological or historical explanations are able to account adequately for this "portability." The idea seems to be that unlocking, as it were, the secret of this "portability" would simultaneously reveal the "power" Bulleh Shāh exercises in the imagination of different religious communities—it also implies, although Rinehart does not say so, that this will help in moving beyond simply labeling him as a syncretic figure.[34]

Nevertheless, Rinehart's critical probing is not able to point the way forward or propose an alternative; it can only accept the need for this.[35] It turns out that the study of Bulleh Shāh has been largely a matter of analyzing discourses or discrediting historical sources as not "authentic" enough. Rinehart focuses only on the polemical writings ("discourses"), who are critical of interpretations that do not correspond to their own religious identities,[36] on account of which she aptly describes the study of this figure as facing an "interpretative impasse."[37] Such an approach is inadequate in examining a figure like Bulleh Shāh, whose poetry is the only historical source available—as Rinehart has herself points out. This saint is part of a living tradition with a complex oral history in which the metaphysics of presence is very much at the heart of it. To say it in other words, there is a close relationship between ritual and poetry in Bulleh Shah's case, as in

34 *Ibid.*, 77: "Yet the very existence of so many portrayals makes clear Bullhe Shah's importance in Punjabi literature. Why is he claimed by Punjabis of different religious communities?"

35 *Ibid.:* "To answer such questions requires developing new strategies for studying Bullhe Shah, his poetry, and his interpreters." However, she does not say what these "new" strategies could look like.

36 *Ibid.*, 72: "But at a deeper level, even those interpreters who challenge the merit of these analyses of Bullhe Shah do not question the very nature of the study itself—they make the same kinds of claims about categories, using concepts such as influence etc.—and present analyses which are grounded in the very same assumptions about the relationship between biography and poetry, religious identity, and authorship albeit with less communal hyperbole."

37 *Ibid.*, 59: "The problem with his method is that it creates a circular process of interpretation. Many of the details that provide the raw material for Bullhe Shah's biographies have been gleaned from his poetry; these details—now instantiated "facts"—in turn are used to explain other aspects of his poetry. Thus interpreters use Bullhe Shah's life to explain his poetry, and his poetry to explain his life. Yet seldom can either strategy be grounded on solid historical evidence."

Indus Sufism more widely.[38] As mentioned above, the spoken word of the saint is performed at the *dargāh* where it is a source of spiritual connection or communion, and its audition is a form of worship (*'ibadat*). Moreover, and this is critical, the Sufi *dargāh* is historically also a sacred place visited by devotees from different religious communities.[39] Yet Rinehart completely overlooks this important dimension of Sufism in the Panjab.

I argue here that the ritual use of poetry at the *dargāh* in devotion to Bulleh Shāh and the ritual participation of Hindus, Sikhs, and Muslims are better able to explain the appeal of this mystic across these religious communities. So, instead of the disagreements among scholars that we see in the writings on Bulleh Shāh, the ritual space of the Sufi shrine allows us to witness or observe the power of this mystic in the *lived* experience of his devotees. Here we can see that this devotional connection with Bulleh Shāh is not the preserve of any one religious tradition; rather it is shared among Muslims, Hindus, and Sikhs.

Moreover, the unique hagiographic identity of Bulleh Shāh—which I will discuss below—is at the heart of the devotional celebration that connects him to the people. For his followers, his rejection of formal or prescriptive ritual and his theopoetics of love (*'ishq*) that are the core of the cultivation of a personal relationship with the Divine are what especially endear him to those who cross his sacred threshold. It was love of God that caused Bulleh Shāh to dance in the streets, among the lower strata of society, the *mirasīs* and the *kanjars*, and in the process this is what enabled him to find his Lord. In short, Bulleh Shāh's defiance of ritual and consequent critique of orthodoxy, together with his insistence on the supremacy of the love of God and his iconoclastic practice, are

38 For a good discussion, see Michael Frishkopf, "Authorship in Sufi Poetry," *Alif: Journal of Comparative Poetics* 23 (2003): 78–108.

39 As argued in this dissertation, especially chapter 3, despite the state intervention in regulating the pilgrimage to Sufi shrines, the popular devotion or practice does circumvent this reformist attempt of the state, and visitors continue to perform *ḥāzirī* by which they enter into the living presence of Bulleh Shāh. It demonstrates, for one, how ideological interests can and are evaded on the ground below, to challenge the textual categorization of figures like Bulleh Shāh in its aim to construct a monolithic identity of this mystic. In the absence of textual sources, believed to be essential to the practice of "critical historiography"—the preferential methodology for historians for most of its disciplinary practice—anthropological inquiry can indeed account for alternative modes of *histories*, such as in the case of Indus Sufism, that may not have a voice in the literary field responsible for the textual production of knowledge. While we may not have anything substantial on Bulleh Shāh insofar textual documentation is concerned and failing to conceptualize the absence thereof in the textual realm is perhaps to miss out on the significance of Bulleh Shāh, especially when his shrine is a place of pilgrimage, and hence his identity as well.

all part and parcel of his mystic identity and ethos, which have won him a special place in Indus Sufism.

Since, as noted above, the Sufi shrine of Bulleh Shah's tomb is a place for worship for different religious communities, this ritual function is able to explain how multiple constructions of him can exist without necessarily being in conflict with the notion of one "true" Bulleh Shāh. The experience of seeing the "self" as the locus of Divine manifestation transcends communal identities and their historical boundaries. Such mystical experience also points the way to the Divine *unity* behind the colorful multiplicity observed in this world. In the eyes of Bulleh Shāh, one can be a Hindu, Muslim, Sikh, Christian, or Jew and fully experience this Divine unity by searching for its essence, or presence, in the heart—beyond religious institutions, rituals, and reified categories of what it means to be religious.[40]

How does the above explanation help with the question of category? If we take seriously how devotees see Bulleh Shāh, then the mysticism of Bulleh Shāh is not at all about the supremacy of any one religious tradition over another. Rather, primacy is given to *experience itself* as a means of *tasting* the truth, over formal or reified notions of religious performance treated in the books of law or ritual. In advocating the need to look beyond formal ritual, Bulleh Shāh's critique would be applicable to any type of religious orthodoxy, that is, any institutionalization of religious experience in canonical books or prescribed ritual practice. In any case, one can see that a considerable synchronicity developed in the Panjab among the different mystical traditions and teachings in the poetics of the sacred "self," as evidenced in the writings of Dara Shikoh, to take the best-known example.[41] The latter's work shows that there was a strong recognition on the part of the spiritual seekers at the time of the unitive dimension of "truth" at higher levels of experience or consciousness beyond communal identities.

In light of this, in the remainder of this chapter, I will delve in greater detail into Bulleh Shāh's metaphysics of Presence.

---

40 See Bulleh Shāh's *kāfī* which negates all religious identities and speaks of mystical enchantment and the *ineffability* of that experience, Taufiq Rafat (trans.), *Bulleh Shah,* 70–73: *Kī jānāñ meiñ kaon*" (what do I know, who am I?), see above for another version of the translation of this *kāfī* by Christopher Shackle.

41 Mahfuz-al-Haq (ed. and trans,), *Majmu'a-ul-Bahrain or The Mingling of the Two Oceans by Prince Dārā Shikūh* (Calcutta: The Asiatic Society, 1929; 2nd edition reprint in 1998).

4.3 ‘Ishq

The central dimension of Bulleh Shāh’s mysticism of *‘ishq* is his dance, which is expressed in *thaiyā thaiyā*, a rhythmic movement of the feet in Indian classical dance. However, despite this formal reference, the accompanying state (*hāl*) of Bulleh Shāh in the dance is one of dishevelment, desire, and longing. It would not be far-fetched to conclude that Bulleh Shāh learned and cultivated the practice of classical dance as he left his Sayyīd status to join the profession of singers (*mirasīs* and *kanjars*). The *shar‘iah* as such would not be applicable to them because of their liminal, and non-Muslim, social identity—although some could be Muslims but restricted to their lower-caste profession. Hence, we see that Bulleh Shāh’s inclusion among the *mirasīs* and the *kanjars* would signal a disruption of social order, a physical crossing over that would invite tremendous censure (*malāmat*) of a Sayyid man. Bulleh Shāh’s crossing of the social space would be provocative on two accounts, first on the basis of religious identification, and second on account of socio-cultural gender ambiguity that characterized the class of *kanjars* in South Asian societies. Last but not the least, the *‘ishq* of Bulleh Shāh was not an abstract one, but rather involved winning the love of his *murshid*, Shāh ‘Ināyat of Lahore of the *‘Arain* caste and a gardener by profession.

The practice of finding God in a living person, as an *insān-i kamīl*, has a long history in Islamic mysticism. The notion of *‘ishq* in Bulleh Shāh’s *kalām*—like in many other traditions—is a tumultuous force that overwhelms, and more often than not, on the path of blame. It has the power to induce ecstasy but also exposes one to peril, which is its contradictory nature. In fact, Bulleh Shāh recounts a number of examples from the past about those who faced dire tribulations in their quest, and in this capacity of *‘ishq* to put the lover in state of grief, danger, or a test, yet there is nothing sweeter than *‘ishq*. He depicts *‘ishq* as a kind of law unto itself:

prem nagar dey aulṭey jāley
khūnī naen howey khūs ḥaley[42]

Indeed, the ways of love are topsy-turvy,
However, your blood-shot eyes remain content, ever the same.

In Bulleh Shāh’s case, *‘ishq* made him dance, and abandon his Sayyīd status, and embrace notoriety in the process. This is evident in a number of his *kāfīs*, and one of the most well-known has been discussed earlier:

---

42 Maqbūl, *Terey ‘Ishq Nachāyā*, 70.

terey 'ishq nachāyā kar ke thayyā thayyā
terey 'ishq ne ḍerā merey andar kitā
bhar ke zeḥr piyālah meiñ tāñ āpey pītā
jahbdey bawhoṛīñ wey ṭabībā nahiñ tāñ meiñ mar gaīy ā[43]

Your love had made me dance to a fast beat!
Your love had taken abode within my heart!
This cup of poison I drank all by myself
Come, come, O physician, or else I breathe my last![44]

In another *kāfī*, whosoever receives the excellence/virtues of *'ishq* begins to dance without step or rhythm. As love enters the body (*tan*), it compels a person to dance without any concern to technique or form: *Jis tan lageyā 'ishq kamāl / nāchey bey sur tey bey tāl*[45] (whomsoever is infected with perfect love/dances without rhythm or beat).[46]

Several verses draw upon the example of those who found themselves on this path and underwent its tribulations. In his remarkable *rah rahwey* Bulleh Shāh tells names of those who were subjected to the path of love, and questions the possibility of surviving it, because no one else seems to have done so: "*rah rahwey 'ishqā mārayā aīy / kah kis nūñ pār autārayā aī.*"[47] (Again and again, oh love, you have killed me/say, whom have you ever spared)

'isā nūñ bin bāp jamāyā
nūḥey par ṭūfān mangayā
nāl paīwu dey puthar laṛāyā
ḍub ūḥanāñ nūñ mārayā aiy[48]

You caused Jesus to be born without a father; you
Summoned the flood over Nuh. You set the son
Quarrelling with the father; you drowned and killed them.[49]

43 *Ibid.*, 151.
44 Translation in English by J. R. Puri & T. R. Shangari, *Bulleh Shah: The Love-intoxicated Iconoclast* (Dera Baba Jamal Singh, Dist. Amritsar: Radha Soami Satsang Beas, 1986), 191.
45 *Ibid.*, 160.
46 Translation by J. R. Puri & T. R. Shangari, *Bulleh Shah*, 303:
"Whosoever is stung by perfect love,/He
dances out of tune,/He dances out of step!"
47 Maqbūl, *Terey 'Ishq Nachāyā*, 186.
48 *Ibid.*, 186.
49 Translation by Christopher Shackle, *Bullhe Shah: Sufi Lyrics*, 243.

Bulleh Shāh goes on to add that Jesus was born without a father—reference to accusations flung towards his mother, Maryām, and the stigma that Jesus as a baby without a father must have had to bear. He adds that Noah was made to face a mighty storm, and amidst this chaos God created a rift between father and son, killing the son in the end. Such then are the ways of love; whom has he ever spared? Another verse along the same lines is the following:

'ishq dī nawīwañ nawīñ bahār
jāñ men sabaq dā paṛheyā, masjid kolūñ jīwoṛā ḍaryā
ḍarey jā ṭhākur dey waṛyā, jittahey wajdey nād hazār
'ishq dī nawiwaṇ nawin bahār[50]

The spring of love is ever new.
When I studied the lesson of love, my heart became afraid
of the mosque. I went to enter the temple of the lord,
where a thousand conches are blown.[51]

In these passages, Bulleh Shāh's imagery of *'ishq* combines corporeal sensuality, blame (*malāmat*), and divine ecstasy. The capacity of *'ishq* to disrupt social conventions, to transgress, is found again and again in Bulleh Shāh's verses, almost as a law of sorts. He mixes the profane and the sacred with abandon in the process, subjecting religious rituals to violent imagery that exhorts to breaking, burning, and throwing away. The coarseness of his verses is an important part of Bulleh Shāh's poetic resonance, and a major reason for his vernacular appeal.

Another critical aspect of Bulleh Shāh's *kalām* is his depiction of his beloved's (*ḍholan māhī*) habit of communicating through secret signs and mysterious ways, residing as he does behind the veil of the letter *mim* (م), which is a play on *Aḥmad*, the Prophet's name: if the *mim* in *Aḥmad* is removed it becomes *aḥad*, one. Bulleh Shāh goes on to say that his Beloved made Manṣūr declare *anā al-hāq*, which He Himself told him; and then He had him put on the gibbet for doing so, all the while standing next to him when he is on the gibbet, enraptured and laughing:

bey ḥad ramzāñ dasdā nī,
ḍholan māhī
mīm dey ūley wasdā nī,
ḍholañ māhī
auliyā' Manṣūr kahāwey
ramaz anā l'ḥāq āp batāwey

50 Maqbūl, *Tere 'Ishq* Nachāyā, 215.
51 Translation by Christopher Shackle, *Bullhe Shah: Sufi Lyrics*, 29.

āpey āp nūñ dār chaṛhāwey
tey kol khalo key hasdā nī,
ḍholañ māhī[52]

He tells of countless mysteries, my dear beloved.
He dwells concealed in *mīm*, my dear beloved.
He calls himself saint Mansur, and it is he who reveals the secret of *I am God*. He puts himself upon the gallows, and standing nearby he laughs, my dear beloved.[53]

These verses disclose his Beloved's ruthlessness with His lovers, and just as He bequeaths ecstasy, compelling the lover to reveal his secret, he also likes to watch the execution in glee. This view of his Beloved conforms to the *malāmtī* temperament that views the divine as enacting a cosmic drama, indulging in deception, and seeming to flout His own rules. The rhythm and style of this exquisite *kafī* capture the imagery of divine paradox to include both the macabre and otherworldly ecstasy: "*rah rah wey 'ishqā mārayā aī / kah kah kis nūñ pār autārayā aīy*" (Again and again, oh love, you have killed me/say, whom have you ever spared).

The beloved's ways are strange, says Bulleh Shāh. He enters the mosque to pray, and then afterwards, and without any reservations or hesitation, he enters the idol temple:

"*wich masit namāz guzārey, but khāney jā waṛdā*"[54] (spends time praying in a mosque/in the next moment, enters a idol temple). As for himself, Bulleh Shāh declares that he has won the wager or game (*bazī*) by breaking the shari'ah and by losing his honor, perhaps because he was not a masterful player, an amateur at the game: "*toṛ shar'a nūñ jit la'ī bāzī, phir dī nak waḍḍhā key / men ān jānī khaiḍū kachchī ān, khaiḍāñ men ākey bākey*"[55] ("I have broken the law and won the game. I go around in disgrace/I was a silly girl who was lost in her game. I played at making things out of mud").[56]

52 Maqbūl, *Terey 'Ishq Nachāyā*, 121.
53 Translation by Christopher Shackle, *Bullhe Shah: Sufi Lyrics*, 189.
54 Maqbūl, *Terey 'Ishq Nachāyā.*, 254.
55 *Ibid.*, 278.
56 Translation by Shackle, *Bullhe Shah*, 195. In an explanatory footnote, Shackle describes "*akey bakey*" as "a children's game in which figures are made out of mud" (407).

Experiential Knowledge: *Nuqta* and *Alif*

A very critical, and popular, dimension of Bulleh Shāh's verses is his denunciation of discursive knowledge (*'ilm*) and learning in favor of the dot (*nuqṭa*) or just the first letter of the alphabet (*ḥarf*, pl. *ḥurūf*), namely, *alif*. The verses extoll the virtues of the dot and the letter *alif* over and above all other letters, forms of knowledge, and texts. Its reference indicates Bulleh Shāh's familiarity with *alif's* older mystical and cosmological associations in Sufi thought. Bulleh Shāh advocates abandoning all other kinds of knowledge in order to grasp the significance of *alif*, which contains the essence of everything in existence: "*aik nuqṭey wich gal mukdī hey / phaṛ nuqteh chauṛ ḥisābāñ nūñ*[57] ("The whole thing ends in a dot/Grasp the dot and quit counting.")[58] In this verse, Bulleh Shāh says that all matters of intellectual thought and dispute alike are resolved in one dot (*nuqta*), so best to leave aside tedious calculations (*ḥisābāṇ*) and speculations to study or read (*phaṛ*) only this single dot. In another *kafī*, Bulleh Shāh says that his *yār* (an informal way to address the beloved friend) has taught him how to read only the dot: *Aik nuqtah yār paṛhāyā hey*[59] (my friend taught me to read the dot).

The bold rejection of rational knowledge is a reference to the science of letters (*'ilm-i ḥurūf*) in Islamic mystical tradition, although Bulleh Shāh seems content with only one letter: "*alif Allāh ratā dil merā / meinūñ bey dī khabar nah kāī*"[60] ("With 'A', my heart is steeped in love for the Lord / I have no knowledge whatsoever of 'B').[61] In this *kafī*, Bulleh again sings that the letter *alif* of *Allāh* has dyed my heart in his color, or inflamed it, such that he has no knowledge of the letter *bey* (ب): "*Aik alif phaṛho chuṭkārā hey*"[62] ("Deliverance lies in learning about one *Alif*").[63] The verse holds that reading the *alif* saves one (*chuṭkārā*) from all earthly and godly difficulties. Also, "*'ilmūn bas kariñ aū yār / aikū alif terey darkār*"[64] ("That is enough of learning, friend / but all you need is *alif*").[65]

The verses below discuss the sequential quality of *alif*, by the addition of which we get astronomical figures. The essence of all things is the *alif*, which

---

57 *Ibid.*, 88.
58 Trans. By Shackle, *Bullhe Shah*, 37.
59 Maqbūl, *Terey 'Ishq Nachāyā*, 90.
60 *Ibid.*, 69.
61 Trans. by J. R. Puri & T. R. Shangari, *Bulleh Shah*, 196.
62 *Ibid.*, 84.
63 Trans. by J. R. Puri & T. R. Shangari, *Bulleh Shah*, 288.
64 Maqbūl, *Terey 'Ishq Nachāya*, 217.
65 Trans. by Shackle, *Bullhe Shah*, 39.

is present in all numerical values, and which bestows existence upon numerical multiplicity:

aik alifoñ do tin char huwey
phir lakh karoṛ hazār huwey
phir ūthoñ bāhijh shumār huwey
hik alif dā nuqtah niyārā hey[66]

One *Alif* became two, three and four
The very same one multiplied into millions,
And then it grew into countless multitudes.
The mystery of one *Alif* is unique![67]

The *kāfīs* below also serve as a potent way for common, or unlettered, folk to criticize the vanity and arrogance of learned scholars. Not only do the verses point towards mystical matters that are considered beyond rational learning, the *kafīs* are scornful of overbearing scholarly types, who seem to resemble beasts of burden, as they are described with grotesque expressions, but they are captured thus in what is *essentially a comical performance* from Bulleh Shāh's cosmological perspective. It is this double-entendre play in Bulleh Shāh verses that is achieved by skillfully revealing the monstrous as farcical. It is also a swift dismissal of the very seriousness of the authority being projected by the stern scholar.

Why do you read cartload of books,
And carry on your head a bundle of troubles?
You look awful, like an executioner;
The journey ahead is most arduous.[68]

In these verses, Bulleh Shāh questions the logic of readings a stack of books, and he asks why one carries such a frightful load, which will lead to the abyss, and why the face of such a scholar becomes like that of an executioner. It is a caricature of a scholar with a serious and somber disposition, evidenced by a permanent scowl on his face that gives him a terrifying disposition: "*kiyoñ paṛhanāiñ gaḍ kitābān dī / sir chānāiñ panḍ 'azābān dī / kiyoñ hoeyān shakal jalādāñ dī*[69]

66 Maqbūl, *Terey 'Ishq Nachāyā*, 84.
67 Trans. by J. R. Puri & T. R. Shangari, *Bulleh Shah*, 288.
68 Trans. by J. R. Puri & T. R. Shangari, *Bulleh Shah*, 288:
69 Maqbūl, *Terey 'Ishq Nachāyā*, 84.

("Why do you read cartload of books / And carry on your head a bundle of troubles? / You look awful, like an executioner").[70]

Another *kāfī* dismisses the one who recites the Qur'ān in order to memorize it, and with a special emphasis on correct pronunciation, merely as a phonetic exercise for the tongue: "*ban ḥafiz ḥifz qurān kareñ / paṛh paṛh ke ṣāf zabān kareñ.*[71] ("You may remember the Qur'an by heart and become a Hafiz, / and purify your tongue with continual recitation").[72] In a similar *kafī*, Bulleh Shāh again says how the transmitted knowledge of Islamic books (*Qadūrī*, a foundational source in Ḥanafī *fiqh* by Ahmad ibn Muḥammad al-Quduri, and the second is *Kanz al-Daqā'iq* by Abdullah ibn Muḥammad Nasafī) began to slip from his mind. He questions the rationale for observing Islamic ritual obligations for those who have imbibed wine and are already in a state of ecstatic intoxication: "*manṭaq, ma'aney kanz qadūrī phaṛayā 'ilm gawatā / namāz rozah aūs kihey karnah, jis madh pītī mātā*"[73] ("I have studied logic and semantics, the *Kanz* and the *Qudūrī*, but all this learning has destroyed me / What use are prayer and fasting to those who have drained the flask of love").[74]

As discussed earlier, one of the most popular aspects of Bulleh Shāh's poetry is his unrelenting criticism of religious rituals. *'Ishq* has made him forget all religious formalities, such as *namāz* and *rozah*. Rules of logic and grammar were also erased from his mind as he heard, or came under the spell of, the single note of unity, played by his beloved:

rozey ḥaj nāmaz nī māey, meinūṇ piyā ney ān bhulāey
jāñ piyā diyāṇ khabarāṇ paeyāṇ
manṭaq naḥw sabhey bhol gaeāñ
aus anhad tār bajāey[75]

My fasts, my pilgrimages, my prayers O mother,
My beloved has made me forget them all!
As soon as I realized the beloved,
My logic, my grammar were all forgotten!
Such was the unstruck melody that He played![76]

---

70 Trans. by J. R. Puri & T. R. Shangari, *Bulleh Shah*, 288
71 Maqbūl, *Terey 'Ishq Nachāyā*, 84.
72 Trans. by Shackle, *Bullhe Shah*, 35.
73 Maqbūl, *Terey 'Ishq Nachāyā*, 343.
74 Trans. by Shackle, *Bullhe Shah*, 29.
75 Maqbūl, *Terey 'Ishq Nachāya*, 185.
76 Trans. by J. R. Puri & T. R. Shangari, *Bulleh Shah*, 281.

Pushing his satire further, Bulleh Shāh takes a stab at religious rituals such as *namāz* (ritual-prayer), and links it to mere exhibitionism by displaying a visible mark on the forehead produced by constant prostration (*sajdah*)—and reciting the *kalima* as a rote custom, that is, without any concern for its meaning or understanding. He asks if the truth always remains hidden. Then, in the refrain, the idea that truth makes itself evident, despite the practice of hypocrisy, is broached in a mockingly tone and ultimately constitutes the point of the entire *kafī:*

> *aiwaiñ mattha zamin ghasāidā*
> *pālammā meḥrāb wakhāidā*
> *paṛh kalmah lok hasāidā*
> *dil andar samajh nah layāidā*
> *kadī bāt sachchī bhī lukdī hey*
> *aik nuqṭey wich gal mukdī hey*[77]
>
> In vain you rub your forehead on earth,
> And show a long arch formed on it.
> Your false note in the recital of *kalma,*
> Makes you a laughing stock of people.
> You do not know the art at all,
> By which to bring the *Kalma* in the heart
> Can Truth be kept hidden for ever?
> Wisdom is contained in a single point![78]

In this *kafī*, Bulleh Shāh defines prostration (*sajdah*) as a futile exertion that consists of rubbing the forehead on the ground, and after successfully acquiring the mark/scar of pious distinction, it is exhibited for everyone to see, without any inner transformation or comprehension of its virtues being affected. In such a display, it does not occur to these hypocrites that their true condition is apparent to everyone, for such is the nature of truth.

We also have a critique of bodily ascetic practices of his day in the following verses in the same *kafī:* "*aik jangal baḥraiñ jāṇdey naiñ / aik dāna rozey khāṇdey naiñ / bey samajh wujūd thakāndey naiñ.*[79] ("Some retire to forests / Some journey across the oceans / Some live on a grain a day / The fools tire their bodies senselessly").[80] Here we see ascetics depart for jungles and oceans, eating only a grain of barley every day, and without any understanding, they exhaust their bodies. A number of *kāfīs* consistently confirm Bulleh Shāh's dismissal of bodily asceticism evident in this verse. In his view, the practice of asceticism needlessly

77 Maqbūl, *Terey 'Ishq Nachāyā*, 88.
78 Trans. by J. R. Puri & T. R. Shangari, *Bulleh Shah*, 350.
79 Maqbūl, *Terey 'Ishq Nachāyā*, 88.
80 Trans. by J. R. Puri & T. R. Shangari, *Bulleh Shah*, 351.

inflicts pain on the body without the intended spiritual benefits to its weary practitioners. The critique of piety is an important aspect of the *malāmatī* ethos, whether the piety takes the form of religious rituals or pious asceticism because it involves personal effort in trying to win God's love—a ridiculous proposition by *malāmatī* standards. The *malāmatī* seeks to avoid taking any conscious credit for her ability to achieve closeness to God through her actions; for God is the only true agent.

Another *kafī* offers a further example of Bulleh Shāh's love for his *yār*, seeing whom is for him equivalent to visiting the house of God in Mecca. In other words, his visit to the beloved is *hajj; "ḥājī log makkey nūñ jāndey, sāḍā hun tūñ haj aū yār."*[81] (Pilgrims go to Mecca / Our visit to the friend is pilgrimage). Moreover, Bulleh Shāh openingly speaks of making his prostration before his *yār*, just as other people prostrate before the Ka'ba: "*lokī sajdah ka'bey nūñ kar dey, sāḍā sajdah yār payārey nūñ.*"[82] (People prostrate towards the Ka'ba / Our prostration is for the beloved friend).

The *kāfī* below is one the most tantalizing and provocative *kāfīs*, and a quintessential statement of his *malāmatī* statement. Many modern exegetes of his verses would have immense difficulty in handling these lines, which invite to drinking wine and eating meat, to ignite the molten fires of lust, to steal, to violently break into the house of God, to hoodwink or hustle the grand thug of all thugs—the word *ṭhag* is the same as the word "thug" in English. The verses mix the profane and the sacred to create a powerful, lyrical effect:

> *pī sharāb tey kahā kabāb, haiṭ bāl haḍḍañ dī ag*
> *chorī kartey, bhan ghar rab dā, aūs ṭhagāṇ dey ṭhag nūñ ṭhag*[83]
>
> Drink wine and eat kebabs, burning your bones
> as the fuel to cook them with. Do burglary and break into the lord's house, to rob that robber of robbers.[84]

These verses above do not receive the attention they deserve in modern interpretations of Bulleh Shāh. Writers who wish to emphasis Bulleh Shāh's social aspect see this as his rebellion against the *'ulamā'* or any priestly class persons who trick people in the name of God. But I would argue that it is difficult to portray Bulleh Shāh as advocating any kind of physical violence or any form of social activism. Bulleh Shāh's notion of political is enmeshed with his metaphysical

81 Maqbūl, *Terey 'Ishq Nachāyā*, 221.
82 Maqbūl, *Terey 'Ishq Nachāyā*, 319.
83 *Ibid.*, 448.
84 Trans. by Shackle, *Bullhe Shah*, 347.

view of how God is the real trickster behind the façade of multiplicity—as argued in this section. Hence, a materialist reading of his poetry offers a very selective and narrow view of Bulleh Shāh's identity and message.

4.4 Metaphysics of Presence

One of his most powerful *kāfīs* openly speaks about how God has come, *deceptively*, to the world as a human being (*ādmī*) to enact all kinds of paradoxes.

ḍholā ādmī ban āyā
āpey āhū āpey chīta āpey māran dahāyā
āpey ṣāḥib āpey bardā āpey mul wikāyā
ḍholā ādmī ban āyā
kadī hāthī tey aswār hōyā, kadī ṭhuṭhā ḍāng bhawāyā
kadī rāwal jogī bhogī hogey sāngī sang banāyā
ḍholā ādmī ban āyā
bāzī gar kihya bāzī khaiḍī, mainūñ putlī wāng nachāyā
meiñ aus phaṛtāley nachnāñ hañ
jis gat yār likhāyā
dholā admī ban āyā
hābīl qābīl ādam jāiya ādam
kis dā jāyā
bullehā auhanūñ tuñ āgey āhā
dādā gūd khaḍāyā [85]

The beloved has come as man. How wonderfully he had come concealed.
It is he who is the deer, he who is the cheetah. It is he who moves fast to hunt them down. It is he who is the master, he who is the slave, he who lets himself be sold.
Sometimes he is mounted on an elephant, sometimes he is sent around begging with earthen cup and staff.
Sometimes he is a master yogi, sometimes a hedonist; he is an actor who assumes different guises.
What a show the Juggler has mounted; he has made me dance like a puppet. I must dance according to his measure, in the style that my beloved has shown me.
Abel and Cain were the sons of Adam, but who was Adam born from? Bullha, even before them, the Grandfather was dandled in his lap.[86]

85 Maqbūl, *Terey 'Ishq Nachāyā*, 179.
86 Trans. by Shackle, *Bullhe Shah*, 145–147.

Bulleh Shāh's image of a Sufi is a synthesis of Islamic and non-Islamic imagery—comparable to the borrowing of Christian imagery in Persian Sufism. For example, he says that one of the signs of the mysterious Beloved or charismatic stranger (*jogī*) is that he wears earrings and beads around his neck, and in beauty is a second Joseph (*Yūsuf*), made in the nature of *Aḥmad*. Here again we see how deception plays a part again, because

ais jogī dī aih nashānī
kan wich mundarāñ gal wich gānī
ṣūrat aus dī yūsuf sānī
aus alifūṇ aḥmad banāyā[87]

What are the marks of this Yogi?
He wear rings in his ears and beads around his neck.
His visage reflects the beauty of Yusuf.
He has turned *Alif* into *Ahad.*[88]

In addition to Bulleh Shāh's image of God as a trickster, who tricks people through his multiplicity, He (*sāīn*) is depicted as visiting the world quietly to watch His creatures' drama (*tamasha*) unfold as well. Despite his apparent absence, God stands among his creation in anonymity and is, therefore, very much present. The consistent focus of Bulleh Shāh is on God's immanence, and yet His active involvement in the mundane sphere is also a paradox, a matter of divine deception.

sāīn chup tamashey nūñ āyā[89]

The Lord has come to the fair in disguise![90]

He goes on to say that God first prohibited Adam from eating the grain of wheat and then encouraged the devil to go after him—in order to tempt and misguide him. So here the real culprit behind the entire drama is essentially God.

ādam kanakūñ man'a karāyā, āpey magar shaiṭan duṛāyā[91]

"You had Adam banned from the wheat; it was you who
Set Satan running after him."[92]

87 Maqbūl, *Terey 'Ishq Nachāyā*, 182.
88 Trans. by J. R. Puri & T. R. Shangari, *Bulleh Shah*, 170.
89 Maqbūl, *Terey 'Ishq Nachāyā*, 198.
90 Trans. by J. R. Puri & T. R. Shangari, *Bulleh Shah*, 368.
91 Maqbūl, *Terey 'Ishq Nachāyā*, 186.
92 Trans. by Shackle, *Bullhe Shah*, 243.

In the same *kafī*, Bulleh Shāh blames God for making Musa climb the mountain and also Ismail to lie down under Abraham's knife. Yet such are the ways of God. The entire *kāfī* speaks about all the people whom he has put in peril or difficulty.

> *mūsā nūñ koh-e ṭūr chaṛahāyū, ismā'īl nūñ zibaḥ karāyo*[93]
>
> "You made Moses climb Mount Sinai, you had Ismail slaughtered".[94]

Referring to this divine stranger, Bulleh Shāh wants to know this mysterious person, wearing such-and-such attire and prods us to ask him who he is with all sincerity and politeness. Perhaps Bulleh Shāh seems to suggest that this stranger will reveal his true identity if asked in all earnest:

> *kaon āyā phan libās kuṛey*
> *tusī phuchū nāl ikhlāṣ kuṛey*[95]
>
> Who has come clad in a new attire?
> Ask him in all frankness, O girls![96]

Naḥnū Aqribū (Self)

An important key to understanding Bulleh Shāh is what we may call his theopoetics of the self (or inner experience). The following *kāfī* gives a better insight into the metaphysics of presence, how the Divine dwells within the human body:

> *saiyyū hun meiñ sājan pāiyu aey*
> *har har dewich samāiyu aey.*[97]
>
> Oh girlfriends: Now I have found you, beloved. You
> Are contained in each and everything [everyone].[98]

The *kāfī* begins with Bulleh saying that the lover (*sājan*) he has found (*pāiyu*) is one who dwells (*samāiyu*) within everyone (*har har*):

---

93 Maqbūl, *Terey 'Ishq Nachāyā, 186.*
94 Trans. by Shackle, *Bullhe Shah*, 243.
95 Maqbūl, *Terey 'Ishq Nachāyā*, 240.
96 Trans. by J. R. Puri & T. R Shangari, *Bulleh Shah*, 251.
97 Maqbūl (ed.), *Terey 'Ishq Nachāyā*, 209.
98 Trans. by Shackle, *Bullhe Shah*, 45.

anā al-aḥad dā gīt sunāiyu
anā al-aḥmad hun phir farmāiyu
anā al-'arab bey 'aiyn bātāiyu
phir nām rasūl dharāiyu aey[99]

You recited the song of *I am Ahad the One*. Then you decreed, *I am Ahmad*. Then you explained, *I am an Arab without the A*. Then you took the name of Apostle.[100]

Entering by way of a song (*gīt*): *an al-ḥaq* (as famously uttered by Mansūr Ḥallaj), and then having this repeated by (Muḥammad) in saying (*famāyu*) *an al-Aḥmad*. A (more accurate) understanding of the relationship between *an al-aḥad* and *an al-aḥmad* is gained once we remove the *aiyn* (ع) of *an al-'arab*, (*rab!*), and then (the Divine) called, or named him (*nām*), Prophet (*rasūl*).[101]

fathammā wajh allāh nūr tusādā
har har wich ẓahūr tusādā
al-insān mazkūr tusāda
aithey apnā sar lukāiyu aey[102]

Then there is the face of God reveals your light. You are Manifest in each and everything. The word *Man* refers to you. Here you have hidden your face.[103]

And the face of God, as light (*nūr*), is what has been manifest (*zahūr*) everywhere and in everything (*Fathammā wajh allāh*), the Divine face hidden in *al-insān* (the human being).

tūñ āiyun tey meiñ na āiy
ganj makhfī ney murlī bajāī
ākh alast gawāī chāhī
authey qālū balā batāiyā aey[104]

---

99 Maqbūl (ed.), *Terey 'Ishq Nachāyā*, 209.
100 Trans. by Shackle, *Bullhe Shah*, 45.
101 Bulleh Shāh, "*ju rang rangayā gūhṛā rangayā*" in *Tere 'ishq nachāyā: Sharḥ Kalām Ḥazrat Bullhe Shāh*, Miyān Ẓafar Maqbūl (Lāhaūr: Maktabā Dāniyāl), 166:
Kun fayakūna āvāzah hoyā dīttī mīm vakhā lī ao yār
Aḥad kolūn aḥmad hoyā bismil mīm nakālī ao yār
102 Maqbūl (ed.), *Terey 'Ishq Nachāyā*, 209.
103 Trans. by Shackle, *Bullhe Shah*, 45.
104 Maqbūl (ed.), *Terey 'Ishq Nachāyā*, 209.

> You came, and I did not come. You played the flute of a hidden treasure. You said, *Am I not?* And desired testimony. There you announced, *They said, "Yes."*[105]

When God (you) manifested then what place is there for me. The hidden treasure (*gunj makhafī*), God, played the flute (*murlī*) to the tune of "Am I not your Lord" (*alast*), and all the creation (as the Divine manifestation) affirmed by saying (*qālū*) "*balā*" (yes!).

> *pargat ho kar nūr sadāiyu*
> *aḥmad tūn muwjūd karāiyu*
> *nabudūn kar būd dekhāiyu*
> *wa nafakhat fihī sunāiyu aey*[106]
>
> Becoming manifest, you called yourself light. You made it present through Ahmad. From nonbeing you made being appear. You proclaimed, *And I breathed into him.*[107]

This manifestation (*pargaṭ*) of God is known because of its light (*nūr*), as it emanates from *Aḥmad* (Prophet Muḥammad), from non-existence (*nabūd*) to existence (*būd*), and (God) blew Divine breath (*fa-anfukhu*, Qur'ān 3.49:18) into the body—or entered into that body:[108]

> *nahnū 'aqrabu likh dittu aey*
> *huwa ma'akum dā sabaq dittu aey*
> *wa fī anfūsakum dā hukm kītu aey*
> *phir keyhā ghunghat pāyu aey*[109]
>
> You wrote *We are nearer.* You taught the lesson of *He is with you.* You gave the command *And in your own selves.* So what is this veil that you have put on?[110]

Thus, it was written (*likh*) (in the Qur'ān): "We are closer to you then the jugular vein (*naḥnū 'aqrabū*), imparting the lesson (*sabaq*) that God is always with you

105 Trans. by Shackle, *Bullhe Shah*, 45.
106 Maqbūl (ed.), *Terey 'Ishq Nachāyā*, 209.
107 Trans. by Shackle, *Bullhe Shah*, 45.
108 Maqbūl (ed.), "*sīḥarfī (vav)*" in *Terey 'Ishq Nachāyā*, 403:
Vāhavā āp muḥammad apnī ādam shakal banāvey
Āpey roz azal dā malik āpey shafī' ho āvey
109 Maqbūl (ed.), *Terey 'Ishq Nachāyā*, 45.
110 Trans. by Shackle, *Bullhe Shah*, 45.

(*huwu ma'akum*), and just in the very moment of delivering that command, the Divine veiled (*ghunghaṭ*) itself.

bhar key waḥdat jām pilāiyu
Mansūr nūn chā mast karāiyu
Aus tun an al-ḥaq āp kahāiyu
Phir sulī pakaṛ chaṛāhaiyu aey[111]

You filled the cup of oneness and gave it him to drink. You
Made Mansur drunk. You were the one who made him
Say, *I am God*. Then you seized him and set him on the gallows.[112]

Filling the goblet with the wine of Oneness (*waḥdat jām*), gave it to Mansūr, intoxicating (*mast*) him, and then in that state of inebriety made him say: "I am the Truth," then put him on the gibbet (for such a transgression).

Wa-laqad karramanā banāyā
Lā ilāh dā pardah lāhayu
Illa al-lah kahu jhātī pāyu
Phir Bullahā nām dharāyu aey[113]

You made him remember *And we have honored them.*
You removed the veil of *There is no God*. You said *but*
God, and peeped out. Then you assumed the name of Bullha.[114]

In the final *kāfī* verses it says lifting the veil of *la ilāh* ("no God") is what allowed the seeing or witnessing of *ilā allāh* ("except/but God"), that is, the Divine Lord. And, so it is, that "Bulleh" was the name given, the mystic-poet Bulleh says about himself. In this last verse of the *kāfī*, we find that "Bulleh" is also one of the names of God.

Do the lovers of God die? To this Bulleh Shāh offers the following verses:

karāñ naṣihat waḍdī jey koī sun kar dil tey lāwey gā
mūey tāñ roz ḥashr nūn aṭhan 'āshiq na mar jāwey gā[115]

I give you sound advice,
If only you were to heed it.

111 Maqbūl (ed.), *Terey 'Ishq Nachāyā*, 210.
112 Trans. by Shackle, *Bullhe Shah*, 45.
113 Maqbūl (ed.), *Terey 'Ishq Nachāyā*, 210.
114 Trans. by Shackle, *Bullhe Shah*, 47.
115 Maqbūl (ed.), *Terey 'Ishq Nachāyā*, 174.

> The dead will rise on the day of judgement,
> But the lover will never die.[116]

Were I to give you weighty advice, if you are willing to take it to heart, Bulleh says, the dead bodies (*tān*) will rise only on the Day of Resurrection (*roz ḥashr*) but the lover never dies. So in contrast to those who have to wait to be resurrected on the Day of Resurrection, the lovers are resurrected on their deaths. In another *kāfī*, Bulleh clarifies this in the following manner:

> *"jey tūn marein maran tūn agey marney dā mul pāwein gā*
> *ḥijāb karein darwishī kolūn kad tak ḥukm chalāwein gā"*[117]

> If you die before your death,
> You'll reap the fruit of death.
> Loathe to live a saintly life,
> How long will you issue commands?[118]

If you "die before you die" only then will you find out the real worth (*mul*) of your death. In abstaining from the mystical way (*darwishī*), till how long will you persist in your resolve, [or literally, rule according to your own authority (*ḥukm*)]. Clearly, the answer is, not very long because death will finally end your dominion over life.

In the modern period, the understanding of such a relationship of love for the Divine has presented a number of challenges. Through the emergence of religio-nationalist identities in the 20th century the contours of sacred and the profane were more firmly established. The role of the state in particular came to exercise a profound role in the defining as well as the production of the Islamic public sphere. The nationalization of the *awqāf* (religious endowments), which included Sufi shrines, was one such strategy to employ Sufi culture in modernizing its ritual life. A new phase was opened in the complex historical relationship between Sufism and political power, which had always sought to co-opt Sufi charisma in service of the state. The Indus Valley Sufi culture faces a number of challenges, and modern notions of religious identity are one of the significant ones, among other developments, and a serious threat to its traditional religious imagination and the lived traditions of love-mysticism and irenic pluralism. Nevertheless, as long as the *dargahs* remain part of a sacred geography, the devo-

116 Trans. by J. R. Puri & T. R. Shangari, *Bulleh Shah*, 317.
117 Maqbūl (ed.), *Terey 'Ishq Nachāyā*, 174.
118 Trans. by J. R. Puri & T. R. Shangari, *Bulleh Shah*, 317

tional culture of Sufi music and love continues to persist in the Indus Valley region.

The dawn of modernity in South Asia transformed South Asian geography with the rise of nation-states in the region. Alongside the emergence of socio-religious reformist movements across the broad religious spectrum in India, nationalism was rapidly changing political life and, consequently, inventing modern identities in response to British colonialism. Pre-modern societies were undergoing an unprecedented change across the economic and social spectrums. Religion had to reconcile with scientific rationalism and refashion itself along scientific lines if it were to remain relevant in modern times, or so argued modernist scholars of Islam. Traditional reformist movements struggled to agree on hermeneutical approaches to the sacred sources—Qur'ān and Ḥadīth—and on approaches to pre-modern legal interpretation. Amidst much of the religious polemics of the 19th century, the role of Sufi shrine culture became one of the significant issues of debate and disagreement. The Wahhabī movement in the Arabian Peninsula came to exert a new, opposing force on Sufi shrines, and its extra-Arabian influence on religious discourse gained momentum with the creation of Pakistan in 1947. The Barelwī *'ulamā'*, predominantly Ḥanafī in their legal orientation, were the only traditional group of scholars who defended *ziyāra* to Sufi shrines and the celebration of *'urs*. However, their reformist bent sought also to reconcile Islamic law and Sufi culture in response to the also primarily Ḥanafī Deobandīs and the (*ghair muqallid*) Ahl-Ḥadīs, by emphasizing the cosmological and mythological importance of the Prophet (*nūr-i Muḥammad*). Their belief in the sanctity of the Prophet above all creation allowed the Barelwīs legally to defend festivals that celebrated the birth of the Prophet, and by association of *nūr-i Muḥammad*, the Muhammadan light, also the sanctity of the 'friends of God,' *'awlīya Allāh*. The Prophet, the greatest *wasīla or link* to God, could hear and see all who came to his grave, and so could the dead saints. An important aspect of this new role of Sufism in Barelwī discourse revolved around the notion of *fatīha-khwānī*—as mentioned in the introduction. Religious acts (*'ibadāt*), as *sawāb*, were offered to the souls of the saints in exchange for God's grace and to benefit from the *faiz* emanating from the grave. The Deobandīs also accept the permissibility of reciting *fatīha* for the dead, but only through extremely circumscribed rituals (as discussed above in chapter 1).

Sufism and Sufi poetry have become an extremely problematic issue in Pakistan and elsewhere in the modern Muslim world. In Pakistan, the poetic sensibility of mystical thought has been detached from its cosmological context and sacred space, namely the Sufi shrine. From the perspective of state discourse, the Sufism of the shrines needed to be Islamized and modernized through the extermination of its medieval elements. The critical debate over Sufism has been framed by this overarching concern to reform shrine culture in Pakistan almost from the inception of the new state.

The issue of language also has played a pivotal role in determining how shrine-based Sufism was to be treated at the local level of politics. The dominance of Urdu provoked ethnic sentiments in many areas. In response, several intellectuals began to emphasize their ethnic identities and seek a revival of their native regional languages. These modern writers and scholars began assiduously cultivating their linguistic heritages. Still, they were always uncomfortable with the devotional aspects of vernacular Sufi poetry, specifically the intimate connection between Sufi *kalām* and the tombs of the saints.[119] These groups were keen to interpret Sufis as revolutionary social and political reformers who wished to raise the level of consciousness of their respective peoples and instill a sense of political activism. Among the religious groups, the only acceptable act of religious contemplation and devotional practice (*'ibādat*) were the legally recognized shared rituals of Islam: *namāz*, *rozā* and *ḥajj*. The nationalists were largely anti-religious and generally disdainful of anything resembling mystical contemplation. Hence, in the view of, both, Islamic activism and ethno-nationalist consciousness, Sufi contemplative traditions and imagination are redundant and useless remnants of a bygone age. For the Islamists, as for the anti-shrine *'ulamā'* and the modernists, Sufism had long lost its connection with its roots in the Golden Age of early Islam and taken on non-Islamic practices and superstitions. Such an understanding of Sufism then allowed for modern, literate Muslims selectively to appreciate its poetical tradition without paying attention to its living and lived component in contemporary times.

In Pakistan, Urdu as the official language, and English as the administrative one, replaced Persian, which had already been displaced by English under the British. Vernacular languages were ignored within the educational institutions, thereby cutting off the majority of the population from the region's indigenous literary heritage. The marginalization of vernacular literary traditions had sever-

---

119 The trend by elite households to hold Qawallī sessions at their homes for private gatherings —just as one would invite any professional singer by paying an agreed upon monetary sum—was one way by which modern folks could enjoy its experience without necessarily believing in its metaphysical beliefs. Traditionally, the *qawals* won acclaim, and monetary rewards, by voluntary participation at *'urs* festival at shrines. It was the masses attending the *samā'* gathering that determined their popularity and the type of emotional response the performers could invoke from the devotional audience. All the legendary folk musicians in Pakistan earned their name by participating in such festivals of saints, and only later came to be invited to perform on national radio and television, and eventually international recognition. However, many a folk performer died without receiving any royalty from such repeated broadcasts.

al repercussions that have been discussed too narrowly. Besides political concerns over religious intolerance in the country, the large-scale secularization of Sufi culture on one hand and the marginalization of vernacular languages on the other were seriously jeopardizing the opportunity and means by which the urban educated classes could even remotely understand the religious experience of devotees at Sufi shrines in any meaningful way. Sufi folklore became too strongly associated with feudalism, magic, superstition and strange as well as dangerous customs of healing. The space of the shrine came to be textually and visually construed as a dark and dangerous site of sexual deviancy and violence. A variety of social ills, such as child abuse, begging, sorcery, and sexual exploitation were seen as part of the shrine culture in modern literature. It is no wonder that the carnivalesque atmosphere at Sufi festivals quickly came to be seen as evidence of the kind of dangers that lurk in the vicinity of the shrine.[120] Barelwī Islam played a crucial role in the normalization of shrine-based Sufism through its constant emphasis on legally recognizable practices such as *du'ā*, prayer, and fasting, all of which are seen as the accurate expression of religious love (*'ishq*), which is a spiritual currency offered to the soul of the saint for manifold return by God. Needless to say, the point here is that much of the folk aesthetics of religious devotion were replaced by the Barelwī style of religious devotion. Today's urban class is most familiar with Barelwī Sufism, and other vernacular traditions of *ziyāra* seem to have lost their literary voice.

"The set of measures towards a theocratic state sought to inculcate obscurantist views and induced a narrowing of the human mind. It involved the suspension of the sensibility of love and reason underlying the religious tradition signified in Pakistan's folk culture."[121] This quotation is symptomatic of the unconscious contradiction at work when dealing with Sufism by conflating "the religious" with "folk culture." "The religious tradition" in this citation is an implicit reference to the plural aspects of "Sufism"—as a counterfoil to religious conservatism—while the category of "folk culture" is appreciated only for its sec-

---

120 The disciplined and very much sober gathering of *tablighi jama'at is* opposed to the type of *'urs* festivities in Pakistan, which in fact, become the largest religious gathering in the Muslim world after *ḥajj*, both in Bangladesh and Pakistan. Such gatherings do have psychological and political dimensions that cannot be overlooked and need comparative work. The annual meetings of the *jama'at* are criticized for breeding religious conservatism, but such massive gatherings rarely invoke this kind of anxiety.

121 Akmal Hussain, "Pakistan's Economy in Historical Perspective: Growth, Power, and Poverty," in *Pakistan: The Struggle Within* edited by John Wilson (India: Dorling Kindersley Pvt. Ltd., 2009) 51.

ular, cultural dimension. Therefore, Sufism in the service of state discourse is both secular and religious at the same time.

A figure like Bulleh Shāh represents a mystical tradition grounded in the thorough mixing up of the sacred and the profane. His decision to forsake his masculine position of social and religious authority for an effeminate and lower social status; his use of dance as an expression of *'ibadat* (worship); and his *malāmatī* ethos of blame, anonymity, and trickery to win divine friendship—all of these cannot be adequately interpreted without seeing him in that tradition of lovers (*'ashiqān*) who were made eternally alive by surrender to the love of the Divine. The metaphoric image of the heart as a way of seeing and experiencing reality is not taken as a legitimate epistemology in modernity, but it is the only lens through which piety such as that exemplified by and associated with the veneration of saints such as Bulleh Shāh.

Chapter Five
Festival Ritual

The *'urs* of Sufi saints is, first and foremost, a celebration rather than an occasion for renewed grief, as one outside the tradition might expect. It marks the transcendence of death by the saint, who is a friend of God (*walī Allāh*). Thus, it is not the saint's death but rather his/her marriage to God that inspires the gathering and collective joy. The special immortality of the Sufi saint allows for a kind of communion between him or her and the visitors. In the vernacular strand of Sufism examined here, one enters a sacred, performative space to be in the living *presence* of the saint.[1] This is described as *ḥazirī*, the meaning of which I discussed in chapter 1.

A Sufi saint is believed to possess a bodily essence—produced by an alchemical transformation of the corporeal state—that is different from that of an ordinary person.[2] The saint most effectively conveys this embodied alterity in the existential overcoming of death.[3] The dead saint is experienced as, in fact *alive* and capable of extraordinary agency in tending to the needs and heartaches of those who visit.[4] The extraordinary transcendence of death by the saint is, therefore,

1 Sufism in much of the Indus, especially in Punjab and Sindh, is given to various forms of performative traditions. Several folk musicians are associated with Sufi shrines, having performed only Sufi verses throughout their careers. During the *Qawwālī* sessions at the time of the *'urs* of the Sufi saint, the verses performed are specifically directed towards that particular Sufi. Hence the *mehfil-i samā'* is structured around the mention of the Sufi whose *'urs* is being celebrated.
2 See Martin Lings, *Symbol and Archetype: A Study of the Meaning of Existence* (Lahore: Suhail Academy, 2000), 23: "The harmonious reunion of these two poles [masculine and feminine] of the psychic substance is termed in alchemy the chemical marriage".
3 Besides the belief that the Sufi saint is an exceptional person, gifted with wonderful powers—commonly read as 'supernatural' abilities according to scientific view, one of the most (literally) creative ideas is that the friend of God can give life as well, like *Īsa* (Jesus), who is a an active inspiration in Islamic mysticism: see Sara Sviri, "*Kun*-The Existence-Bestowing Word Power of Language," in Sergio La Porta and David Dean Shulman, eds., *The Poetics of Grammar and the Metaphysics of Sound and Sign* (Leiden: Brill, 2007), 35–67.
4 The grave in Islam plays a vital role in defining the Muslim religious experience, informing Islamic ethics, and impressing upon everyone the importance of a pious life. Literature is abundant on the condition of the soul after death, describing the kind of tribulations and examinations that await every individual in the grave and the kind of privileges that will be granted to the pious soul. The questioning of the deceased by the angels in the grave may be a uniquely Islamic intervention that is described in fascinating detail: for a sense of the gravity of the process, see Christian Lange, "Barzakh," in Kate Fleet, Gudrun Krämer, Denis Matringe, John Nawas, Everett Rowson, eds., *Encyclopedia of Islam, Three* (Leiden: Brill, 2011). The very mention of the

https://doi.org/10.1515/9783110781496-007

an exceptional feat. Still, in Sufi folklore it is also ultimately an act of Divine grace toward the friends of God.[5] Moreover, the mystical body of the saint, occupying an interactive space between the saint in the tomb and the devotees and depicted in a range of metaphoric imagery, is *experienced* in the performative culture of Sufi shrines.[6]

There are several ways to interpret the emotive and celebratory aspect of *'urs*. It can be studied in its political context, as a ritual, and also in sociological perspective by examining the meanings and functions that inform its practice. The aim in this chapter is to examine the celebration of *'urs* in its theopoetic dimensions (see below), an approach that has not been comprehensively undertaken in contemporary study of *'urs* thus far. Here I shall argue that the notion of *'urs*, as union between the Sufi saint and God, is inherently a formulation of the theopoetic imagination in celebration of that mysterious union.

In considering the *'urs* celebration as a ritual phenomenon, I will consider the theopoetic use of the word *melā* in vernacular Sufism first by examining three *kāfīs* in total, one by Bulleh Shāh (d. 1757) and the other one performed by Ḥamid 'Alī Bela (d. 2001), a modern virtuoso folk musician who dedicated himself to a great mystic of Lahore, Shāh Ḥussain (d. 1599), and who is known for having sung only his verses. I will then consider a third *kāfī*, also by Bulleh Shāh, which celebrates the Hindu festival of Holī in the context of the Sufi metaphysics of *nūr muḥammadī*, "the light of Muḥammad". This third *kāfī* is a particularly good example of sacred play at work in vernacular Sufism, and I will discuss it with particular reference to Bulleh Shāh's other mystical verses on the notions of divine play and reflexivity.

The Sufi theopoetics to be examined below in the vernacular *kāfīs* offers a very different style of religious engagement than the one dictated by the con-

grave provokes anxiety among the believers because of the harsh interrogation of the soul that is said to begin immediately therein. Hence, against this frightening procedure described in the literature authored by religious experts to alert Muslims of the perils of the grave, the Sufi saint offers a striking contrast because the Sufi has transcended this stage altogether to possess intercessory powers for the living and even the ordinary dead.

5 In vernacular Sufi folklore and *qawallīs*, it is common to come across notions of how the Sufi will intercede on behalf of the devotee in the grave, or how giving the name of the Sufi to the angels will avoid the harsh interrogation by the angels who descend into the grave at death. Finding a burial space next to the tomb of a Sufi saint is considered a blessing because of his/her intercessory powers.

6 See Amira Mittermaier, "Reimagining Space: Dreams and Saint Shrines in Egypt," in George Stauth and Samuli Schielke, eds., *Dimensions of Locality: Muslim Saints, Their Place and Space* (Bielefeld: Transcript Verlag, 2008), 47.

cerns of *īṣāl-i ṣawāb* for the dead.[7] In general, scholarship on Sufi shrines has not paid sufficient attention to poetic aesthetics (theopoetics) in drawing a connection between devotional practices on the one hand and modes of religious knowing on the other. Therefore, it is essential first to expand the definition of what happens in the ritual space at Sufi shrines and to include modes of theopoetic thought and aesthetics as integral parts of the religious experience at shrines—aspects that have not been examined previously in scholarship on Sufi shrines in the subcontinent.[8]

In vernacular Sufism of the Indus region, *'urs* is commonly described as a *melā*. This expresses the festive nature of the social gathering that takes place in the vicinity of the Sufi shrine. While there are many aspects to shrine celebrations that are almost impossible to classify because of the clear ambiguities and contradictions at play, nevertheless the central role and symbolism of the Sufi saint are widely acknowledged in the scholarly literature.[9] The figure of the Sufi saint has continued to be of vital importance in the devotional shrine culture over the centuries, and this importance has persisted widely in the modern era, not least in Pakistan. Meanwhile, the notion of *baraka* has been the singularly most important conceptual term used in scholarship to describe the per-

---

7 Deobandis reject any action that may give a hint of agency to the saint, as discussed in an earlier chapter, and they correspondingly allow only *īṣāl-i ṣawāb* to gain access to the *baraka* of the saint's tomb. The debates between the reformers over Sufi tombs and rituals intensified in the course of the 19th century, which had consequences for the religious culture of these sites. Indisputably, the space of the Sufi shrine has changed dramatically from its pre-modern social ecology. In Pakistan, the devotional space of Sufi shrines was caught between three interconnected historical forces: reformism, the state, and modernity. Each of these elements participated in regulating Sufi shrines, first under the aegis of British colonialism and later under the nation state. The overlap of these three factors is critical to understanding the transformation of Sufi shrines in modern Pakistan. In Pakistan, at least, Sufi shrines have lost much of the social dynamics of religious pluralism that previously marked the character of their festivals. However, Sufi shrines are still very much places for confluence different religious traditions and meeting grounds for Muslim elites and ordinary Muslims. It cannot, however, be denied that religious and political developments have put tremendous pressure on the earlier social environment of Sufi shrines. Such pressure is especially evident in how sources of patronage have come to be controlled by the government and in the political impact of religious reformism (discussed above in Chapter 3).

8 There has been some progress towards overcoming this gap in scholarship. See Abha Chauhan (ed.), *Understanding Culture and Society in India: A Study of Sufis, Saints and Deities in Jammu Region* (Singapore: Springer 2021). The work was recommended by an anonymous reviewer, for which I am thankful.

9 For an important, and the only extensive anthropological study of Sufi festivals see Samuli J. Scheilke, *The Perils of Joy: Contesting Mulid Festivals in Contemporary Egypt* (Syracuse: Syracuse University Press, 2012).

ceived religious efficacy and sacred nature of Sufi shrines and the material objects associated with them. Yet the saint's living quality or presence is equally relevant for the experiencing devotee.[10] The approach taken here, in contrast to Qamar-ul Huda's effort literally to bind ritual praxis at Sufi shrines to an 11th-century Sufi treatise on mysticism in examining the 'functional' presence of the saint, does not subscribe to reading the devotional practices at *dargahs* as having a direct correspondence with texts or manuals of Sufi devotions.

Secondly, to state the obvious, there are several socio-cultural layers to the ecology of Sufi shrines in Pakistan (as in other regions), some overlapping and others contradictory. Despite the ambiguous nature of *'urs* in its sacred-profane dimensions, the devotional core of Sufi shrines is its religious—in particular, as I emphasize here—its theopoetic *habitus.*[11]

For the devotees, just as the word *ḥāẓirī* carries a meaning that differs from the "official" meaning given to *ziyāra* to Sufi shrines, the use of the word *melā* to describe the *'urs* of a saint is not without significance—as I will show in the discussion of the *kāfīs* below. The *melā* should be seen as occasion for a religious experience in relation to the living presence of the Sufi saint, whose sacred body at the tomb is the primary reason for the ritual performance of the festival. The *kāfīs* studied here help us understand the theopoetics of the celebration (*'urs*), and the associated notions of *sacred time* that the poetry of Sufi saints (such as Bulleh Shāh) is able to impress upon its audience. At the center of the *melā* is the resurrected body of the saint, which is central to both performance of ritual (*ḥaẓirī*) and recitation of poetry (in *samā'*) in the vicinity of the *dargāh*. Therefore, the *kāfīs* recited/sung at the *melās*, as the spoken words of saints, enable us to grasp the "metaphysics of presence," which, I argue, is essential to understanding the ways of experiencing the sacred on the occasion of *'urs*. The *'urs*, in its anthropological dimension, is a celebration of colorful multiplicity, which makes it what it is: a *melā*. The *melā* is open to all religious communities and social classes to mingle and mix, to sit in a circle of friendship to experience the sacred in its manifold colors. The present chapter offers a small contribution to a phenomenology of the typical *melā* of Sufi saints of the Indus.

---

10 See Qamar-ul Huda, "Celebrating Death and Engaging in Texts at Dātā Ganj Bakhsh's *'Urs*," *The Muslim World* 90 3/4 (2000), 377–394.

11 I follow the use of this term (*habitus*) in the above-cited recent study by Schielke on celebration at Sufi shrines in Egypt, *The Perils of Joy*, 33: "To speak of habitus in relation to festive practice means, first, to indicate that certain dispositions of the body (in a wider sense including dress, speech, movement, and the like) are specific to the festive context and contribute to its atmosphere."

In what follows, I examine the vernacular meaning of *melā* in Indus Sufism and then turn to its usage in the spoken word (*kāfī*) of the Indus saints to explore the phenomenological dimensions of the festival. The chapter aims to argue how the experiential dynamics of the *melā* as a celebration of the saint's *'urs* (mystical union with the One) is simultaneously about deriving joy from the manifest multiplicity at the *melā* as a sacred *hierophany*. The fanfare and the tumult, the social and religious diversity, the music, and the ecstasy is for the devotees a celebration of the living presence of the saint through shared rejoicing and lighting of lamps at the graves of the resurrected. In other words, its description as *melā* (in the *kāfīs* examined) has a metaphysical significance. It expresses how devotees experience ritual space and time at the Sufi tomb. The *kāfīs* illustrate how the metaphysics of presence (*ḥāẓirī*) informs the theopoetics of ritual experience at the *melā* (*'urs*) of saints.

Next, the chapter briefly discusses the treatment of the festival in earlier scholarship, focusing on the contribution of two anthropologists whose valuable analyses of celebration—Werbner on *'urs* and Schiekle on *mulid* (birthday celebration)—offer two different perspectives on the phenomenon. The aim and objectives of these two studies vary, but both attempt to understand the performative dynamics at work in the celebration.

5.1 *Melā* in Indus Sufism

In vernacular Sufism of the Indus valley region, the *'urs* (wedding) of Sufi saints are, as noted earlier, commonly described as a *melā*. The meanings of the word *melā* given in the *Punjābī Urdū Lughat* are given in the following order: *fuqarā kā jalsa* (a gathering of *faqīrs*), *'urs*; *tamashā*, *tamāshāgāh* (a spectacle or place of spectacle); *rawnaq* (festivity), *malāp* (meeting), *waṣal* (union); and *hajūm* (crowd). *Melā* derives from the word "*mel*" that means to meet, and it implies a rejoicing in this coming together (*atiḥād*).[12] The notion of union is implicit in the meaning of *melā* and it signals a confluence of diverse personages that creates the dynamics of the *melā*; even more importantly, it signals a loss of individual identity and differentiation.[13] In this sense, the etymological force of the

12 Tanwīr Bukharī, *Panjābī Urdū Lughat* (Lāhaur: Urdū Sā'ins Būrḍ, 1989), 1440.

13 It is also used on occasion as a pejorative by the urban folks to mean unstructured or unorganized state of affairs, and generally associated with rural life. That is, to call a structured or formal event as a *melā* is a derogatory way to describe an unruly state of affairs in urbanized culture.

word *melā* is not easily gleaned from its commonly used English equivalent "festival".[14]

In the case of Indus Sufism, the *melā* of saints is expressive of a particular kind of collective joy associated with visiting a friend of God to celebrate the saint's special transcendence of death. The union implied in the meaning of *'urs* as a wedding translates into a *melā* in the public sphere and in theory cuts across differences of gender, ethnicity, and religious identity.

Predictably, reformist movements have been extremely critical of this visible diversity of collective gatherings at the *melā*s of Sufi saints. Reformist interventions seek to turn these assemblies into more regimented, ritually structured social and religious gatherings. Two forces have been most active in this: the state, through its nationalization program beginning in the 1960s, which has given the control of major shrines to the Department of *Awqāf*; and the Barelwīs and their religious network of scholars and students across Pakistan. The latter, in particular, has been heavily involved in regulating shrines and participating in their devotional culture (albeit according to their own interpretive guidelines). The Barelwīs believe in the intercessional role of saints through the *īṣāl-i ṣawāb*, which in their interpretation provides the only Islamic legal defense of *ziyāra*. The present study examines those modes of devotional activity that extend well beyond the legal sphere alone. Such modes have remained unmapped mainly because of their performative and oral mediums of expression. They need to be seen and described in the context of lived social practice involving meanings beyond those from the discursive textual domain of Muslim religious authorities.

For example, the 'urs of Shāh Ḥussain is well known in Lahore as a carnival for the lower classes, one held outside the formal vicinity of the *dargah*. However, even better known is the *melā chirāghān* (festival of lights), which is an integral part of religious/devotional ritual at Shah Hussayn's tomb. The *melā* is, first and foremost, a religious festival that celebrates Shāh Ḥussain as a saint. According to Gaborieau, there are two aspects of the *'urs*, one sacred and the other profane. [15] The "sacred" realm of the *'urs* is defined in the following way:

---

14 On the use of the term "festival" more generally, Beverly J. Stoeltje remarks that the term *festival* itself is rarely used, whether speaking of events such as Mardi Gras (Catholic), Sukkot (Jewish), Holi (Hindu), Shalako (Zuni), or the Feast of Fools (French), in "Festival," Richard Bauman (ed.), *Folklore, Cultural Performances, and Popular Entertainments. A Communications-centered Handbook* (Oxford: Oxford University Press, 1992).

15 Marc Gaborieau, "The Cult of Saints in Nepal and Northern India," in Stephen Wilson, *Saints and their Cults: Studies in Religious Sociology, Folklore and History* (Cambridge: Cambridge University Press, 1983), 297.

The rituals involved are both individual and collective. As in the occasional rites, the faithful go around the tomb, touch it and leave offerings individually. But the annual festivals also involve a collective cult, which always includes recitation from the Koran and canonical prayers. Where music is allowed, there is also devotional and mystical chanting. Lastly, animals, usually goats, are sacrificed to God in the name of the saint and are offered up to him symbolically together with the food prepared for the faithful. The latter then consume both in a banquet which, though it may seem like a picnic, is in the context of the festival a sort of sacrificial meal: the faithful eat the remainder of the food which has been blessed and offered up in the name of the saint.[16]

As for the "profane" realm, Gaborieau continues:

After this, the rest of the festival becomes entirely profane: a sort of fair lasting all night (the ceremonies having begun at sunset) and the rest of the following. *This is called melā, the name given to the same type of gathering at the shrines of Hindu gods* [emphasis mine]. One also finds there the same mixture of commercial activity and entertainment. It is a great occasion for traders and restaurateurs who set out their wares by the shrines and do brisk business. There are all kinds of entertainments: Muslim fakirs do turns, such as walking through fire. There are a large number of jugglers, tumblers and exhibitors of animals, all of whom are often members of Sufi orders. There are musicians, dancing girls and prostitutes. The faithful sing and dance too. The fair quickly becomes a carnival, one of whose functions is to allow a certain license. Such then are the general lines along which the cult of saints is organized.[17]

5.2 *Melā* in *Kāfī* Verse

As noted above, the *'urs* of saints is called a *melā* in the vernacular idiom. The term *melā* is usually seen as a generic word to describe the ludic aspect of any collective event. Yet in vernacular culture, the *'urs* of Sufi saints being labeled a *melā* should not be understood as marking it as "secular" and without religious significance. In this section, the aim is primarily to examine three vernacular Panjabī *kāfīs* to trace the religious meanings of *melā* in them, which can give a good sense of the religious inflections or shades of meaning that are to be found in the performative—and imaginative—space of Sufi poetics and devotion.

It will be essential throughout this discussion to keep in mind the performative dimension of these *kāfīs*. These verses are meant to be sung, and as tends to be the case in performance, they elicit and condition the participatory expe-

16 *Ibid.*
17 *Ibid.*, 298.

rience of the audience.[18] Performances in devotional gatherings such as *samā'* do not follow a tightly structured format, and there is much room for improvisation and spontaneity. In a way, each sacred performance in a religious gathering is unique, unrepeatable.[19] The performers at Sufi shrines are aware of the theological significance of the verses and the meanings, which the skilled musician will exploit by modulation of musical scale, tone, and rhythm. It is for this reason that performers in sacred gatherings can stay with a single verse for some time, repeating the stanza with variations and in differing cadences, until they sense that the audience is ready to move on.

The present section will focus only on the structure and content of the three selected *kāfīs* to analyze their theological and aesthetic features. In particular, I want to show how the meaning of the *melā* in these Panjabī verses contributes or compares to the understanding of *'urs* as a ritual. Also, it is equally important to clearly state the limits of the analysis at the outset: an investigation of the underlying musical composition or structure will not be attempted, as such analysis falls beyond the scope of this study.

Of the three *kāfīs* to be considered, two are by Bulleh Shāh and the third one performed by the famous folk singer, Ḥamid 'Alī Bela, whose author is not known. The first by Bulleh Shah is a variation on a popular folk *kāfī* that deals entirely with religious motifs and narratives. The second *kāfī* of Bulleh Shāh and the one performed by Ḥamid 'Alī Bela, are more directly concerned with the theme of *melā*. The *kāfīs* by Bulleh Shāh is an example of literary and theological pluralism, and therefore adds an important complementary dimension to the various expressions of *melā*. Specifically, this *kāfī* of Bulleh Shāh, "*horī khelūṇ kī kaheh kar bismillah*", conveys the sacred playfulness at work in sanctifying the popular Hindu festival of *ḥolī* with the invocation of the *basmalah* (saying "in the name of God, the merciful compassionate One") and the creedal affirmation of *lā ilāha il-allāh* ("there is no god but God"). This *kāfī* is about the significance of play and offers a contrast to ritual regimen-

18 Needless to say, the dynamics of the performance of these *kāfīs* in a devotional gathering are different from those in a secular concert, wherein the audience would not share in the emotive associations integral to the composition of the *kāfīs* (the next chapter will consider the issue of religious experience in the performative space of Indus Sufism).

19 I have personally attended numerous *samā'* performances of the internationally renowned *Qawwāl*, Nuṣrat Fateh 'Alī Khan (d. 1997) at a Sufi shrine in *Mazang* cemetery, in Lahore, which were much less formally organized. The *qawwalīs* would include verses from other compositions (what can be described in colloquial terms as *joṛ toṛ*—literally: to join and to break) in order to sustain the religious experience or mood (*ḥāl*) of the moment, than his concerts in secular or commercial settings, either in within the country or at international avenues.

tation at Sufi shrines under the legalistic influence of reformism in Pakistan. In the context of reformism, rituals of *ziyāra* are legitimized through the idea of sacred exchange (*īṣāl-i ṣawāb*). However, in these *kafī*s the metaphysics is intricately interwoven with a mode of aesthetics that speaks of an entirely different kind of devotional engagement and imaginative thinking. Also, in examining the structure and context of the *kāfī*s, we have to pay attention to the hermeneutics of theological intercourse. In two of the *kāfī*s on the theme of *melā*, the very act of gathering (*melā*) embodies a form of religious witnessing through seeing, playing, hearing, and experiencing. These embodied aesthetics and the bodily gestures accompanying the *kāfī*s are not only an integral part of vernacular theopoetics but also hold clues to understanding religious experience. Therefore, to a large degree, the *kāfī*s surveyed can be understood as representative of the hermeneutics involved in Indus Sufism, especially how classical religious symbols and motifs—Islamic, Sikh, or Hindu—are translated into the vernacular idiom.[20]

The *kāfī*s examined here interpret the *melā* in the context of a much larger narrative of events. In each case, the event manifests as a spectacle, both in worldly and otherworldly senses. Although all of the examples here require different interpretations, the *kāfī*s together provide a sense of how the notion of *melā* is employed towards an entirely religious end. The word *melā* is recognized across South Asia and is used to convey several events, but here the aim is to examine its particular application in the context of Sufi shrine culture—the *melā* as an expression of vernacular versions of Sufi theosophy.

***Kāfī* by Bulleh Shāh**

The complete *kāfī* is presented below and will be discussed in detail. In offering, as comprehensively as possible, a translation of the *kāfī*, I have had to take recourse to providing two versions. The first line of English translation (in parentheses) is my attempt to offer as literal a translation as possible, without concern for poetic form or meter. The second line of English translation is that by Chris-

20 A valuable work in better understanding the hermeneutic dynamics shared between different religious communities, see Yogish Snehi, *Spatializing Popular Sufi Shrines in Punjab: Dreams, Memories, Territoriality* (London: Routledge, 2019), 16: "Punjab had a significant presence of Nath practices and repository of shrines associated with Nath *yogis*...so predominant was the metaphor of *jogi* in the Indus region that the yogi became a popular figure in poetry written in Punjabi and Sindhi."

topher Shackle, a leading authority on Panjabī literature, history, and poetics.[21] In adducing two different translations of this *kāfī*, the aim is to provide deeper and better insight into the poetics and meaning of the *kāfī*. In Panjabī verse, the combination of simplicity and striking wit is difficult to convey in a single (literal or free) English translation.

Āo faqīro meylley chaliye 'ārif kā sun bājā rey (1)
(Come *faqīr* to the *melā*, listen to the gnostic's music.)
Come, fakirs, let us go to the fair, and listen to the music of the adept.

anḥad sabad sunū bahū rungī tajaiye bhikh biyājā rey (2)
(Listen to the multi-colored 'unstruck' word, cast aside your selfish interests)
Listen to the many-hued unstruck music and abandon
your robe of falseness.

anḥad wājā sarab malā pī nirvairī sarnājā rey (3)
(That unstruck musical instrument is undifferentiated in its music/emanation)
The unstruck instrument is friendly to all, the pipe has no hostility.

meley bāḥajhūṇ melā awtar ruṛh gayā mol weḥājā rey (4)
(Without such union/meeting at the festival, the *melā* is devoid of all merit)
Without union the fair is no good; both capital and
interest are squandered.

kaṭhan faqīrī rasta 'ashiq qa'im karo man bājā rey (5)
(Adopt the difficult path of *faqīrī*, the self is a musical instrument)
Fakirhood is difficult, but it is the path of the lover, so
make your restless mind settle.

bandā rab bhiyu aik bullhā paṛā jaḥan barājā rey (6)
(The human and God as one in Bulleh, sitting in Divine kingship.)
The creature and the lord become one, Bullha, while the world
is ungoverned.

The title of Bulleh Shāh's *kāfī*, *āo faqīro melley chaliye,* is a slightly playful exhortation to go (*chaliye*) to participate in a festive gathering (*melā*). The desire to go to the *melā* is triggered by sacred music. It is hearing a Divine sound/melody by Bulleh Shāh (*'ārif kā sun bājā rey*) that is a sign (*ishāra*) for the mystic to go towards the *melā*, the gathering or celebration.[22] This sacred music is none other

---

21 Bullhe Shah, *Sufi Lyrics*, ed. and trans. Christopher Shackle, *Murty Classical Library of India* (Cambridge: Harvard University Press, 2015), 23–25.

22 In Sufism, mystical knowledge is expressed with the term *ma'rifa*, and the one who possesses such knowledge is an *'ārif*, and "although there are alternative opinions, most Sufi authors argue for the superiority of *ma'rifa* over *'ilm* (discursive knowledge)," John Renard, *Historical Dictionary of Sufism* (Oxford: The Scarecrow Press, Inc., 2005), 139; Christopher Shackle in his

than one sounded by the *'Ārif* (knower or even gnostic), whom Dārā Shikūh, in his treatise on Hindu-Islamic mysticism, defines as the knower of God. For him,*'ārifs* are those "who are immune from senselessness or thoughtlessness both in this and the next world."[23] In their state of sincerity and innocence, they come to be in tune with the Divine. The *bājā* of the *'Arif* is in tune with the Divine sound, which refers to the harmony and rhythm that a knower of the sacred possesses. This can manifest itself in verse or musical compositions. Therefore, this *'ārif kā bājā* is a reference to the Divine nature or revelation of the words, hymns and sounds heard in sacred spaces.[24] It is the *'ārif* who becomes a conduit for transmitting the Divine sound to humanity. This creative synergy or rhythm in the *'ārif kā bājā* with cosmic roots in the unstruck sound (*anḥad sabad*) can be said to spring forth in/at the *melā*, a sacred gathering or celebration. That is, the *melā* is directly associated with, or even a manifestation of, the mystic's hymns, verses, songs, and sacred musical compositions (*'ārif kā bājā*): *anḥad sabad sunū bahū rungī tajaiye bhikh biyājā rey* (Listen to the multi-colored 'unstruck' word, cast aside your selfish interests)

The second verse of the *kāfī* employs an important theological concept in Sikh theology and in Nāth and Haṭha-yoga. According to Mandair's explanation in his note on "*anhad sabda*", the term means literally "unbeaten or unstruck," i.e., when the source of a sound effaces itself in the moment of being sounded. In the *Ādi Granth* the words *anhad/anāhad* are variants of the Sanskrit *anahata*. The term *sabda [śabda?]* often occurs within the *Ādi Granth* in combination with the term *anhad* (or its variant *anāhad*)."[25] Bulleh Shāh invites the listener to pay

---

translation in *Sufi Lyrics* defines it thus: "the word "adept" (Ar. *'ārif*) is a Sufi term, like "fakir" (Ar. *faqīr*). But the *kāfī* also draws on the language of yoga and is written in the characteristic Hindi of the yogis", 379; for a more detailed meaning of the term in Ibn 'Arabī's work see Toshihiko Izutzu, *Sufism and Taoism: A Comparative Study of Key Philosophical Concepts* (California: University of California Press, 1983).

23 Dārā Shikūh, *Majma ul-Baḥrain or The Mingling of the Two Oceans*," translated by M. Mahfuz ul-Haq ([first published in Calcutta in 1929] Karachi: Royal Book Company, 1990), 69.

24 The sacred nature of the Sufi verses and poetry has become more divisive issue in the modern period as reformist voices seek to restrict the parameters of Divine revelation to the text of the Qur'ān either in a literalist manner, or as the traditional *'ulamā'* for who accept their selection of commentaries as the most legitimate way of accessing the meaning of the Divine Revelation in the Qur'ān.

25 Arvind-Pal S. Mandair, *Religion and the Specter of the West: Sikhism, India, Postcoloniality, and the Politics of Translation* (New York: Columbia University Press, 2009), 467, n. 50, see S. S. K. "Sabad (Punjabi)," *Encyclopaedia of Indian Literature, Volume IV*, ed. Mohan Lal (New Delhi: Sahitya Akademi, 1991): "The sabad in the *Adi Granth* pertains to the sound vibrations of the word uttered by the Guru (Guru-shabad), the 'mantra' given by the Guru to the seeker...if the subad is recited continuously, its sound will lead the aspirant to eternal bliss...[and]

heed to the *anḥad sabad* and to forget about this-worldly cares and economic interests.[26] The second stanza of the verse is a call to forgo materialist concerns, *tajaiye bhikh biyājā rey*, and to concentrate on listening to this sacred music or sound. This can also be read as transcending religious or theological boundaries to focus on the Eternal sound or word, the primordial unity, which gives color to all the diversity in the universe.

On anḥad sabad or sabd

The terms *anḥad sabd* here accord with the command of God in the Qur'ān expressed in the word *kun* (Be!), to which Dārā Shikūh in his eighth discourse discussing the significance of sound ("*Bayān Āvāz*") refers directly: "Sound emanates from the same breath of the Merciful which came out with the word *Kun*, (or, Be), at the time of the creation (of the universe). The Indian divines call that sound *Sarastī*, which, (they say), is the source of all other sounds, voices, and vibrations."[27] It is worth looking at Dārā Shikūh's full description (*nād*):

> *Anāhat*, which has been in Eternity Past, is so, at Present, and will be so in Future. The *Ṣūfīs* name this sound, *Āwāz-i-Muṭlaḳ* (or, the sound of the Absolute), or *Sulṭān-ul-Adhkār* (i.e the Sulṭān of all devotional exercises). This (sound) is eternal and is (also) the source of the perception of *mahā akas;* but this second is inaudible to all except the great saints of both the communities. *Secondly, Āhat*, or the sound which originates from the striking of one thing against another, without its (i.e sound's) combination into words. *Thirdly, sabd*, or the sound which emanates together with its formation into words.[28]

Similarly, "the Hindavī Sufi poets translate the divine command in Arabic (*kun!*) into the local Hindavī word for the *nāda* or sound that sets the universe in mo-

---

in the words of Guru Amar Das, the sabda is quintessence of the saint whose utterances are the truth," 3722.

26 W. J. Johnson, "anāhata nāda" *A Dictionary of Hinduism* (Oxford: Oxford University Press, 2009): (anāhata sabda) ('a sound produced other than by beating') An 'unstruck sound' heard by saints and yogins as the immediate prelude to enlightenment or samādhi, sometimes equated with om. Its manipulation is a common element in haṭha-yoga practice."

27 Dārā Shikūh, *Majma ul-Baḥrain, or, The Mingling of the Two Oceans*, edited in the original Persian with English translation, notes and variants by M. Mahfuz al-Haq (Calcutta: Asiatic Society of Bengal, 1929), 47.

28 Dārā Shikūh, *Majma ul-Baḥrain, tr. by M. Mahfuz al-Haq*, 48. "*Sāmi'a*, (or, the sense of hearing), is connected with 'the great element' ('*Unṣur-i-A'ẓam*), namely *mahā akas*, through whose instrumentality we hear sounds. And it is through the sense of hearing that the real essence of *mahā akas* is manifested to the religious devotees" 41.

tion."[29] Traditionally, the metaphysical notion of the unstruck sound, *anḥad shabad*, is associated with yogic practices and exercises, a metaphysical concept, but here it is being given an interpretation that leans towards listening to a Divine sound in the context of a *melā*, that is, in a social gathering of *faqirs*.

Bulleh Shāh's verses are thoroughly grounded in praxis rather than abstract metaphysical speculations. This accords with the overall tenor of his words, which place more emphasis on the heart rather than the form of ascetic exercises. It is the finding of truth amidst other seekers of truth; therefore, the unstruck sound is heard only by those sitting among the *faqirs*. There is a connection here between the *samā'* as the mystic's music at shrines and the cosmic, sacred, eternal word as *anḥad sabad*. This verse of the *kāfī* refers to not any abstract notion of music, but to its sacred performance in gatherings of mystics at a *melā*. It is a gathering that is intimately connected to the *samā'* of mystical verse as sacred music—a ubiquitous lived practice at Indus Sufi shrines.

The act of going to a religious gathering for a mystical experience is construed as the way of *faqirī* itself. The following *kāfī* by Bulleh Shāh defines the meaning of celebrating *'urs* as going to a *melā*, the invitation to which is recognized by the sound of the eternal word, *anḥad wājā*. This *anḥad wājā* does not discriminate; it is an instrument of Divine justice and love. The *kāfī* is also a good example of the linguistic interplay between terms that have different theological contexts and are put together to transcend issues of religious identification. It refers to the purpose of visiting a *melā* as a response to that eternal sound, without which it would not really be a *melā* but instead a waste of precious time (*meley bāḥajhūṇ melā awtar ruṛh gayā mol weḥājā rey*). Therefore, it is best to adopt the difficult path of love and to attune the self to that eternal vibration. The *melā* is a space for peaceful and sacred coming together, focusing on this original sound that unites everything and everyone. In other words, a union or meeting occurs in a gathering where differences are transcended. Otherwise, a gathering that does not focus on the original harmony becomes a place of strife and conflict. Although the path of *faqīrī* is ridden with difficulty and hardship (*kaṭhan rasta faqīrī 'ashiq qa'im karo*), Bulleh Shāh advises one to adopt its ways and nuances. Here the *man bājā rey* means that the self/body/soul is also a musical instrument, which corresponds to the above *'ārif kā bājā* who is a source of Divine secrets and connection. The self needs to be brought in tune with the *'ārif kā bājā* who hears the *anḥad sabad* and plays the melody of Divine unity. The real *melā* is a gathering of *faqīrs* where the chord of Divine

---

29 Aditya Behl, *Love's Subtle Magic: An Indian Islamic Literary Tradition, 1379–1545* (Oxford: Oxford University Press, 2012), 45.

unity is struck. Here Bulleh has become one with his Lord, sitting (as at the *darbār*) in his luminous splendor (*bandā rab bhiyu aik bullhā paṛa jaḥan barājā rey*). Christopher Shackle translates the word *barājā* as "ungoverned," that is, in the sense of involving civil strife and conflict. But if the meaning (of *barājjā*) is taken as *barājamān*, "luminosity," "splendor," "sitting in state of authority or power," "leader of an assembly, or court (*darbār*)," it can be understood to represent Bullhe Shāh's claim to Divine splendor, analogous to Ḥallaj's *āna l-ḥaqq*.

Kāfī (Performed) by Ḥamid 'Alī Bela (author unknown of *kāfī*)

The second *Kāfī* that employs *melā* to express religious sensibilities and interpretation is by an unknown author, performed by a leading 20th -century folk singer and dedicated to the famous 16th-century Lahorī mystic, Shāh Ḥussaīn. Below is the *kāfī* with a literal translation of the verses that certainly does not convey its rhythm but does indicate the kind of vernacular poetics at play. The aim of the literal translation here is to show the imagery—imagery that, for the devotional listener, strikes the heart, drawing a tangible connection between the cognitive element of listening (which is better understood as a form of seeing) and the feeling element experienced in the body.

Auṛak janā mar waye chal meyley nū chalīye
(In the end we are going to die; let's hasten to the *melā*)

Qasmāṇ nū khāndā teyrā ghar waye
(Abandon your home)

Phelā meylā adam kīthā
(The first *melā* was enacted by/in Adam)

Malkaṇ weykh kar sajadā kithā
(Upon seeing which, angels prostrated themselves)

Gayā har aek daye wich waṛ waye
(Entering then into everyone)

chal meyley nū chalīye
(Lets hasten to the *melā*)

Qasmāṇ nū khāndā teyrā ghar waye
(Abandon your home)

Dosra meylā Khalīl minayā zībaḥ karaṇ puthar lāṭayā
(The second *melā* Abraham performed as he laid out his son for sacrifice)

waye chal meyley nū chalīye Qasmāṇ nū khāndā teyrā ghar waye
(Let's hasten to the melā; abandon your home)

Tīsrā meylā rāthī hoyā ān būey taye jibrīl khalotā
(The third *melā* happened at night when Gabriel stood at the door)

Sunā gaya buraq uttaye chaṛ waye
(That lovely one climbed atop Buraq)

Chauthā meylā Ḥussayn dī wārī karbal sari khalqat waykh
(The fourth *melā* was Ḥusain's turn, witnessed/seen by the entire universe)

Gayā sir neyzey uttaye chaṛ waye
(As/how his head ascended the spear)

waye chal meylley nū chalīye
(Let us hasten to the *melā*)

"In the end, we shall die, so let us go to the *melā* and forget our domestic worries," is the message of the opening verses of the *kāfī*. This verse is the refrain for the entire *kāfī*, expressing urgency, or desire, to journey towards a *melā* and to forget about worldly concerns because everything will eventually end anyway in death. But what is this *melā?* Of what is it a celebration? The *kāfī* narrates a sequence of events that is a familiar part of religious knowledge among the elite (*khāss*) and serves as a commentary for ordinary folk (*āmm*) on what this *melā* signifies. In contrast to the silence of death, it is an enticement towards a carnival, an eternal one.

The first *melā* was done (*kīthā*) or enacted by Adam, and upon seeing Adam, angels prostrated themselves. In this making of Adam, God entered all human beings.[30] In this verse, the *melā* is situated in primordial time, when human creation begins in the archetypal figure of Adam. God brought Adam into existence by entering (*waṛ*) into him, signifying an intimate association between the human and God, the theological status of which became a matter of controversy among theologians.

In the first instance, the literal meaning of the verse seems to be in line with the traditional account unless we consider this first verse, *Phelā meylā adam kīthā*, as a cataphatic expression. In effect, this *kāfī* verse is subverting the conventional dualistic chronology in giving agency to Adam, as the one who does *(adam kītha)* the first *melā*. The *melā* is a cosmic event that involves creation

30 In the Qur'ān, Adam was given life by God's breath in the body made of clay and ordered all the Angels to bow to him.

through an archetypical act realized in the participation of Adam in the *melā*. God makes Adam's creation an act of cosmic self-reflexivity.[31]

The narrative next shifts from Adam to the patriarch of monotheism, Abraham. In the following verses, the second *melā* is said to be performed (*minayā*) by al-Khalīl (Abraham).[32] In order to sacrifice his son, he laid him on the ground (or the altar) and sat upon his chest. Here the *melā* is seen as a ritual, the act of sacrifice that Abraham attempts to perform after leading his beloved son up the mountain. The command to sacrifice his son, *Isma'īl*, was given by God as a test of Abraham's faith. In the next verse, Abraham is not mentioned; rather it is again that anonymous one who springs (*chaṛ*) upon Abraham's son's chest. Both *Khalīl* and God are fused in this heart-wrenching performative ritual—otherwise, who else in his sane mind would possess the capacity to lay down their own son for sacrifice? It is a chilling image, to say the least.

The third *melā* happened (*ḥoyā*) at night, in a moment of deep stillness, when Gabriel stood at the door—of the Prophet—and the beloved (*sonā*) Muhammad climbed atop the heavenly steed, *Buraq*. The reference is to the Prophet's *miraj*, ascension to God on a winged horse, accompanied by Gabriel, and finally, an encounter between the Prophet and God. Here in the *kāfī* the journey is expressed entirely without mentioning the name of Muhammad or God; instead, it is a lover mounting a mythical horse, as the messenger Gabriel brings the news of the arranged meeting at night. Here a desire is immediately reciprocated by the Prophet in mounting *buraq*; it is a *melā* enacted in secrecy.

For the fourth *melā* it was the Prophet's grandson Ḥusain's turn (*warī*), and after his death it was witnessed by the entire creation at Karbala, when his head was mounted on a spear. This is the last verse of the *kāfī*, a final affirmation of the breath-taking spectacle. It is interpreted in terms of Ḥusaīn's turn, his destiny, and a *melā* performed in full view of the entire world—angels, prophets, humanity (past, present, and future)—that culminates in Ḥusaīn's head ascending (*chaṛ*) a spear. The spectacle manifests the Divine in the body of Ḥusaīn, witnessed openly as the head that mounts the spear. The *melā* takes place not as a ritual nor in secret but as a spectacle in the world, in accord with God's *Shariah*, establishing the manner of Divine manifestation/truth in this world. This form of violence to God's lovers is found in Bulleh Shāh's verses as well,

---

31 See Leigh N. B. Chipman, "Adam and the Angels: An Examination of the Mythic Elements in Islamic Sources," *Arabica* 49.4 (2002): 429–455; and *idem*, "Mythic Aspects of the Process of Adam's Creation in Judaism and Islam," *Studia Islamica* 93 (2001): 5–25.

32 Abraham is known by the *laqab* as al-*Khalil*, friend of God in *Qur'ān*, an honorific title. See John Renard *All the King's Falcons: Rumi on Prophets and Revelation* (Albany: SUNY Press, 1994), 48.

where it is described by him as the way of love—discussed in the present study in the chapter on Bulleh Shāh's Sufism.

Kāfī by Bulleh Shāh on Sacred Play and Celebration

Horī[Holi] khelūṇ kī kaheh kar bismallah
(Saying *bismillāh* ("in the name of God") I will play *Holi*)
I will play Holi, after saying *bismillah*

Nām nabī kī rattan chaṛhī bund paṛī ilā allāh
Rung rangīlī aūhi khilavey jo sakhī hoey fanā fī allāh
(The color of the *nabī* 's reality sent forth a droplet (no god) except God; the ecstatic lover playing with color passes away in God)

I wear the name of the Prophet as my jewel, and the
Words *but God* as my pendant. He is the one who
Operates this colorful show, from which the lesson of
annihilation in God is learned

Horī khelūṇ kī kaheh kar bismillah
(*Saying bismillah* I will play *Holi*)
I will play Holi, after saying *bismillah*

alastu birabikum[33] *pītham boley sabh sakhiān ney ghughaṭ kholey*
qālu balā hī yūn kar boley, lā ilāh ilā allāh
(Am I not your Lord? asked the Beloved, and all lovers unveiled in affirmation: there is no god except God)
When the beloved said, *Am I not your lord?* The girls all removed their veils.
With the words *They said, "Yes,"*
they said, *There is no god but God.*

Horī khelūṇ kī kaheh kar bismallah
(*Saying bismillah* I will play *Holi*)

Naḥnu aqarab kī banī bajā'ī man 'arif nafsahu kī kūk sunā'ī
Fathuma vajahu allahu kī duhūm machā'ī darbar-i rasūl allāh
(Playing the flute of We are closer (to you than your jugular vein), and the cry of know your-

33 Qur'ānic reference to "*alastu birabbikum*", translated as Am I not your Lord? See in Qur'ān, *Sūrat al-A'rāf* (The Heights) (7:172): *wa-iz akhaza rabbuka min banī ādam min ẓuhūrihim zurriyyatahum wa-ashhadahum 'alā anfusihim alastu birabbikum qālū balā shahidna an taqūlū yawma al-qiyāmati innā kunnā 'an hāza ghāfinlīna* (And when your Lord took from the children of Adam from their loins their descendants and made them testify regarding themselves, "Am I not your Lord?" They said "yes we testify [to this]" Lest you say on the Day of Resurrection indeed, we were unaware about this), see also Gerhard Böwering, "Time in Persian Sufism," *Iran* vol. 30 (1992), 77–79.

self; thus a fanfare in the court of Prophet: wherever you look you will find the face of God)[34]

Horī khelūṇ kī kaheh kar bismallah
(*Saying bismillah* I will play *Holi*)

Hāth joṛ kar pā'on paṛon kī 'ājiz ho kar bīntī karon kī
Jhaghṛā kar bhar jholī lon kī nūr muḥammad ṣalalāh
(With all my being, I drop at your feet and clasp my hands to place my request; (Or else) quarrel with you to take mine share of blessings, with the grace of the light (*nūr*) of Muhammad)[35]

Horī khelūṇ kī kaheh kar bismallah
(*Saying bismillah* I will play *Holi*)

Fā zikrunī kī ḥorī banā'on wa shukrulī piyā ko rijhā'on
Aisay piyā kay main bal bal jā'on kaisā piyā subhān allāh
(So if you remember me (I will remember you), and with my gratitude I will please; such a beloved whom I unceasingly adore, all praise to Him (God))

Horī khelūṇ kī kaheh kar bismallah
(*Saying bismillah* I will play *Holi*)

ṣibgat al-allāh kī bhar pichkārī allāh al-ṣamad piyā mūnh par mārī
nūr nabī dā ḥaq say jārī nūr muḥammad ṣalalah
(Collecting the dye of God, the Eternal One, flung it at the face of the beloved; In/from the truth of the Prophecy manifests the light of Muhammad)

Bulleh Shāvh dī duhūm machī hay lā ilā ilālāh
(Bulleh Shāh's fame has spread far and wide: there is no existence except God)

Horī khelūṇ kī kaheh kar bismallah
(*Saying bismillah* I will play *Holi*)

---

34 Here we see references to the Qur'ān and *ḥadīth: naḥnu aqarab* and *fathuma vajahu* refer to God's presence within the human body (i.e., "closer than the jugular vein"), and the "face of God," respectively. Ordinary folks are familiar with these Qur'ānic verses. As for *'arif nafsahu,* this is a famous Divine Saying (*man 'arafa nafsahu faqad 'arafa rabbahu:* "who knows himself knows his Lord."

35 See Annemarie Schimmel, *And Muhammad is His Messenger: The Veneration of the Prophet in Islamic Piety* (Chapel Hill: The University of North Carolina Press, 1985), Uri Rubin, "Pre-Existence and Light: Aspects of the Concept of Nūr Muḥammad," *Israel Oriental Studies* 5 (1975), and *idem.*, "More Light on Muḥammad's Pre-Existence: Qur'ānic and Post-Qur'ānic Perspectives..." Andrew Rippin and Roberto Tottoli (editors), *Books and Written Culture of Islamic World Studies Presented to Claude Gilliot on the Occasion of his 75th Birthday* (Leiden: Brill, 2014): 288–311; Ayesha Irani, "The Prophetic Principle of Light and Love: *Nūr Muḥammad* in Early Modern Bengali Literature," *History of Religions* 55.4 (2016): 391–428.

This third *kāfī* above is an example of the how Bulleh Shāh effortlessly blends different religious aesthetics into a seamless whole. More than with any other vernacular Sufi in the Indus region, the extent of Bulleh Shāh's use of Hindu and Sikh vocabulary is remarkable. In this *kāfī* the obvious reference is to the Hindu festival of Holi associated with Krishna.[36] It is a sacred festival that involves a dimension of play. In addition to direct reference to a sacred festival of Hindus, the *kāfī* evokes the theosophical idea of *nūr-i muḥammadī* as an altogether legitimate pretext for engaging in the play of Holi. Furthermore, Bulleh Shāh makes a direct reference to *Allah*'s dye or color (*ṣibghatallāh*), and other Qur'ānic references to express the notion of Divine play poetically.

The refrain of the *kāfī* emphatically declares that one is to engage in the festival of *Holi* with the evocation of *bismillah.*[37] Muslims use the Qur'ānic formula before commencing any task, small or big, believing that it bequeaths sacred efficacy to human actions.

> *Nām nabī kī rattan chaṛhī bund paṛī ilā allāh*
> *Rung rangīlī aūhi khilavey jo sakhī hoey fanā fī allāh*
> (The name of the *nabī* is the emanation of color sending forth a droplet of (no god) but God; only that lover in union with Divine playfully brings forth such diversity of/in color)[38]

The *kāfī* verses above that come next delve into the idea of the Muhammadan Reality, which then in the rest of the verses lays the theosophical basis for seeing in the Hindu festival the cosmology of *nūr Muḥammad.* These opening verses point to Bulleh Shāh's fidelity to Sufi theosophic belief in the Muḥammadan reality, that is, creation as (or in) the name (or person) of the *Nabī* (Muḥammad), the color of reality as the affirmation of the sole lordship of God (*būnd paṛī ilā-allāh*) in the last part of the *shahāda.*[39] Irrespective of the religious context of Holī in Hindu mythology, Bulleh Shāh sees the celebration of this Hindu festival as a manifestation of the Divine. The second verse links the Sufi metaphysics of *fana fī-allāh, "extinction/absorption in God"*, with the notion of playing (*khilavey*)

---

36 A. Whitney Sanford, *Sacred Play: Ritual Levity and Humor in South Asian Religions* (Albany: State University of New York Press, 2010), 37. "Visitors observing the Holi festivities might be struck by the raucousness of the festival, as water fights and the throwing of colored powders, some of the most obvious manifestations of Holi, are par for the course," 37.

37 See William A. Graham, "Basmala," in Jane Dammen McAuliffe et al., eds., *Encyclopaedia of the Qur'ān* (Leiden: Brill, 2001), 1:207–212.

38 Christopher Shackle, *Sufi Lyrics: Bullhe Shāh 1680–1758* (Cambridge: Harvard University Press, 2015).

39 The doctrine of the Reality of Muhammad is an integral aspect of Sufi metaphysics, and its foremost exponent is the famous Mystic Ibn 'Arabī.

by the one who is multi-colored (*rung rangīlī*). This plurality of color may be taken to encompass the entire creation as well, but more specifically here, the colorful festival of Holi becomes an expression of *nur Muḥammad*. It is equally important to take note of Bulleh Shāh's reference to the Prophet as *nām nabī*, because the form of the name of Muhammad itself is a subject of metaphysical exegesis and part of Sufi hermeneutics (see Chapter 4).[40]

> *alastu birabikum pītham boley sabh sakhiān ney ghughaṭ kholey*
> *qālu balā hī yūn kar boley, lā ilāh ilā allāh*
> (Am I not your Lord? asks the beloved, at which all lovers unveiled in affirmation: responding with there is no god except God)

The next *kāfī* refers to Divine revelation regarding *alastu bi-rabbikum*, which speaks of the primordial covenant in the Qur'ān. It is difficult to understate the importance of this covenant in the history of Sufi thought and practice. Known as the Day of Covenant (*yawm al-mīthāq* or *rūz alast*), the motif is interpreted as a time of primordial Unity, employed effectively by Rumī, as also by Ḥafiẓ, as the "*sāqi* of the day of *alast*," the day of intoxication.[41]

> Ḥāfiẓ also boasts of being famed as a drunkard from the very first day of the pre-Eternal Covenant (*rūz-i alast*), and rails against the ascetic who cannot understand that his intoxication with human beauty is a necessary consequence of his vow in pre-Eternity.[42]

The Covenant is taken here as an act of love, a primordial bond of love, and thus it is common to depict the relationship between God and humans in erotic imagery.[43] The reference is an example of a metaphysics of presence, and the imagery of lifting up of their veils (just as an adorned bride does for her husband on their wedding night) signals union. Thus, the *sakhaīn* proclaims: "there is no

---

40 See Carl W. Ernst, "Muḥammad as the Pole of Existence," in Jonathan E. Brockopp, ed., *The Cambridge Companion to Muhammad* (Cambridge University Press, 2010), 126: "Never has anyone departed from the M of Muḥammad, and no one entered the Ḥ. [As for] his Ḥ, the second M, the D, and the M at the beginning: the D is his permanence [*dawām*], the M is his rank [*maḥall*], the Ḥ is his spiritual state [*ḥāl*], and the second M is his speech [*maqāl*]."
41 Annemarie Schimmel, *The Triumphal Sun: A Study of the Works of Jalāloddin Rumi* (Albany: State University of New York Press, 1978), 152.
42 Husayn Ilahi-Ghomshei, "The Principles of the Religion of Love in Classical Persian Poetry," in Leonard Lewisohn, ed., *Hafiz and the Religion of Love in Classical Persian Poetry* (New York: I. B. Taurus, 2010), p. 98, On Islamic law and theology of the covenant, see Bernard Weiss, *The Spirit of Islamic Law* (Athens: The University of Georgia Press, 2006), 31–34.
43 See Lewisohn (ed.), *Hafiz and the Religion of Love in Classical Persian Poetry* (New York: I. B. Tauris, 2010).

god but God."[44] However, the *sakhaīn* refers to the *gopīs*' interlude with Krishna who "played Holī with the cowherd boys and with *Rādhā* and other *gopīs*," which is the pretext for energetic celebration and amorous play, aptly described as the "festival of love."[45]

These respective verses relate to the anthropomorphic depiction of God in the Qur'ān, that is, in God's asking of the following question: "am I not your Lord?" This dialogical question privileges the relationship between God and humanity as a primordial relationship (in creation), and as one of Divine union. The directive for such participation in the festival of the Hindus, according to Bulleh Shāh, is justified in the Qur'ān in the Divine verses that speak of God's proximity to human beings. It is this intimate relationship with the Divine, as understood within a metaphysics of being, which calls for celebrating this preexistent privilege.

> *Naḥnu aqarab kī bansī bajā'ī man 'arif nafsahu kī kūk sunā'ī*
> *Fathuma vajahu allahu kī duhūm machā'ī darbar-i rasūl allāh*
> (Playing the flute/tune of "we are closer" (to you than your jugular vein), and the cry of "know yourself"; a splendor in/of the court of Prophet: "wherever you look you will find the face of God")

In this *kāfī* verse, Bulleh Shāh juxtaposes first the Qur'ānic verse 50:16 with the playing of the flute (*bansī*), which in the Indian context immediately suggests the flute of Krishna. The flute symbol is an important trope in Mawlana Rumī's poetry, with which someone like Bulleh Shāh would have been well acquainted. Nevertheless, the mention of the flute (*bansī*) fits wholly within the context of the Indian religious landscape. After all, Bulleh Shāh is celebrating the festival of Holi in the name of God, and the overlapping of the symbolism associated with Krishna with the Qur'ānic verse in this *kāfī* makes the appropriate connection. Moreover, as in the case of Rumī, the flute here can quite as quickly stand for the human body to symbolize the creation of Adam into whom God breathed His spirit/breath, or *ruḥ* (Qur'ān 15:29 and 38:72). In this manner, the use of the flute as a symbol lends itself to different interpretations, making it difficult in the context of Bulleh Shāh's milieu to prefer one interpretation over the other. Arguably, given Bulleh Shāh's *wujūdī* metaphysics, both readings are equally valid,

44 As it is, the *'urs* means 'wedding with God' to represent Divine union.
45 McKim Marriott, "The Feast of Love," in Milton Singer, ed., *Krishna: Myths, Rites and Attitudes* (Chicago: University of Chicago Press, 1966; first published by East-West Center, University of Hawaii) 207; for a general overview on Holī, including a photo of an 18th-century miniature of "Krishna playing Holi with *Radha* and the *gopis*", see also Shakti M. Gupta, *Festivals, Fairs and Fasts of India* (Delhi: Clarion Books, 1991), 38–42

and *must* be taken together to grasp the devotional effect of the *kāfī*. *Nahnū aqarabu min ḥabli al-warīdi* is the most robust Qur'ānic expression of the proximity of the Divine to the human soul, and it is incredibly potent because of the use of the corporeal image of the jugular vein.[46] The verse depicting God as being closer to the human than his jugular vein is visualized in the *kāfī* as the playing of the flute by the Divine.[47] This is then further complemented in the *kāfī* by a well-known, late *ḥadīth qudsī* "he who knows himself knows his Lord."[48] This too has a broader significance in terms of its meaning in Indian mysticism. In this case, Bulleh Shāh is building on the pre-existing Sufi tradition of emphasizing Divine intimacy through this Qur'ānic verse and then the *ḥadīth qudsī*—completely apart from the fact that these interpretations would be disputed or rejected by other traditionists and theologians.

The next line of the *kāfī* starts with another string of Qur'ānic words: "*fathuma vajahu allahu*," which expresses the omnipresence of God in every direction.[49] However, the full effect of the cosmology of "the Face of God" is realized in the *duhūm machā'ī rasūl Allāh*, as manifested (*tajjalī*) in the person of the Prophet, evident in/by the splendor of his court (*darbar*), which can also mean the tomb of the Prophet. *Darbar* of the Prophet also signifies Divine sovereignty or Kingship.

> *Hāth joṛ kar pā'on paṛon kī 'ājiz ho kar bīntī karon kī*
> *Jhaghṛā kar bhar jholī lon kī nūr muḥammad ṣalalāh*
> (With all my being, I drop at your feet and clasp my hands to place my request; (Or else)

---

46 See Qur'ān, *ṣūra qāf*, 50:16: *wa-laqad khalaqnā al-insāna wa-na'lamu mā tuwaswisu bihi nafsuhu wa-naḥnu aqrabu ilayhi min ḥabli al-warīdi* (And we have already created man and know what his soul whispers to him, and we are closer to him than his jugular vein).

47 Gerhard Böwering, *The Mystical Vision of Existence in Classical Islam: The Qur'ānic Hermeneutics of the Sufi Sahl al-Tustari* (Walter de Gruyter, 1979), 203: "Tustarī understands the Qur'ānic verse, "We are nearer to him than the jugular vein (*ḥabl al-warīd*) (50, 16) as referring to the vein (*'irq*) which is in the depth of the heart (*gauf al-qalb*) and interprets it as meaning that "He (God) is nearer to the heart (*qalb*) than that vein."

48 See Lenn Evan Goodman, *Jewish and Islamic Philosophy: Crosspollinations in the Classic Age* (Edinburgh: Edinburgh University Press, 1999), 16. Ibn al-Arabi collection, *mishkat al-anwar,* 'Attar's *Tadhkirat al-awliya'* [ed. Nicholson, 2 vols., London: Luzac, and Leiden: Brill, 1905–7], vol. II: 291. See also Divine Word, 72, n. 106.

49 See Qur'ān 2:115: *wa-lillāhi al-mashriqu wa-almaghribu fa-aynamā tuwallū fa-thamma wajhu llāhi inna l-lāha wāsi'un 'alīmun* (And to God belongs the east and the west, so wherever you turn so there is the face of God, indeed God is All-encompassing All-Knowing), see also Andrew Rippin, "Desiring the Face of God," in Colin Turner, ed., *The Koran: Critical Concepts in Islamic Studies* (New York: Routledge Curzon, 2004), 265–272.

quarrel with you to take my share of blessings, with the grace of the light (*nūr*) of Muhammad)

This *kāfī* is a good example of Bulleh Shāh's attention to devotional aesthetics and gestures familiar across religious cultures in South Asia. In these verses, the overriding theme of the sacred play is affected by focusing on the role of the supplicant. Bulleh Shāh has her entreating (*bintī*) with gestures of submission and humility, that is, by clasping or pressing her hands together (*hāth joṛ kar*), or if need be, falling at the Lord's feet (*pā'on paṛon kī*). If that fails, as seems to be implicit here, she must move even to quarrel (*jhaghṛā*) with the Divine, and by the grace of *nūr Muḥammad*, in asking to have plentiful blessings, take her share of it (*bhar jholī lon*). The *kāfī* describes both passive and then active methods of getting the beloved to pay heed to the supplicant. Such informal behavior as devotional play to get one's way with the Divine presupposes an intimate relationship between the supplicant and God. The dual play of being passive and active is essentially a mode of coquetry; especially the depiction of quarrelling with the Divine is a playful transgression of boundaries. In a flash, the Divine beloved is rendered passive in accepting the demands of the supplicant, who then helps herself to the eternal flow of Divine blessings granted in the *nūr Muḥammad*. Secondly, the gesture of bringing one's hands together, with attention and courtesy, and then falling on the ground in submission, represented in the prostration (*sujud*), is performed in *ṣalāt* (ritual prayer). Therefore, without direct reference to the *ṣalāt*, Bulleh Shāh prefers to place the gestures of humility and submission in a wider cultural context, thus deconstructing the ritual of prayer as a form of coquetry that is essentially about both humility and submission, but also a hint of playful insolence, or tactful cunning, to steal favor from the Divine beloved. These aspects of quarrelling or stealing, and even tricking the Divine, are found in other *kāfī*s of Bulleh Shāh as well. They are something of a trademark of Bulleh Shāh's Sufism (see above, Chapter 4).[50]

The *kāfī* is written in the feminine gender to express desire, which corresponds to the *virahaṇī*, the motif of an ardent yearning of a woman for her beloved or husband.[51]

---

50 [NB: Such playfulness is also well-known in Hindu popular piety, as for example in Krsna piety (hymns) in Brindavan and elsewhere]

51 See Annemarie Schimmel, *Pain and Grace: A Study of Two Mystical writers of Eighteenth-Century Muslim India* (Leiden" E. J. Brill, 1976), 173, for a more detailed account of the *virahiṇī* tradition see Ali A. Asani, "Sufi Poetry in the Folk Tradition of Indo-Pakistan," *Literature and Religion* vol. 20 1 (1988), 86.

> *Fā zikrunī kī ḥorī banā'on wa shukrulī piyā ko rijhā'on*
> *Aisay piyā kay main bal bal jā'on kaisā piyā subhān allāh*
> (So if you remember me (I will remember you), and with my gratitude I will please; such a beloved whom I unceasingly adore, all praise be to Him)

Fā zikrunī here is a direct reference to the Qur'ānic verse in which the Divine wants to be remembered: such remembrance (*zikr; Ar. dhikr*) is an expression of gratitude and will be reciprocated by God's remembering His servant.[52] Accordingly, Bulleh Shāh sings of pleasing (*rijhā'on*) or delighting the beloved with gratitude. The combination of remembering God and gratitude, or remembrance as a form of gratitude, has been the subject of extensive Muslim theological reflection. Praising God becomes synonymous with gratitude. Abū Ḥamīd Muḥammad al-Ghazālī (d. 505/1111) treated this theme extensively in his *Kitāb al-ṣabr wa-l-shukr* (*"On Patience and Gratefulness"*).[53]

The next verse sings praises of having a beloved who will reciprocate such ardor, and therefore Bulleh Shāh is ready to be sacrificed (*bal bal janā*) for the *piyā* (beloved, sweetheart). Here he evokes the standard Islamic expression of *subḥān Allāh,* "Praise/ glory be to God."[54] The formula itself is a form of gratitude because of its explicit acknowledgment of Divine beneficence.

> *ṣibgat al-allāh kī bhar pichkārī allāh al-ṣamad piyā mūnh par mārī*
> *nūr nabī dā ḥaq say jārī nūr muḥammad ṣalalah*
> (Collecting the dye of God, the Eternal One, fling it at the face of the beloved; In the truth of Prophecy[55] the light of Muhammad is manifest/manifests itself)

These are vexingly difficult verses to translate, and they are open to several interpretations. However, these verses do seem to conform to the theosophical notion of identifying Divine truth (*ḥaq*) with Muhammad.[56] In literal translation,

---

52 See Qur'ān, *Sūrat al-Baqara* (The Cow) 2:152, *fā-udhkurūnī adhkurkum wa-ushkurū lī* (so remember me and I will remember you, and be grateful to me).

53 See Lamin Sanneh, "Gratitude and Ingratitude," *Encyclopaedia of the Qur'ān*, ed. Jane Dammen McAuliffe (Leiden: Brill, 2002), 2: 314–15.

54 See "Glorification of God," Mustansir Mir, *Encyclopaedia of the Qur'ān*, edited by Jane Dammen McAuliffe (Leiden: Brill, 2002), 314–15; "understood in the sense of *tanzīh* (Transcendence) by al-Ṭabarī, al-Zamakhsharī, and al-Rāzī (transcendence) see D. Gimaret, "Subḥān," *Encyclopaedia of Islam, Second edition,* edited by P. Bearman and Th. Bianquis *et al.*

55 In theological discourse and in the Qur'ān, the Prophet is called both as a *nabī* and as a *rasūl* (messenger), the former applies to all Prophets, while the later to Prophets who are given a book (*kitāb*).

56 Such a view is not unique with Bulleh Shāh, as it constitutes an entire tradition of Sufi theosophy revolving around the Muḥammadan Reality.

the *kāfī* speaks about throwing the color of God (*ṣibgat al-lāh*) at the face of the beloved who is the Eternal One, designated *al-ṣamad* in the Qur'ān.[57] Muḥammad is the beloved immersed in the color of God, and *al-ṣamad*. This specific word belongs to the shortest Qur'anic chapter, called the *Sūrat al-Ikhlās* (the Chapter of Sincerity), more traditionally known as the *ṣura* of *tawḥīd* (unity), because of its entire emphasis on God's unity. The second couplet identifies the emanating light of the Prophet (*nūr nabī dā ḥaq*) as an emanation of truth.

The *kāfī* celebrates the Hindu festival of Holi, and the use of this specific word signifies the color of God (during Holi everyone is playfully sprayed with color to mark the occasion). It is a compelling theological motif used to explain how the manifold colors (diversity) of the *melā* are, in reality, an emanation of God. It transforms the metaphysical doctrine of *nūr Muḥammadī* into a playful gesture of throwing color (*bhar pichkārī*), and in the process simultaneously celebrates the Hindu Holi in praise of the Muhammadan Reality.[58] The literary or experiential synthesis is such that it would be difficult—or represent a callous insensitivity to its poetical genius—to identify any single metaphysical belief as possessing more or less weight in the *kāfī*.

> *Bulleh Shāvh dī duhūm machī hay lā ilā ilāllāh*
> (Bulleh Shāh's fame has spread far and wide: there is no god except God)

The last verse ends in an affirmation of Divine unicity in *lā ilāha illā-llāh*. This is the first part of the Muslim declaration of faith, which in Sufi lore is seen as af-

---

57 See in Qur'ān, 2:138, the entire verse: *ṣib'ghat allāhi wa man 'aḥsanu min allāhi ṣib'ghatan*" (Allah's color or dye, and who is better than God at coloring), *ṣūra al-baqara* (The Cow); Annemarie Schimmel, Cf Ali S. Asani, "Sufi Poetry in the Folk Tradition of Pakistan," *Religion and Literature* 20.1 (1988), 81–94, who also translates *ṣibghat Allāh* as the 'coloring of God,' *Pain and Grace: A Study of Two Mystical Writers of Eighteenth-Century Muslim India* (Leiden: E. J. Brill, 1976), 180; however according to James A. Bellamy, "The word *ṣibgha* refers to the Christian baptism (q.v.), so the exegetes were obliged to redefine it. They take it to mean *dīn* or *īmān*, or they equate it with the *millat* of *Ibrāhīm*, in Q 2:135, which they take to mean Islam," "Textual Criticism" *Encyclopaedia of the Qur'ān Volume Five*, ed. Jane Dammen McAuliffe (Leiden: Brill, 2006), 243; for *al-ṣamad* (The Eternal, the Absolute) in *ṣūra al'ikḥlas* (112:2), for a fuller examination of the word, and its post-Qur'ānic usage among the theologians, see Daud Rahbar, *God of Justice* (E. J. Brill: Leiden, 1960), 37–39; furthermore, in Chishtī *mehfil-i samā'* the performance of *rang* (color) concludes the ceremony, in which the names of all the Sufi masters in continuum, spiritual genealogy, are recognized as being immersed in the same color, *ṣibghat 'Allah*; also the performance of *rang* is associated with Amīr Khusraw, the famous disciple of Khawaja Nizām al-Dīn *'Awliyā*, who is said to have composed it upon being accepted by Khawaja Nizām's as his disciple.

58 See chapter 4 on how such mixing, in the case of Bulleh Shāh, refers to seeing two different expressions of the divine as a manifestation of Divine unity.

firmation of God as the only existent. In other words, quiddity belongs only to God. Here, in essence, it is a declaration of Bulleh Shāh's own state (*ḥāl*) of experiencing Divine union, and thus not finally different from Ḥallaj's "*āna l-ḥaqq*" ("I am the [Divine] Truth"). *Duhūm* (fame) refers to the tumult over Bulleh Shāh, as talk about him, which could potentially signal danger, or positively, recognition of his glorious experiential state. In any case, like a number of his other verses, this *kāfī* too has a strongly rapturous feel, concise yet loaded with *kaifiyyāt*, which is a hallmark of Bulleh Shāh's vernacular poetry in general. It is the economical, *haiku* quality or precision of his verses that lends itself to easy memorization and performance. He has a seemingly effortless capacity for firing the listener's imagination. For this reason, it is common among performers of Sufi music to employ a verse or an expression from Bulleh Shāh's *kalām* in the middle of another composition in order suddenly to arouse passion in the audience. In sum, the *kāfī* sacralizes the gesture of throwing color in Holī and signifies sacred playfulness in tune with the metaphysical reality of Muhammad.

5.3 *'Urs* as Festival (Celebration)

The polysemic aspect of the *mawlid, or mulid* (as a celebration of the saint's birthday), is aptly described by Schielke in the following manner:

> there is not just one habitus of celebrating the mulid, but many. A mulid's main embodied emotional tone is that of love and joy, but that tone is mixed with moments of grotesque laughter, solemn commemoration, voluntary hardship, sexual tension, and sometimes aggression.[59]

Such a complicated image of the *mawlid* (and the same can be said of *'urs*) is not a novel one; it is affirmed in other studies as well and was likely what made the *'urs* an object of curious fascination in early accounts by European travellers and orientalists, who saw it as standing in opposition to 'official' and more austere Muslim practices regulated by the religious scholars (*'ulamā'*). For example, Lane, in Egypt, observes that, "at the funerals of saints or holy men, the usual laments were replaced by hymns of joy."[60] Similarly, Garcin De Tassy reports that in India, "these gatherings comprise the *faqirs*, devotees of all classes,

59 Schielke, *The Perils of Joy*, 40.
60 Cf Stephen Wilson, *Saints and their Cults: Studies in Religious Sociology, Folklore and History* (Cambridge: Cambridge University Press, 1983), 41.

musicians and libertines, rascals and thieves." The exact words De Tassy uses to describe such gatherings are "half-mundane" and "half-religious." He describes at length the carnival-like atmosphere and flow of activity at the shrine of Salar Mas'ud Ghazi as follows:

> Tents and merchants' stalls are erected all around; sweets of all kinds and colours are artistically spread out.... There are musicians who play on different instruments, jugglers who play skillful tricks, dancers from the Deccan with astonishing suppleness.... In the middle of the ravishing spectacle, intoxicating liquor made of exudation of hemp flowers is passed everywhere.... However everyone goes to the tomb and pays homage by offering flowers and sweets. Singers and musicians express their devotion in their own way. Thousands of candles and lanterns shed their brightness on the flowers of lotus and cypress. This continues till sunset.[61]

The ambiguous or ambivalent nature of the social and ritual space at Sufi shrines has been amply noted and well documented. The ritual space of the Sufi shrine is a meeting ground between the extraordinary alterity of the image and body of the Sufi saint as a friend of God, and the crowds of ordinary folk that throng the shrine's threshold.[62]

Hava Lazarus-Yafeh has argued that "there are few festivals in Islam, and Muslim festivals seem to lack colour or meaning"; yet she claims that Muslim festivals have "inherent pagan mythological" characteristics "expressed in dramatic rituals."[63] According to Catherine Bell, the term "festival" as applied to

---

61 Garcin De Tassy, *Muslim Festivals in India and Other essays*, trans. and ed. M. Waseem (Delhi: Oxford University Press, 1995), 43.

62 Although rulers and religious scholars also visit Sufi shrines, in the present as in the past, out of motives ranging from the spiritual to the political, two aspects of the Sufi shrine in particular are central to its ecology: first, its role as a place for supplication (*du'ā*), and second, its theopoetic qualities in the popular practice. Both of these features are part of the religious culture and aesthetics at Sufi shrines, and even the diverse forms of entertainment and pleasure at shrine celebrations can be read not just as an 'accidental' attribute of the shrine space (providing a opportunity only to have fun), but in some measure as part of the culture of hospitality extended to all who visit the Sufi saint. Nevertheless, this liminality of Sufi shrines has much to do with the political function of Sufi shrines, insofar as they serve as a kind of intersection where rulers, religious scholars, and ordinary folk (*āmm*)—along with non-Muslims—interact with each other (see Chapter 3).

63 Hava Lazarus-Yafeh, "*Muslim Festivals*," in *The Development of Islamic Ritual*, ed. Gerald Hawting (Burlington: Ashgate Variorum, 2006), 317. She gives the following as a possible explanation for the lack of festivals in Islam: "The meagerness of Muslim festivals is very difficult to explain, but it may be connected with the severely monotheistic character of Islam and its constant fight against pagan phenomena and remnants of the pagan past. Muslim—just like Jewish —festivals are pre-Monotheistic heritage, and no suitable way may have been found to Islamize them as, for example, the Ḥadjdj rituals were Islamized."

the two official Islamic celebrations "may be a bit strong for both of them."[64] By contrast, Lazarus-Yafeh argues that in Christianity festivals have a "human, personal aspect," while in the case of Judaism the "reenactment of historical events" organize Jewish festivals.[65] She goes on to argue that Islamic monotheism as such did not allow for any cultivation of festivals, and that the popularity of *mawlids* could be explained as fulfilling the (Sunni) need for "human mediation," similar to that in Christianity.[66] The practice of visiting shrines and celebrating the *mawlid* of the Prophet is seen as fulfilling a basic psychological or emotional need that is not met in the severe "orthodox" monotheism of Islam.[67]

Pnina Werbner describes several features of *ziyārat* to the shrine of Ghamkol Sharif in Pakistan. She classifies the sum of these activities as the "*symbolic complex of blessing.*" She then goes on to point out "subtle differences in modes of saintly blessing" in order to analyze the notion of *baraka* so commonly deployed to explain the charisma of the saint.[68] Werbner defines the *'urs* as both a ritual and a popular religious festival, but concedes that it is not a tightly organized affair."[69] However, she qualifies her definition by adding that "despite this apparent lack of structure, the *'urs* is nevertheless a structured ritual moved forward by the power of blessing."[70] According to Werbner, these ritual acts are "ways of reaching out materially to the saint's grace."[71] She identifies eight stages or movements that define the *'urs:* visiting sacred sites *julus, langar, mulaqāt, zikr, shajara, taqrīr*, and finally, *du'a.*

> *Du'a* means supplication, benediction and blessings. Any person can say *du'a* on behalf of the congregation. In this sense *du'a,* like *baraka,* is a general Islamic term. But the *du'a* of a *pir,* said in the final day of the *'urs* after all *zikr* recitations and *langar* feedings of the mass-

64 *Ritual: Perspectives and Dimensions* (Oxford: Oxford University Press, 1997), 125.
65 "Muslim Festivals," 326.
66 In Lazarus-Yafeh's opinion, the following attributes best describe the official festivals, that give them their strictly monotheistic character:

In sum, the main features of the two Muslim official festivals are prayer and charity, to which are added various popular customs, such a festive meal, festive dress, visits to cemeteries and pilgrimages to the tombs of saints. Yet the festivals lack special ritual observances and specific associations with seasons of the year or with historical or religious events.

67 *Ibid.:* "The most striking expressions of this development are various *Mawlids, i.e* birthdays and memorial days of prophets and local saints, which have mushroomed throughout the Muslim world, their customs and rejoicings spilling over on the official festivals."
68 Pnina Werbner, *Pilgrims of Love: The Anthropology of a Global Sufi Cult* (Oxford: Oxford University Press, 2003), 250.
69 *Ibid.*, 249.
70 *Ibid.*, 250.
71 *Ibid.*

> es, is enormously powerful. It is believed that at that moment the soul of dead saint which the *'urs* commemorates, and the souls of all the *auliya* and the prophets, gather over the congregation. Their combined spirituality is directed towards the saint's appeal to appeal for blessing and healing. That is why no one goes home before the final *du'a*. It is the whole point of the *'urs* ritual."[72]

Werbner further argues that "together these [eight stages] make up the ritual process through which space and person are sacralised during the *'urs*. Reciprocally, *'urs* is the moment in which the charisma of a living or dead saint, embodied in his lodge or grave, is actualized before the whole congregation."[73]

The final *du'a* is a standard feature which concludes the *'urs*, and a ritual activity that is a bone of contention between Deobandīs and Barelwīs. As discussed above in chapter [1], the Barelwīs have supported this ritual of *du'ā* as a form of *iṣāl-i sawāb* (*fatiḥa*), and they have seen the role of saints as a *wasila*, with the *du'ā* being directed towards God. In Werbner's opinion, the Barelwī network perpetuates its ideology through the ritual of *'urs* because it provides the *'ulamā'* with a venue, and at the same time, she does notice their ambivalent intermediary role between the *pīr* and his followers, that is, "while the *maulvi* preaches, the *pir* blesses".[74] She notes that the Barelwī "endow the ritual practices and Sufi beliefs associated with saints and shrines with wider political and public legitimacy."[75]

In this study, I argue the need to rethink Werbner's treatment and definition of *'urs*. It is not through the ritual of *'urs* that "the Barelvi movement is interpolated into the saintly shrine system in South Asia," rather it is through legal and theological discourse on *īṣāl-i ṣawāb*, which Werbner completely overlooks in her study.[76] Second, I argue the need to approach the event of *'urs* in broader perspective by taking into account vernacular interpretations in which the distinction between the event's status as a ritual, on one hand, and as a festival, on the other, is not clear cut. My aim is to rethink the ritual space of the Sufi shrine in the context of vernacular performative aesthetics and mimesis, over and above the ritual of sacred exchange. In fact, the saint's festival moves beyond ritual towards a much larger cosmological or metaphysical expression of transcendental immanence. Werbner is unable to make much of the saint's transcendence of death, it would seem, because of her interest in the power of bless-

72 *Ibid.*, 252.
73 *Ibid.*, 258.
74 *Ibid.*, 257.
75 *Ibid.*
76 *Ibid.*, 242.

ing (and ironically, without understanding the primary function, or even the legal signification, of the ritual in the first place, as the only means by which the saint's charisma is experienced, and as the final purpose of the entire event). Hence, she can say that "the declaration that the saint is alive and powerful in the grave, whether told as an explanation to the anthropologist or as a didactic lecture by a *maulvi* to the congregation attending the *'urs*, is something that may be said of any Sufi saint, anywhere."[77]

The overall weakness of Werbner's study stems from not contextualizing the saint's role with reference to the issue of religious authority in Islam, which also hampers an understanding of its political significance and, most critically, the kind of stakes involved in the process. She only briefly considers aesthetics by including a few devotional verses without analyzing their theosophical significance; nor does she seem to appreciate the role such modes of poetic thought play in informing religious experience beyond the ritual structure of sacred exchange. The aesthetics in which Werbner is interested are entirely those woven into the dynamics of her eight-fold scheme to understand *'urs* as a ritual.[78] I agree with Van der Veer's insight that "ritual practice refracts a variety of social relations as well" and also that "meanings can be seen as communication about ritual as much as communication in ritual."[79] In this the performative aspects and aesthetics of 'the doing and becoming' possesses a kind of autonomous agency as well.

The purpose in this section has been to highlight the literary performative functions of *melā* as a trope in Panjabi mystical verse. Each of the three *kāfī*s treated provides us with an opportunity to gain insight into the nature of vernacular hermeneutics and its engagement with theological ideas. We should bear in mind that these *kāfī*s are most effective when performed, especially in sacred spaces or devotional gatherings. The themes, metaphors, and symbols in these *kāfī*s are experienced through the performative media at Sufi shrines and also stimulate ways of talking, imagining, and thinking about the Divine. On the occasion of *'urs*, conversations among Sufi friends are also modes of mystical experience and knowing shared by observant participants knowledgeable in Sufi

---

77 *Ibid.*, 280. Despite her own experience of having recounted this several times, she does not recognize its existential importance or relevance for her study.

78 In addition to focus on sacred exchange, Werbner does not qualify the Sirhindī metaphysics in relation to its negation of *wujūdī* metaphysics of Ibn 'Arabī, and how such difference would matter in understanding religious experience, apart from mentioning its distinction in the passage printed in full from her interlocutor's narrative.

79 Peter Van Der Veer, "Playing or Praying: A Sufi Saint's Day in Surat," *Journal of Asian Studies* 51.3 (1992), 548.

lore. This is easily seen in how when the name of Ḥusaīn is mentioned, the tone and rhythm of speech changes; or when a tear wells up in an eye of another companion, in remembrance of or sadness at an event (personal or collective) or listening to stories about the extraordinary compassion of saints.[80]

80 In writing about the significance of the saint's festival of *mulid*, I think Michael Gilsenan (*Sufi Saint*, 63) expressed it best: "He [the Sufi saint] is the main topic of conversation, and from the lips of those who actually knew him they hear of his miracles and his teachings. The Saint is, as it were, diffused into the world of men."

Chapter Six
Ḥāẓirī and the Ritual of *Samā‘*

In modern scholarship, Islam has generally been viewed as a fundamentally logocentric religious tradition, not least because of its conception of divine revelation.[1] The science of Qur'ānic exegesis (*tafsīr*) and Ḥadīth as the collected words and practice of the Prophet, sit at the center of its authoritative apparatus. Further, Islamic jurisprudence (*usūl al-fiqh*) developed as a reflective engagement with the two sources of Qur'ān and Ḥadīth in the effort to discover God's law, or *sharī‘a*. However, if the immortal bodies of Sufi saints in Indus folklore are incompatible with textual Islam and other variants of Sufism that emphasize the primacy of *sharī'a*, then such a construction would have it be at odds, ultimately, with the sacred texts of Islam, the Qur'ān and Ḥadīth. The question necessarily arises as to how one defines the relationship between the fundamental importance of the immortal bodies of Muslim saints (as argued here), on the one hand, and, on the other, the centrality of Muslim sacred texts, preeminently the Qur'an (since, if anything in this creation can be said to embody God, it would be His Speech or Word, namely, the Qur'ān).[2] The answer to this question has a bearing on the state of scholarship in the field and how studies have situated the role and legitimacy of Sufi thought and piety in Islam.

Now there are many different aspects to the performative space of the Sufi shrine, with also many ways to interpret its diverse rituals and performances of *samā‘*. In this study, the focus has been on understanding why people journey so determinedly and enthusiastically to the *dargāh*s of Sufi saints. What kind of experience is derived from such a ritual practice? Even more simply put, what is the nature of the religious experience of *ziyāra?* The question is whether we can really claim to examine this phenomenon without understanding the "meaning" that *ziyāra* holds for the devotees. The lack of engagement on the part of scholars with this issue of meaning in studying the performative aspects of worship at the *dargāh* divorces the persons worshipping from the emotional content of their experience.[3] Therefore, I have focused on the theopoetic dimensions insofar as they may be present and discernable in the devotional realm. Moreover, under-

1 See Chapter 3 for the discussion on the political significance of textual Islam, how the authority of the traditional *‘ulamā* is connected with scripture through the use of the science of ḥadīth.
2 It is on the basis of this difference that some scholars have argued that the body of Christ, as understood in the Christian tradition, not the Bible, is comparable to the text of the Qur'ān.
3 It is these very ritual actions and performative dynamics that anthropologists and ethnomusicologists want to learn much about.

https://doi.org/10.1515/9783110781496-008

standing the role of the emotions can serve as a much-needed corrective to what has been so far an over-emphasis on "technique" and "discipline" that has ironically led to a rationalizing and reduction to formalism of the very phenomena that ethnographers aim to study—see discussion in Chapter 2.

I began the first chapter of the dissertation with a discussion of these two ritual expressions—*ḥāẓirī* and *fātiḥa-khwānī*—in order to contrast them to the recitation of *fātiḥa* at the grave of Sufi saints as a practice claimed to be mandated in Islamic law. The purpose was to reveal the incongruity between these two modes of ritual practice, with *ḥāẓirī* considered as indigenous to folk devotion of the Indus, and the recitation of *fātiḥa* as a scriptural, or textual, intervention to replace the theopoetic experience of sacred presence at the *dargāh*s of Sufi saints. The effort here has been to consider the implications of this ritual incongruity, especially insofar as it opens up a new perspective on the dialectics of worship (*'ibadah*)—specifically considering the parallels between many kinds of ritual worship (including *ḥāẓirī*) and scriptural piety—in Islam over the course of history. This incongruity I have claimed is indicative of a deep-seated division and tension, even an epistemological divide that lies at the heart of religious authority in Islam. This epistemological rift is that between, on the one hand, the thinking of the custodians of textual Islam, cultivated through their production of pious scholarly commentaries and glosses as a mediating presence between the words of scripture and ordinary folk; and, on the other hand, the devotees' faith in the miracle-working saint, whose very body and breath is believed to be suffused with the divine and thus serves to connect ordinary persons of faith *directly* with God through their presence at his/her tomb.[4]

After death the body of each saint becomes a pilgrimage goal, hence the celebration of *ḥāẓirī* or *'urs*. In popular imagination, the ritual itself signifies the

4 See chapter 3 for discussion of the issue of religious authority in Islam, especially how different reformist movements sought to reconstruct the authority of law schools in the aftermath of political loss of power in 19th-century India. Scholars of different persuasions argued the hermeneutics of scripture, and in fact ever since all efforts to reform have been centrally preoccupied with regaining "Islam" as an authentic experience, one that can be seen purely rooted in the teachings of the Qur'ān (and ḥadīth) and the strictures of developed *Sharī'a*. Vernacular Sufism associated with the Sufi shrines has been the target of reform or critique, or both, in the view of almost all reformist projects, which has led to a marginalization of its performative dimensions, later to be appropriated by the Pakistani state as "folk culture." The notion that Sufi practices of the *dargāh* could possibly at all also be a form of living memory and thereby an epistemology for gaining sacred knowledge has been unthinkable, something utterly "unthought."

transcendence of death achieved by Sufi saints through divine grace.[5] In this regard, the *dargāh* is analogous to the Qur'ān in its role as the Word of God through which the faithful in his or her recitation participates in the Divine.[6] The identification of the *dargāh* as analogous to scripture is virtually *unsayable* from the common standpoint of textually oriented religious scholars. It is as if the limits of the textual, of *logos* as reason ends, and where the realm of popular religious practices, of religion as experiential participation, begins. However, if we are attentive, we can see that what transpires in the ritual and performative space of the *dargāh*, especially through *samā'*, is very much analogous to the experience of Divine proximity sought in more "orthodox," 'ulamā-approved practices such as Qur'ān recitation in ṣalāt or recitation of the Fātiḥa at saints' tombs.[7] This function of the *dargāh* as a site of human-Divine interaction is es-

5 Since most studies have not examined *'urs* in terms of ritual practice, seeking to construct what it means only through texts, its relevance as a source of mystical knowledge and power has not been satisfactorily treated in academic scholarship.

6 Patrick Laude, *Singing the Way: Insights in Poetry and Spiritual Transformation* (Bloomington: World Wisdom, 2005), 49: "In Islam, the contemplative poetry of mystics is both a commentary upon the Qur'ān and a kind of prolongation of the sacred text; the integration of Sufi poetry—along with Qur'ānic passages—within the mystics' ceremonial gatherings bear witness to this fact."

7 The Ḥadīth science of transmitter biography, *'ilm al-rijāl*. upon which the entire edifice of the religious authority of the traditional *'ulamā* rests, mainly because it was considered a means for maintaining an epistemological connection with the Prophet, was intended to represent the record of "living experience" of the community, through the chain of (person-to-person) transmission. Its development paved the way for the formation of four schools of law, especially after its textual canonization. For this reason, ḥadīth is sometimes seen as the queen of religious sciences in Islam. Ḥadīth for the *'ulamā* helps in understanding scripture because it is based on the Prophet's knowledge of God's word, and a synchronicity between the experience of the Prophet and the Qur'ān. In the case of the *dargāh*, the same role may be assigned to the performative art of *samā'* as it, through the ritual means of *samā'*, brings the devotees to experience the sacred presence of their saint, whose body *is* the cosmological link with the Prophet, and indeed, God. This means that the Prophet is also a Sufi saint, and his "sainthood" (*wilāya*) constitutes the esoteric knowledge of scripture. Sufis have clung to the superiority of the body of the Prophet Muḥammad, his sacrality and immortal presence. Hence the debates over *nubuwwa* and *wilāya*, and this is also the reason that the *'ulamā* have to press the superiority of *nabuwwa* over *wilāya* in their attempts to reform and regulate the Sufi shrines. *Fātiḥa* for the Prophet or for the saints is a means for the *'ulamā* to preserve their textual authority by signifying them both as holy, but dead. Yet this was not how devotion to the Prophet or the saints has been practiced ritually—as discussed previously. In understanding the question of intercession, and how it has been problematic or the paradox it has presented, I draw on the comparative work of a scholar from outside the field of Islamic Studies, Kimberley C. Patton, "God and His Angels Pray for the Prophet": A Qur'ānic Paradigm," chapter 9 of *Religion of the Gods: Ritual, Paradox, and Reflexivity* (Oxford: Oxford University Press, 2009).

tablished not in the ontological space of the text, but through the embodied ritual performance of *ḥāẓirī* in the precincts of the *dargah.*

The nature of the tension becomes apparent when we compare this experiential function of *samā‘* with the textual authority of the *ulamā‘*. In Sufism, the connection between the devotees and the *dargāh* has been a ritual means for the ordinary person of faith to connect with the power and grace of God, and this ontological relationship is both expressed in and expressive of the aesthetics of Indus Sufism.[8] Devotees experience miracles (*karamāt*) of the Sufi saints in their daily lives and speak of the peace and sense of intimacy gained in visiting one of the *dargāhs*. Worshippers also feel ecstasy when attending the *samā‘* performances and listening to the Sufi poetry, either regularly during the week, or on *'urs*. It is important to note that such experiences are received as *gifts* by the devotees, confirming what Jean During has so perceptively written—and to what anthropologists or ethnomusicologists pay too often too little heed—in his study on the significance of hearing in Sufism: "it is the ecstasy that leads to the dance, and not vice versa."[9] This has important consequences for understanding the transmission of mystical knowledge in Islam in ways that has received too little attention up to now, something for which this study seeks redress. Therefore, the purpose is to see how we can study the phenomenon of mystical experience in Indus Sufism at the *dargāh*, by attending to the truth of the saint's grace and power that devotees feel and affirm so convincingly in their bodies.[10]

It is ultimately to supplant the embodied relationship signified in *ḥāẓirī* that the textual orthodoxy seeks to inscribe its practice of supplication (*du'ā*) through the recitation of *fātiḥa* – in order to put limits on the possible metaphysical interpretations of the experience at the *dargāh*. The danger for the *'ulamā* of the ritual act of *ḥāẓirī* is that it potentially undermines their textual authority over

8 Studies that view Sufism only through the filters of its institutional mechanisms and texts have paid no attention to the effectiveness of ritual as a means of connection with the living presence of saints: see chapter 2 above.

9 Jean During *et al.*, "Hearing and Understanding in Islamic Gnosis," *The World of Music* 39.2 (1997): 132. This work as a whole is very useful for understanding the importance of hearing in Islam.

10 Two studies that offer a study of the aesthetics are Seema Golestaneh, "Listening, non-knowledge and the auditory body: Understanding Sufi Zikr Ritual and Sama as Sites of Aesthetic experience," edited by Pedram Khosronejad, *Saints and their Pilgrims in Iran and Neighbouring Countries* (UK: Sean Kingston Publishing, 2012) and Paulo G. Pinto, "The Anthropologist and the Initiated: Reflections on the Ethnography of Mystical Experience among the Sufis of Aleppo, Syria," *Social Compass* 57.4 (2010), 464–478.

scripture, as formulated in schools of law.[11] The recitation of *fātiḥa*, as a ritual activity either at the grave, or prior to distribution of food (*langar*), or at the end and beginning of *samāʿ* performances, marks the saint as a pious, but dead person. Nevertheless, despite the efforts of the *ʿulamāʿ* to enforce this as the "correct" method of performing *ziyāra*, the devotees continue to cultivate an active, intimate, personal relationship with Sufi saints through *ḥāẓirī, one* rooted in the experience of the living presence of God's friends at their *dargāhs*.

6.1 Scripture and Sufi Poetry

In what follows, I want to discuss first some theoretical concerns in the study of Sufism. Second, I review the scholarship that can help us conceptualize the relationship between scripture and Sufi poetry. Third, building on the work of William Graham and Ali Asani, I will make a case for seeing the devotees' experience of the Sufi *dargāh* as analogous to the experience of scripture as God speaking, thus breaking from traditional emphasis on the Qur'ān as effectively or functionally the sole Islamic scriptural "text."[12]

The religious devotion (involving connection, or *nisbat*, in the ritual of *ḥāẓirī*) of the people at the Sufi *dargāhs* plays a major role in shifting the central Islamic focus on scripture from its presence as a recited text, to seeing the sacred body of the Sufi saint as a powerful locus of experience of the Divine analogous to that

11 See chapter 3 for discussion on how the preoccupation with hermeneutics of reform in the modern period led to a reinforcement of scripture as authoritative divine text, which thus became the concentrated locus of attention in efforts at "recovery of authentic Islam." This led to marginalization of the oral-aural aspects of devotion at Sufi shrines. Eventually, the merger of scripture and the body politic, Pakistan, as an "Islamic" state (a scripturalist Islamic state that has never existed previously), has ever since ensured the epistemological domination of scripture as authority. Cf. Clifford Geertz, *Islam Observed: Religious Development in Morocco and Indonesia* (Chicago: The University of Chicago, 1971 edition), 69: "... in the Islamic case the stepping backward seems often to have been taken for the leap itself, and what began as a rediscovery of the scriptures ended a kind of deification of them." The ideology of Pakistan is a merger or synthesis between two types of bodies: the religious body (of the scholars) and the political body (of the state). This is mainly because despite the divisions among the various religious movements, all view the religious and political spheres as one in Islam. As a consequence of the concern with reforming the corrupt practices of Sufism, the *dargāhs* of Sufis in Pakistan have been controlled by the state through their management by its Ministry of Religious Affairs (Department of *Awqāf*).

12 For purposes here, Ali S. Asani is among the few scholars to have studied vernacular Sufi poetry and forms of devotional experiences, and William Graham has dealt specifically with both scripture (especially *Qur'ān* and Ḥadīth) and ritual.

realized through scriptural recitation and contemplation. As stated above, such a conception of human-Divine encounter is anathema to Islamic orthodoxy, unthinkable in most reformist quarters, as it suggests that an encounter with a dead saint might compare to an encounter with the Divine word of God, which is tantamount to blasphemy.[13] To make my counter-argument, I turn to the work of two Islamic-studies scholars, Ali Asani and William Graham, whose contributions are critical here.

I begin with Asani's work on the role of Sindhi poetry before turning to Graham's treatment of the category of scripture. The use of qur'ānic verses in vernacular poetry, known as *iqtibās*, has served as a means by which scripture enters the oral, performative culture,[14] how "liturgical use enhanced the status of the poetry and its authors. As the reciters came to understand poetry as embodying spiritual truths, it began to play the role of a religious scripture."[15] Hence, we have a number of examples of a transformation of poetic genres into scriptures in which the poetry becomes effectively scripture,[16] such as the *Risālo*, which is "regarded as commentary on the Qur'ān in Sindhi and has been called the "Hindu's Gita."[17] Similarly, this is also true for other vernacular traditions, such as the *Adī Granth* for the Sikhs and *Dadū Banī* for the Dādūpanth.[18] Asani explains that such poetry in the Islamic case attains its scriptural status because Muslims regard it as standing in for the Qur'ān, a point similar to that made by Tony Stewart.[19] Hence, verses attributed to Bulleh Shāh (d. 1754 C. E.) and other Sufi poets,

---

13 The late Fazlur Rahman was accused by the Pakistani *'ulamā* of collapsing such a distinction in his work on the Qur'ān.

14 Ali S. Asani, "South Asian Literatures and the Qur'ān," in *Encyclopedia of the Qur'ān:* "The incorporation of a qur'ānic verse into a vernacular text served several purposes. First, it sanctified the text for both the author and the audience, thus making it more sublime. Second, the skill with which the Arabic sacred text was woven into the fabric of the vernacular demonstrated the author's literary prowess. Third, the verse could also serve as a proof text validating the author's religious beliefs and convictions."

15 Ali S. Asani, "Sindhi Literary Culture," in *Literary Cultures in History: Reconstructions from South Asia*, edited by Sheldon Pollock (Berkeley: University of California Press, 2003) 634.

16 *Ibid.* 635.

17 *Ibid.*

18 *Ibid.*

19 In referring to Tony Stewart's essay, "In Search of Equivalence: Conceiving the Muslim-Hindu Encounter Through Translation Theory," *History of Religions* 40.3 (2001): 260–87, Asani writes that "the fact that communities who recite and sing vernacular religious poems frequently regard them as text which encapsulate the teachings of the Arabic Qur'ān," in "South Asian Literatures and the Qur'ān."

like panegyrics (*gināns*), are all "commonly regarded as spiritual commentaries on qur'ānic verses...embodying the inner signification of the Qur'ān."[20]

In this process of sacralization of Sufi poetry, Asani refers to Paul Nwyia's work on devotional practice describing how the early Sufis, being "constantly preoccupied with the Qur'ān as the word of God (q.v.), their memories were eventually 'qur'ānized'."[21] It is held that such an intense preoccupation led to seeing everything in their lives in the light of the Qur'ān, a devotional relationship which still holds centuries later for the vernacular poets of Gujrat, Panjab, and Sindh.[22]

But how are ordinary folk drawn into hearing Sufi poetry as equivalent or analogous to the Qur'ān? In good part because "Sindhi literary culture has been overwhelmingly poetic in nature for most its history, primarily functioning in aural and oral modes."[23] Also, it is hard to overestimate the role of music in the vernacular cultures of both Panjab and Sindhi: "Sindhi poetry is so fused with musical tradition that to recite a verse inevitably means to sing and perform it, frequently with a musical accompaniment."[24] That is to say, "liturgical use enhanced the status of the poetry and its authors. As reciters came to understand poetry as embodying spiritual truths, it began to play the role of a religious scripture."[25] Finally, I want to emphasize a further point that is highlighted in Asani's discussion, one that greatly expands the interpretative horizon of scripture:

> As in the case of much medieval north Indian vernacular poetry, there is a stress on an *interiorized* [emphasis mine] form of religion involving the search for salvation under the

---

20 *Ibid.* See also Shemeem Burney Abbas, *The Female Voice in Sufi Ritual: Devotional Practices of Pakistan and India* (University of Texas Press, 2002), 20: "Among the musician communities in the Sind, Shah's *Risālo* is treated with great respect because of its sacred Quranic context, and the book is not allowed to lie around on the floor."

21 *Ibid.*

22 It is important mention here that while Qur'ān, as the Divine Word of God, is treated with great reverence; we must not forget that the Indus mystics speak incessantly, in their poetry, about the need to purify the heart (see Chapter 4). In a way then the heart takes precedence over the "text".

23 Ali S. Asani, "Sindhi Literary Culture," 631.

24 *Ibid.* According to Shemeem Burney Abbas, *The Female Voice in Sufi Ritual*, 167, fn. 42: "A Sufi epic based on three books. In Sindhi *risālo* means "the message." The *Risālo* narrative is based on thirty sections, each of which is built around a *sūr* that is a form or mode of singing, a rhythm, or melody. Ten of the the *sūrs* in the *Risālo* are built around female heroines (*surmise*) and thus are called *Sūr* Sohni, *Sūr* Marvi, *Sūr* Mumal-Rano, *Sūr* Leela-Cenesar, *Sūr* Sassi-Ma'a-zuri, and *Sūr* Sassi Desi."

25 *Ibid.*, 634.

> guidance of a master. In this personal search, formal learning is not only unnecessary but may in fact be a hindrance.[26]

While search or quest for a *pīr* by the individual, who it is anticipated will be able to help the traveller (*salik*) in his or her journey to personally experience or taste divine truth, is certainly part of a well-established cultural norm—and one that cuts across several religious traditions and practices in South Asia—the role of the Sufi shrine where devotees can spiritually connect, and establish communion, with the living presence of saints must not be overlooked.[27] As is the case in the devotional culture of Indus Sufism, listening to the poetry (in *samā'*) of Sufi saints, like Bulleh Shāh, is very much a core part of the ritual of process and experience of *ḥāẓirī*.[28] Most importantly, then, oral poetry is received as the inner essence ("*batin* of the *batin*") of what the Qur'an contains, a process by which the listeners are drawn into a continuous *ta'wil* as an inner devotional experience.[29] In sum, the Sufi poetry sung or performed by *qawwāls* at the *dargāh* receives the attention that it does because of the devotional bond (*nisbat*) that the visitors to the shrine have established with the saint.

I fully concur with Asani's description of the oral-aural character of the vernacular Sufism of the Indus, and the creative role of its liturgical function. In understanding the scriptural status of mystical verses of Sufi saints like Bulleh Shāh and many other mystic-poets in the wider region, Asani's work opens the way for a closer look into the confluence between the language of scripture, Arabic, and vernacular languages such as Sindhi and Panjabī. In Asani's view,

---

26 Ali S. Asani, "Sindhi Literary Culture," 629–30.

27 See a recent "Foreword" by Ali S. Asani in Michael Frishkopf & Federico Spinetti (editors), *Music, Sound, and Architecture in Islam* (Austin: University of Texas Press, 2018).

28 Shemeem Burney Abbas, *The Female Voice in Sufi Ritual*, 59: "Listening to *qawwālī* repertoire or samā' and *sufiāna-kalām* for devotional purposes forms part of the rituals that devotees undertake during a *hazrī* or *salām* at the shrine. The *qawwāls* share the same worldview whether they are men or women. They look upon themselves as devotees of the Sufi saint or poet. Lahori supports his claim when he uses the very term "*hazrī*" in connection with *qawwāls* who sang at the Sufi shrines in Lahore before the partition of 1947 and after it. He mentions notable musicians such as Ustad Bare Ghulam Ali, Ustad Chote Ghulam Ali, Niaz Hussain Shami, Fateh Ali Mubarik, and Dina Qawwal, who performed *hazrī* in terms of singing mystical texts at the *'urs* of Data Ganj Bakhsh in Lahore. He also mentions women musicians who performed *hazrī* at the Sufi shrines in Lahore, especially at the time of the *'urs*."

29 Asani links this to the poetry of Rumi in discussing how the poetry of Sufis is believed to be the inner meaning of the Qur'ān, but the phrase "*batin* of the *batin*" was shared in a personal conversation of the author with Ali S. Asani on his work on vernacular poetry in Sindh.

this transition takes place because of the importance accorded to acquisition of knowledge through a process of interiorization.

Sufi poetry gives form or voice to this *interiorized* or esoteric knowledge of the Qur'ān, achieved first and foremost by the Sufi saints of the region. In other words, the saint's poetry becomes scripture as a consequence of its profound engagement with the text of scripture, a development that can be seen in the performative space of Sufi rituals such as *ḥāẓirī*, where the poetry/song is heard and received as the inner meaning of the Qur'ān.[30]

In my analysis, it is important to pay closer attention to what specific ritual or ritual form has such an epistemological validity, or "equivalence" in power, so that it is able to carry the inimitable (*mu'jiz*) Word of God and to enter into the hearts of the people in this untranslatable form? In the case of sufi Islam, this was achieved through the performance of *samā'*, which received its sacred, archetypical, status as a means of worship in the vicinity of the *dargāh*. Therefore, as this dissertation argues, the question of how Sufi poetry comes to function analogously to scripture can only be answered once we understand that the oral-aural process can best be grasped in the ritual context of the Sufi *dargāh*.[31] Vernacular poetry, at least in the case of Indus Sufism, achieves its scriptural status in the context of the ritual function of *samā'* performed at the *dargāh*.

Finally, I want to turn to William Graham's reconceptualization of "scripture", which I feel is helpful in thinking about and visualizing how poetry in festivals and ritual is able to take on scriptural properties and effectively function analogously to scripture. In particular, Graham's focus on the experiential dimension of scripture's oral/aural function as an epistemological force emphasizes how texts reveal themselves functionally in practice, that is, in lived context. This contextualization allows for a better disclosure of the "inner" dimensions of religious devotion, through sensual modes of experience. Effectively, then, this means putting the "body" back into how we conceptualize experience in

30 Patrick Laude, *Singing the Way*, 56: "Sufi poetry can be characterized as an overflowing or outpouring of grace through the channel of words. There is no trace of virtuosity in such a practice of poetry: the poet [Sufi] is too overwhelmed by the flow of images and words to be able to manipulate them in a technical way."

31 It is absolutely essential to see the role of the Sufi *dargāh* to understand how Sufi poetry achieves it scriptural status, from its orthodox fixity as a text to experiencing it in the vernacular as the esoteric meaning of the Qur'ān.

terms of its epistemic value, or what we can accurately call a phenomenology of experience.[32]

Beyond the Written Word makes a strong case for considering the oral and aural dimensions of scripture, and offers examples of devotional activities that demonstrate these functions for religious communities. This greatly expands the conceptual meaning of scripture since it involves taking into account its active oral and aural uses that have been historically the primary channels through which scripture has flowed most directly into the lives and affected religious persons in tangible, physical fashion. It is this that we as scholars must attempt to study if we wish to understand the functional significance of scripture in human experience. Scripture's function in practice can take on a wide spectrum of meanings, rhythms, tonalities, and forms that lead us "*beyond the written word*" to the world of embodied experience, and to the body itself. To start with, Graham provides a concise description of how we view or think of scripture today: "the term 'scripture' is usually reserved for religious texts that have been committed to the written or printed page, as the word itself and its common equivalents [indicate]".[33] However, upon careful inspection of the way in which the term "scripture" has come to signify for the modern mind *only* a material, written or printed text, are we able to see what it means to go beyond the written text? The common modern notion of scripture as only a tangible, physical reality, is what is shaken once we open our vision towards the horizon of possibilities that the varied content of actual scriptural texts encompasses: "Myth and legend, historical narratives, ritual books, legal codes, ecstatic or mystical poetry, apocalyptic visions, utterances of prophets and teachers, divine revelations, and hymns and prayers to a deity have all found a place in one or another scripture."[34] Scripture is not just a physical book, a printed word in a bound folio, as we as scholars in today's world in particular typically treat it, because "ours is not only a literate, but a book and print culture".[35] In other words, "so tied are we to the written or printed pages that we have lost any awareness of the essential orality of language, let alone reading."[36] Then there is the issue of

---

32 In his "Islam in the Mirror of Ritual," Graham makes a case for looking at ritual itself to gain a better understanding of its meaning rather than reading it from its categorization in texts, or imposing speculative theories on ritual data. See chapter 2 for my discussion of this article and its relevance for focusing on *ḥāẓirī*.

33 William A. Graham, *Beyond the Written Word: Oral Aspects of Scripture in the History of Religion* (Cambridge: Cambridge University Press, 1987), 4.

34 *Ibid.*, 3.

35 *Ibid.*, 9.

36 *Ibid.*

ambiguity with texts "that might seem peripherally "religious", yet whose functions still can be classified as "scriptural" in many cultural contexts."[37] In consideration of these above factors "we are finally left with few if any external criteria for separating 'primary' and 'secondary' sacred texts across the wide range of scriptural traditions";[38]

> A book is only "scripture" insofar as a group of persons perceive it to be sacred or holy, powerful and portentous, possessed of an exalted authority, and in some fashion transcendent of, and hence distinct from, all other speech and writing.... A text becomes "scripture" in active subjective relationship to persons, and as part of a cumulative communal tradition. No text, written oral, is sacred or authoritative in isolation from a community.[39]

Similarly, in the case of Indus Sufism, focusing on the ritual of *ḥāẓirī* at the *dargāh* of the saint with a view to what it can tell us about the sacred space and its role in Indus Muslim piety, we are better able to attend to its "meaning" in the eyes of the devotee. This has been the entire point of the dissertation. My argument is that the immortal saint, whose sacred body is the site for celebration, is the locus where devotional poetry acquires a sacred status. The performance of verses in the ritual space of Sufi shrines transforms the poetry so that it becomes part of the *lived* experience of the saints and their grace. That is, the living presence of saints at their tombs is what devotees most fully experience in the performance of devotional poetry. Conversely, the latter's devotional *nisbat* with the saint of the shrine (*sahib-i mazār*) is the context in which oral tradition becomes sacred, *analogous* to the Divine word. This context places both the performer and devotee in an interactional or symbiotic relationship, bringing them together in veneration of the saints at the *dargāhs*. *Samā‘* is not so much about arousing "historical memory" as a means of remembrance as it is about invoking or creating the very *presence* that gives *samā‘* its sanctity in the ecstatic experience. The shrines provide the sacred space for the experience of the sacred in the living presence of saints' bodies; it is a major reason why, as discussed in chapter three, the state and the religious scholars want to regulate, delimit, and even expunge its religious significance. Their scripturalist reforms, focused on the Fātiḥa, are part of a political strategy to control the experience of *ḥaẓirī*, to stipulate how its devotional meaning is to be interpreted, and thus eliminate any alternative means by which the textual authority of the religious scholars can be challenged through devotion focused on saints'

37 *Ibid.*, 3.
38 *Ibid.*, 4.
39 *Ibid.*, 5.

bodies. This is the fundamental framework that underlies the debates that advocate the superiority of *sharī'a* over *ṭarīqa*, because in practice it becomes a means through which the *'ulamā* can regulate devotional practices.

In essence, my argument is that the immortal body of the saint with its sacred power is a catalyst for the transformation of vernacular poetry into scripture.[40] This can be appropriately defined as the theopoetics of Sufi shrines. The saint's immortal body is the medium through which the devotional culture at shrines achieves its sacrality. Just as ordinary food (*langar*) becomes sacred food, so too "poetry" performed in devotion to the Sufi saint is experienced as sacred Word.[41] The interaction between the devotees and the performers in the living presence of God's friends is experienced *as equivalent with the listening of the Divine word.* The connection (*nisbat*) of the devotees and the performer with the living presence of the immortal body of the saints is the alchemy at the heart of the "scripturalizing" process.

The shrine has always been noted for the pluralism of its sacred space, as it has been a site of religious devotion for multiple faith communities in wider South Asia. Shrines play a vital role in bridging divides among various communities through the timeless message of seeking the divine within—the "interiorization" of the spiritual in the human, as Ali Asani has rightly emphasized. As a devotional practice, then, Indus shrine Sufism presents us with a sacralization of the saint's human body and the poetic panegyrics devoted to the saint, the experience of which cuts across religious boundaries in popular practice in the Indus region as elsewhere in South Asia.

6.2 Theopoetics of *Nisbat*

From the preceding discussion, through reference to Graham's work, I have tried to make clear how active oral and aural devotional experience as a practice must be taken into account in understanding the meaning of *ḥāẓirī*, *nisbat*, and the

---

40 In chapter 4, I discuss the question of "authorship," in vernacular poetry as raised by Robin Rinehart in terms of the performative culture of Sufism, *"The Portability of Bullhe Shah: Biography, Categorization, and Authorship in the Study of Punjabi Sufi Poetry,"* Numen 46.1 (1999): 53–87.

41 Just as in the case of *fātiḥa*, as *īṣāl-i ṣawāb*, aims at regulating the experience of *ziyāra*, to divert what the reformers believe is a misplaced devotion to the body of saints towards God instead. Its recitation over food prepared on religious occasions is commanded by the same logic (Clifford Geertz's work on feasts in Java treated this phenomenon). In Islam, *fiqh* stands in place of scripture, as far as the *'ulamā'* are concerned.

dargāh. We are now ready to discuss *ḥāẓirī* and *nisbat* as a horizon of meaning in religious devotion to Sufi saints. Communion at the *dargāh* is best understood in connection with the experience of its effects on the devotee's existence.[42] Approaching it from such a holistic perspective helps understand why devotees claim that the states (*kayfiyāt*) of ecstasy (*ḥāl* and *wajd*) are, first and foremost, a gift and, therefore why they strongly object to their being treated as effects emerging from bodily discipline ("technique") and habits. The person gifted with such experiences finds them unquantifiable by usual external standards of assessment.[43] Such experiences of grace are not explicable in a rationalist framework, which gives them their special epistemological status.

The proper context, therefore, for studying such religious experiences must begin with how the devotional relationship with Sufi *dargāh*s has value for the devotees. In particular, the aesthetics of devotion must be viewed in relation to ontology. To grasp the cultural or "social poetics" of Sufism necessitates an attempt to grapple with the kind of miraculous interventions that devotees experience and report. These involve how the Sufi saint's grace breaks into their lives as if out of nowhere. To grasp *nisbat* as a locus for theopoetics means we must be ready to look beyond our positivist notions of reality, to abandon—even if momentarily—our rationalist modes of thinking.[44] As a consequence of a rationalist epistemology, the problem is that in contemporary scholarship, our analytical accounts tend to sideline precisely these kinds of Sufi phenomena.[45] The key

42 We experience death of our loved ones (family and friends) and in time worry about our end as well, becoming familiar with different kinds of suffering through personal experience. People visit saints on account of their personal challenges and dilemmas that present themselves in life. I have witnessed tears of joy and sadness in the eyes of people at the *dargāh*, the sick visiting to be healed, or persons coming to petition on behalf of their parents, friends, and children. Devotees tell their stories of grief and also share news of joy with their saints. So needless to say the *dargāh* is a field of powerful emotions as devotees visit in gratitude and love from far and near. More than anything, *nisbat* is what underlies the structure of experience in *ziyāra* to Sufi shrines.

43 William James writes about this extensively in his *The Varieties of Religious Experiences: A Study in Human Nature* (New York: Penguin Books, 1985; reprint). Mircea Eliade's concept of hierophany is also relevant, concerning how such experiences have an empirical (historical) dimension.

44 In the study of Sufism, the only work that has, as a matter of method, factored in the role of such lived experience is by Katherine P. Ewing, *Arguing Sainthood*. I have discussed the significance of Ewing's work above in chapter 2.

45 Cannell, Fenella. "The Anthropology of Christianity," in *The Anthropology of Christianity*, ed. by Fenalla Cannell (Durham: Duke University Press, 2006), 3: "Religious phenomena in anthropology may be described in detail, but must be explained on the basis that they have no foundation in reality, but are epiphenomena of "real" underlying sociological, political, economic, or other material causes."

to understanding the material culture of Sufi shrines requires that we bring their emotional context into view. To put it in simple terms, the ontological status of the shrine in the eyes of the devotees precedes the devotional aesthetics of the religious culture of Sufi *dargāhs* and characterizes and infuses the performative dimensions of worship at the sacred threshold of the Sufi saint.[46] This is how "ecstasy leads to the dance," in that the *karamāt* of Sufi saints are the source of joy and celebration for which devotees gather and which motivates their actions at the *dargāh.*[47]

As the devotees personally experience this connection, or *nisbat*, this experience brings with it a knowledge or insight about matters of grace. With time, the connection gradually crystalizes into a realization of one's status as a devotee of the loving and compassionate Sufi saint. *Nisbat,* for the devotee, then becomes a source of personal identity by being connected to the *dargāh* of a Sufi saint.[48] Devotees invoke the name of the saint at the *dargāh.* The name is voiced (not shouted) from deep within the core of the body, from the stomach, or it can be whispered silently under one's breath. Calling the name of the Sufi saint is a form of *zikr* (Ar. *dhikr*) or "recollection" ultimately focused on God and the truth (*ḥaq*) that the *zikr* contains or that the saint embodies. Thus, a person is "won over" in his/her concourse with the Sufi saint, a process whereby a relationship of the heart begins to realize its emotive efficacy.

The metaphor of the heart is often employed in Sufi poetry.[49] It has an epistemological function that leads the devotees to the sacred center, to realize the

---

46 By "material culture" I mean the entire spectrum of the sensual dimensions of the *dargāh:* performative arts, ritual practices, architecture, feast (*langar*), and the smell and sounds of the religious ("sensual dimensions").

47 Mikko Viitamäki, "Mystical Gazal in Sufi Practice. Poetics, Music and Metaphysics in the Verses of Sāh Niyāz Aḥmad," *Oriente Moderno* 92.3 (2012), 529: Building on Chittick's work, Viitämaki explains that "The significance of *hayāl* for understanding God has been elaborated on by Ibn 'Arabī and his followers. *Wuğūdī* ideas find a fortuitous confluence with poetics in *gazals*, which at best appeal directly without detouring through the brain. Confusing statements in particular force the audience to resort to their *hayāl* rather than reason (*'aql*). This experience, according to Ibn 'Arabī, is closer to sense experience and thus more immediate than the rational process of stringing concepts together and drawing conclusions."

48 A good number of *kafīs* speak about this connection with saints, of belonging to them. For example, one of the most common is the claim of being *malang* of 'Alī (*'Alī da malang*). Another *kafī* of Shāh Hussaīn goes like this: *"sahib terī bandī ān, changī ān, yā mandī ān,"* (Master, I am yours, good or bad, I am yours)...*mein rung murshid dey rungī ān* (I am dyed in the color of my saint). This meant to say that devotees do not wear red to become mystics; rather it is the Sufi saint who clothes the devotees in red.

49 In a medical sense the heart is no more than an organ, albeit with an important function, in that it pumps blood, supplying it to the entire body. Metaphorically, the heart is identified in our

essence of *‘ibadat* (worship).[50] In Sufism, the heart is considered the seat of experiential knowledge by which devotees gain direct experience of the divine.[51] In order to avoid the Cartesian fallacy, this inner space, I believe, can be examined through the study of ritual practice and thereby coming to an understanding of how the devotional body interacts, kinesthetically, with the Sufi shrine. However, in focusing on the body, care must be taken not to marginalize the role of emotion to prevent our inquiries from becoming disconnected from the phenomenon under study.

---

modern age with the emotive side of our being, and it is accordingly differentiated from the intellectual side that, obviously, stems from the brain. In the history of philosophy, the origins of this dualism, categorized as the mind-body problem, are to be found in our Cartesian inheritance—Descartes' quest for epistemological certainty, which finds its way into our theories on the human body.

50 Qur'ānic verse: *ṣura* 50.16: *naḥnu aqrabu ilayhi min ḥabli l-warīdi* (we are closer to him (*al-insān*) than [his] jugular vein." Also, see chapter 5 for Bulleh Shāh's use of this verse in his own verses. Sufi mystics have made the most of this verse to express our relationship, as human beings, with God. In Islamic intellectual history, broadly speaking, the Mu‘tazilites and the Ahl al-Hadith and Asharites were engaged in polemics over the most appropriate way to explain the anthropomorphic content in the Qur'ān. In Islam, the question of anthropomorphism is irrevocably tied to scripture, the holy Qur'ān, and it fell upon the religious scholars to find the appropriate balance between God's transcendence (*tanzih*) and immanence (*tasbih*). Sufism is bracketed within the Ash'ari camp [sense?] in its transcendence of human reasoning, but little has been done to think about the question of anthropomorphism in view of ritual practices of *samā'* where the notion of the imagery of the divine embodied in the body of saint is often evoked. For example, in Indus Sufism, Bulleh Shāh's poetry does much with Qur'ānic verses on divine proximity:*"maulā aḍmī ban ayā."* (God becoming human (*admī*) has come)—*qawwālī*. In Islamic philosophy, the entire question of anthropomorphism is connected to explaining how the unmoved Mover, God of Aristotle, is linked to an imperfect world of change in a way that does not compromise God's radical transcendence yet does not deny God's (anthropomorphic) attributes.

51 A major study on the Chishtī order looks at the metaphor of a burnt heart to describe its efficacy for the Sufis: Carl W. Ernst and Bruce B. Lawrence, *Sufi Martyrs of Love: The Chishti Order in South Asia and Beyond* (London: Palgrave Macmillan, 2002), a groundbreaking study of Sufism in general, which, apart from referring to an extremely important image and event of a Sufi saint passing away during *samā'* in response to a mystical verse, does not really delve in the emotive aspect of its topic. Another work by Scott Kugle, *Sufis and Saints' Bodies: Mysticism, Corporeality, and Sacred Power in Islam* (Chapel Hill: The University of North Carolina Press, 2007), 289, pursues an in-depth study of embodiment but only through the agency of texts. Also, a study by Michael Sells has a chapter on the heart: "Ibn ‘Arabi's Garden Among the Flames: The Heart Receptive of Every Form," In *Mystical Languages of Unsaying* (Chicago: The University of Chicago Press, 1994).

Ḥāẓirī is equally the realm of gratitude, love,[52] tears, and desires—the theopoetic springs from here, the mindset that devotees bring with them for their conversations with the Sufi saints, to tell these friends of God about their plight, and as well as to receive an answer. Therefore, understanding the importance of the role of grace (*karam*), as I have repeatedly noted, can help us in some measure to attend to this feature of the ineffable in religious devotion. The connection between *ḥāẓirī* and the ritual function of *samā‘* provides the human context for the hearing and experiencing of the divine word at the *dargāh.*

The ritual of *ḥāẓirī* rests upon the recognition of the direct influence and living presence of Sufi saints at the *dargāh*—a form of religious devotion that is incongruent with its prescriptive counterpart in textual Islam. Its foremost effect, as a grace (*karam*), is in opening the devotee's soul to the *hearing* of the divine word in *samā'*. Moreover, *nisbat* with saints becomes the means for the ordinary devotee to experience the fruits of mystical knowledge without the mediation of the textual knowledge of the *ulamā‘*. In particular, the ritual *samā'* offers an immediate means for the ordinary and the unlettered to be initiated in the mysteries of Sufism because of the grace of *nisbat*; all that matters, according to those who regularly visit the *dargāhs*, is the purity/openness of the heart of the visitor.

There is further an interactional dimension to the whole experience of *nisbat* with Sufi saints. One needs to learn, little by little, as one goes along, to recognize and discern how the extraordinary enters into the realm of the ordinary. What our scholarly studies often miss in the study of religious experience is the underlying emotional content and its epistemological value in acquiring mystical certainty.

6.3 *Samā‘* at the *Dargāh*

We are now ready to deal with the issue of "discipline" or "technique." To this I want to add "narrative, which has become a preoccupation in the social sciences and the humanities in attempts to understand the religious experience, and especially states (*kaifiyāt*) of ecstasy (categorized by terms such as *ḥāl*, *wajd* and *mastī*).[53] I do not intend to critique existing studies on the *samā‘*. I want instead

52 See William Chittick, *Divine Love: Islamic Literature and the Path to God* (New Haven: Yale University Press, 2013), Benyamin Abrahamov, *Divine Love in Islamic Mysticism: The Teachings of al-Ghazālī and al-Dabbāgh* (New York: Routledge, 2003).

53 In general, my main problem with ethnographic studies on *samā‘* that have sought to analyze states of ecstasy has primarily to do with their marginalization of its epistemological context. The notion that such states can be described as a gift, or spiritual gift, is freely admitted in

to draw on my field experience to resist the practice of emphasizing performative structure without considering its existential ground for the devotees beyond cursory mention of its emotive dimensions. Such an approach is, I would argue, quite insufficient to characterize the action taking place.

As mentioned above, this chapter is concerned with the ontology of the Sufi *dargāh* and *haziri* as a *real* presence that has a palpable effect on devotees, informing their understanding of the relationship between religion and life.[54] That is to say, the resurrected body of the saint is the center of devotional focus, and it is a body that has transcended death, which informs the poetics of ritual and hospitality and how these two aspects mediate gnosis (*maʿrifa*) or mystical knowledge.[55] So, principally, this coming together of ritual devotion and mystical verse is the reason that I describe the devotional culture of Indus Sufi shrines by the term "theo-poetics" in contrast to "social poetics."[56]

---

participants' accounts but quickly jettisoned in favor of the anthropologist's analysis and explanation of how things "really" happen or work.

54 It is not enough merely to see a formal or structural type of dialectic between ritual and poetry. Aesthetics alone do not tell the entire story if we fail to get a feel for the emotive force that permeates the various cultural manifestations and gives them life. The soul of these two ritual dimensions (*ḥāẓirī* and *samāʿ*) are linked by the devotional element I am identifying as personal *nisbat* with the Sufi saint. After all, the visit (*ziyāra*) to the Sufi shrine is a journey undertaken to be in the living presence (*mawjūd*) of the Sufi saint. Not only the devotee but the performer of *samāʿ* is also bound by *nisbat*, if the performance is to be beneficial for all by the grace of the saint. To clarify further, I am helped by way of Maggie Kast's"Dancing in Sacred Space: Some Reflections on Liturgy and Performance," *Religion and the Arts* 4.2 (2000), 219: "In *The Book of Sacramental Basics*, Tad Guzie describes a human rhythmic in which lived experience is named by stories and framed by festivity. The festivity takes the form of symbolic enactment. This symbolizing, expressive process is at the root of all sacraments. Liturgies name and celebrate experience in this way, but so do other human rituals (birthdays, anniversaries) and certain kinds of performances. Thus, both performance and liturgy are symbolic processes expressing human experience."

55 *Samā'* is not, as a performance, an observance honoring the death anniversary of a Sufi saint; it is rather a form of ritual communion. The fallacy of describing the *qawwālī* as an event marking the death anniversary is all too frequently found in the leading scholarship on the subject, e.g. in Regula B. Qureshi, "*Qawwālī:* Making the Music Happen in the Sufi Assembly," in *Performing Arts in India: Essays on Music, Dance, and Drama*, ed. by Bonnie C. Wade (Berkeley: University of California, 1983): 19: "*Qawwālī* assemblies commemorate the death anniversaries (*ʿurs*) of the numerous Sufi saints at their shrines or wherever their devotees gather for the occasion." Nevertheless, Qureshi's work on the whole is an excellent source for understanding the structural features of *qawwālī* as a musical performance.

56 Richard Wolf employs Michael Herzfeld's term "social poetics" to talk about how the performance of *dhamāl* gets worked from within by the "social actors" to tweak, modify, or alter, rhythmic patterns and movements. "Poetics lies at the nexus of an individual acting in the world—embracing, dissociating from, and deforming categories—and collective forces (or historical tra-

Devotion constitutes the means by which ritual and the realm of poetry are intrinsically linked in experiencing sacred presence. This is the reason for my argument that it is necessary to see ontology and (ritual) aesthetics as inseparable in Indus Sufism.[57] The material culture of Sufi shrines should be viewed in the context of religious devotion, of which one of the chief expressions is encompassed in *ḥāẓirī*. The ritual of *ḥāẓirī* involves listening or hearing (*samā'*) the sacred music in the vicinity of the *dargāh*. To understand the efficacy or function of *samā'* is to understand the metaphysics of *ḥāẓirī*. In both cases, it is an act of communion. This is the reason why listening to *samā'* is considered, or celebrated as, an*'ibadat* (act of worship) in Indus Sufism.[58] Ritual and poetry are woven together in religious devotion at the *dargāh*. It is a spiritual (*ruhanī*) bond that facilitates communication or communion and attunes the senses to the hearing of mystical poetry.[59] This profound devotional relationship between the devotees and the Sufis saints of the *dargāh*, a sacred space of communion, is very much analogous to the experience of the Divine through scripture, whether in recitation, reading, or studying it.

6.4 On Ecstasy (*ḥāl* and *wajd*)

Bulleh Shāh's verses describe the condition of ecstasy as a state that leads to the person dancing un-rhythmically (*be tāl*) and senselessly (*be hāl*), in a state of passionate love (*'ishq*).[60] This dancing without rhythm or beat, and in a manner resembling what would appear to the eye as a frenzy, a stepping out of beat, re-

jectories) that makes these categories seem fully formed," "The Poetics of Sufi Practice: Drumming, Dancing, and Complex Agency at Madho Lāl Husain (and Beyond)," *American Ethnologist* 33.2 (2006): 247.

57 Ritually, as well as epistemologically, Sufi poetry cannot be understood apart from its performative function, as in *samā'*. Sufi verses are meant to be *heard* at Sufi shrines in view of ritual devotion to Sufi saints, and in this role they are taken as *'ibadat* (worship)—a point made earlier as well.

58 There is a large repertoire of *qawwalī* performances extolling the virtues of gazing on the face of the saint as a *namāz* (Ar. *ṣalat*), or *eīd*: "*Ganj shakar wal tak tak ashiq lok manandey eid;*" *mainu yār dī namāz paṛan.*"

59 Jean During et al, "Hearing and Understanding in the Islamic Gnosis," *The World of Music* 39.2 (1997): 127–137.

60 See chapter 5 for the discussion on Bulleh Shāh.

fers to the transcending of usual performative structure and order in the throes of religious ecstasy.[61]

Under the influence of *wajd* and in a state of *hāl,* in *qawwālī* devotees do not dance rhythmically (as it may sometimes appear), in imitation of the beat or the poetic meter of the text (of verses) performed. Ecstasy appears in, or rather descends into, the body, stirring it from within, in a feeling of being pulled back and forth—as if the body were caught up in a strong ocean undertow.[62] The feeling arises within the body and spreads to the limbs, shoulders, neck, and thighs. The force is most powerfully felt in the heart—often described as a piercing—from where it shoots upwards, vertically, to the head, to a different level of consciousness, where it might be said that *hearing* really takes place.[63] In this way, it

61 My view offers a different experiential understanding of *samā'* than the one given by Regula B. Qureshi, "*Qawwālī:* Making the Music Happen in the Sufi Assembly," In *Performing Arts in India: Essays on Music, Dance, and Drama*, ed. by Bonnie C. Wade (Berkeley: University of California, date): 120: "To activate the spiritual link thus formalized requires however that mystical love be cultivated spiritually and aroused emotionally. This is achieved through "listening" to the mystical songs (*sama'*) and repeating the name of God rhythmically (*zikr*). The *qawwālī* assembly is *sama'* set in practice: through the medium of the *qawwālī* performer, the Sufi devotee listens to the mystical poetry rendered musically and enhanced by a powerful rhythm suggesting *zikr*." Nevertheless, I agree with her insightful observation about *samā'* as *zikr.*

62 One of the best descriptions of this, in my view, is given by Patrick Laude, *Singing the Way*, 69: "The "dance" that takes place within the chest of the contemplative is akin to the penetrating infusion of presence that is like breathing in of the divine ether: this presence no one can see or define, for grace eludes any attempt at appropriating it."

63 As Jurgen Wasim Frembgen had the occasion to observe during his fieldwork, in "*Dhamāl* and the Performing Body: Trance Dance in the Devotional Sufi Practice of Pakistan." *Journal of Sufi Studies* 1 (2012), 101: "Having thus gone into *ḥāl* they may increase their trance through rotating their head, pirouetting with their body, jumping and other frenetic and spontaneous movements," and, "A particular ecstatic circling movement which I only observed among *dhamālīs* is rotating the head in frenzy, virtually like the head of a drill in motion. This is commonly known as *sar kā dhamāl* (96). These observations about the bodily movement refer to the practice of *dhamal* but are applicable to other performative modes where the *dhol* is not the dominating instrument. Regula Burckhardt Qureshi has looked at *qawwālī:* "The responses to *qawwālī* fall into two main categories: one contains the range of spontaneous expressions collectively termed "delight" (*kaif*) and directly reflecting the devotee's state of mystical arousal. This state may find—within recognized broad limits—free expression ranging from rhythmic swaying, arm and head movements and exclamations to weeping (*riqqat*), shouting and, rarely, to complete ecstatic abandonment (*hāl, kaifiyat*). Then the devotee will rise and move freely within the central space, walking, gesticulating or shouting if he feels so moved. The culmination of this state is expressed in the turning movement of the Sufi dance of ecstasy (*raqs*). In recognition of the divine gift of this ecstatic state, the leader of the assembly rises and with him all other listeners, until the state subsides." In Regula Burckhardt Qureshi, "*Qawwālī:* Making the Music Happen in the Sufi Assembly," *Asian Studies* [*Performing Arts in India: Essays on*

not just the ear that does the hearing, but the entire body that is transported to a higher state to engage in hearing. Such a condition serves to incite and provoke the un-rhythmic "dance" in the devotee.[64]

Among the few studies that have attempted to inquire into the meaning of mystical moods and states of ecstasy (*kaifiyāt*) in the experience of the Sufi practioners is that by Shemeem Abbas Burney.[65] Burney interviewed several leading performers of Sufi folk ("*sufiāna-kalām*"): "Shaikh Ghulam Hussain, a music director..., used the word *kefīat* as a state of altered consciousness. In my interview with him I understood *kefīat* as a refined state of spiritual awareness, induced through the combination of Sufi poetry and music."[66]

It is also useful for our purpose to consider the relationship between the spoken word and the body of the Sufi folk performer as described by Shemeem Abbas:

> Male singers of *sufiānā*-kalām, such as Alan Faqir, would dance as they held a stringed instrument in one hand. Usually *sufiānā-kalām* is sung to a *caprī* (castanets) or a *tambūr* or *yak-tārā* (stringed instruments). Folk singers like Alam Lohar and his son Arif Lohar would use the *cimṭā* (long tongs) to create the percussion, and their vigorous body language almost resembles a dance. I have seen a male Baluchi musicians dancing to the rhythm of the *rabāb* (a small stringed instrument) while singing mystical poetry.[67]

Music, Dance, and Drama, ed. Bonnie C. Wade (Berkeley: University of California)] 18.2 (1987): 123.

64 Some can experience this state without bodily display, but what is really important here is that ecstasy in any member of the gathering is treated with care and gentleness, understood to be a sacred experience. Ecstasy in any participant has an effect on the entire gathering, and in a sense its experience among the devotees is the fulfillment of the ritual function of *samā'*. But it is also very important to remember that ecstasy is not just about bodily movement, as many participants experience it while remaining still.

65 Shemeem Burney Abbas, *The Female Voice in Sufi Ritual: Devotional Practices of Pakistan and India* (University of Texas Press, 2002). Note, for example, p. 61: "The performers mediate spirituality through inducing a state that they call *kefīat*. They do this all within the context of a performance. They will use their linguistic resources of poetry, myth, syntax, semantics, and tonal and speech patterns together with music to create the *samā*'. They will improvise linguistic constructions at every stage, and they will create rhythmic melodies with drums and other musical instruments to bring about the *kefīat* in the *samā*'. At the grassroots level, *samā*' is also understood as a context or environment, and the aim of every competent performer of Sufi music and discourse is to create the *samā*'."

66 Shemeem Burney Abbas, *The Female Voice in Sufi Ritual*, 13. See also, Mikko Viitamäki, "Mystical Gazal in Sufi Practice. Poetics, Music and Metaphysics in the Verses of Sāh Niyāz Aḥmad," *Oriente Moderno* 92.3 (2012), 534: "the combination of the sonorous soundscape, well-known poetic themes and at times indirect and even illogical language create statements that can be grasped intuitively and that appeal directly, without detouring through the brain."

67 Shemeem Burney Abbas, *The Female Voice in Sufi Ritual*, 9.

A further passage from Abbas, in describing the role of the performer in communicating or transmitting the "experience" (*kaifiyat*, pl. *kaifiyāt*) of the *kāfī*, through the body, greatly advances our understanding of the performative, embodied context of such ecstatic dance and music:[68] "The musician generates body language to engage in the performance. The audience, in turn, complements the performance with strong, appreciative, linguistic clues. The audience additionally communicates through body language."[69]

The most essential component in the ritual performance of *samā‘*, and one that allows for communication or transmission of mystical experience to take place between the performer and the devotee, is that for both sides, it is an act of worship (*'ibadat*) in devotion to the Sufi saint. During the audition, it is the recognition by the performer and the audience of the special nature of the saint whose *'urs* is being celebrated, and by whose grace the ecstasy is to be experienced.[70]

At the the *mela* (*'urs*) of Sufi saints, the performance of *samā‘* takes place at night and is performed over the entire span of darkness until daybreak. It gradually picks up

68 Here I also add Richard Wolf's interview of a famous Sufi drummer in connection with the role of a performer in Lahore, "Pappu called knowledge (*'ilm*) of music the only thing (*kām*) that is really true (*saccā*): When one is playing a rhythmic pattern (*thekāh*), another person must match it. Music admits no falsehood (*jhūt*); one cannot simply 'throw something in.' Pappu compared this musical truth with that of the dhol player playing dhamāl with intoxication (*mastī*). Completely absorbed in spirituality, the drummer matches his own being with that of the *sarkār* and, by extension, with God. The drummer whose performance is colored by his temperament (*tābi'at*) or who plays with his eyes open or with money in mind is lacking such truth. By the same token, Pappu views 'mast qalandar' as a zikr motivated by the saint Shah Jamāl, who enters Pappu's heart during a performance and causes his lips to utter the phrase. This, in turn, causes listeners to recite zikr with the drums. Pappu's remarks creatively link a musical form of truth testing with a spiritual one: playing in unison is morally equivalent to dissolving one's being in the love of God. This is poetics as musical-spiritual ideology," in "The Poetics of 'Sufi' Practice: Drumming, Dancing, and Complex Agency at Madho Lāl Husain (and Beyond)," *American Ethnologist* 33.2 (2006), 256.

69 Shemeem Burney Abbas, *The Female Voice in Sufi Ritual: Devotional Practices of Pakistan and India* (University of Texas Press, 2002), 9.

70 In discussing the case of major shrine in Sind, Frembgen Wasim Jurgen ("*Dhamāl* and the Performing Body," 104) writes that "For common pilgrims, *dhamāl* is a customary, stereotyped and ritualized, but also intensely celebrative form of annually expressing their veneration of the saint, receiving a vision of the great Qalandar, communicating with him and seeking to be close to him. In addition, many devotees and also a number of *dhamālīs* explain that they perform the devotional dance as an offering in connection with a wish and vow (*mannat*) uttered in prayers of supplication."

momentum, and throughout the night it remains a ritual of *ḥāẓirī*. This means that both performer and devotee focus on the living presence of the saint, as a form of *'ibadat*. In this both the performers and the devotees share the bond of *nisbat* with the Sufi saint, and as a consequence, both sides are united in the ritual enactment of *samā'*.[71]

During one of my visits to the shrine of Bābā Bulleh Shāh, in Qasūr, I was invited to a cup of tea by a devotee of the saint who was standing near the grave, a smile on his face. We happened to strike a conversation after I had performed *ḥāẓirī*. Neither did I ask him for his name nor did he inquire about the purpose my visit or ask for personal details. We sat on the floor just outside the vicinity of the tomb in the courtyard for a few minutes. He brought the tea in a small, white cup himself. After I had finished my tea, I thanked him for his hospitality and got up to depart. As we stood facing each other, he had this to say: "What to speak of Bābā Bulleh Shāh?" In the next instance, he performed a graceful, full-turn about (anti-clock direction) on his left foot, striking the ground with his right heel, coming full circle to face me (as if the performance was the explanation to the question). In my interpretation, the gesture was an act of communication, a way to transmit *kaifiyat*, a religious experience or feeling. One can indeed continue to speculate about the mystical significance of the circle, but to me, at that moment, it was a reference to experiential dimensions of Sufism, to the metaphysics of presence: to *find* divine presence with just a turn or a spin.

One of the perennial challenges for the scholar of religion is how to study religious experiences without ignoring or denigrating their experiential value (for their practitioners). How do we correlate this with the methodologies involved in our inquiry that seek empirical evidence, reliability that can be verified by other scholars, if our findings are to be taken seriously? In writing this chap-

71 Shemeem Burney Abbas, *The Female Voice in Sufi Ritual: Devotional Practices of Pakistan and India* (University of Texas Press, 2002), 14: "The term *wajd* (trance) may seem pejorative to orthodox Muslims or to someone in the West. *Kefīat* and *hāl* are states of ecstasy in the same mystical experience, although there is a difference in degrees of expression. *Hāl* is a state of arousal outside the control of self. *Kafīat* is subtle and covert although it involves ecstasy. Even singers are involved in the *kefīat,* and there is no *samā'* (context) without the interactional *kefīat* between listeners and singers. These are subtle dynamics of a performance. The musicians who sing the lyrics or the listeners who attend the speech events perceive the state in its manifold aspects. Among some participants it is desired state that the musicians create. For the musicians it is an indication of the success of their performance: They have succeeded in creating a spiritual ambience through language and music if some among their audience have reached *kefīat* or *hāl*. They are obligated to continue singing until the affected overcome the state of ecstasy."

ter, I am have tried to strike a fine balance between our scholarly concerns and the transcendental claims of Sufi practitioners for their experiences during their audition of sacred music. Many of the excellent ethnographic studies I have used in this chapter to substantiate my claims have, in my view, correctly identified ritual practices which involve ecstasy, but I differ from them as to their neglect of how central a role religious devotion plays in experiencing states of ecstasy. Hence, if we wish to take the "insider's" account of experiencing the various states of ecstasy (*kaifiyāt*), such as *wajd* and *ḥāl*, into consideration, then (as problematic it may sound to us in view of our usual scholarly methodologies) our concern must involve an effort to understand the *meaning* behind the claims of the devotees; namely, that the experience of the presence of the saint does not result from hearing *samā‘*, rather, it is the presence of the saint that allows for ecstasy (*kaifiyāt*) to take hold of the devotees and be expressed in *samā‘*, as a pure gift from the Sufi saints (*naẓar-i karam*).[72]

In the previous chapters (specifically, chapters 1 and 5), I argued for understanding the Sufi saint in his tomb as a special type of mystical body, with its immortality to be understood as a divine grace bestowed on him/her. In transcending death and achieving union with God, precisely what the *‘urs* celebrates, the saint is immediately present (*mawjūd*) to his devotees. In being granted this extraordinary power of immortality, the Sufi saint's sacred body becomes the locus for directly experiencing the divine presence of God at the tomb; it is thus of no minor significance that the vernacular poetry voiced/offered in devotion to the saint is experienced as divine speech/saying.

My claim in this chapter is that the immortality of the saint, as celebrated in the festival of *‘urs*—union or marriage with God—challenges the narrative of Sufism as a predominantly logo-centric religion of the "text" because the Muslim community affirms the Qur'ān as the literal word/speech of God.

Since the primacy of textual hermeneutics in modern scholarly interpretation has often been considered the only fully valid source for constructing meaning in Islam, the oral culture of Sufi shrines has often been treated as merely

72 Referring to the primordial covenant (*alastū rabbīkum*), During observes: "this myth reminds us that the lifting of consciousness from our existential foundations, inseparable from a certain state of grace, takes place directly through the means of musical hearing." Jean During, "Hearing," 129. See also Ian Richard Netton, "The Breath of Felicity; *Adab, Aḥwāl, Maqāmāt* and Abū Najīb al-Suhrawardī," edited by Leonard Lewisohn, *Classical Persian Sufism: From its Origins to Rumi.* (London: Khaniqahi Nimatullah Publications, 1992), 467: "These "stations (*maqāmāt*) and "spiritual states" (*aḥwāl*)—the first "to be effected by the individual's endeavor" and the second to be "*regarded as divine gifts*" [emphasis mine] for they "do not result from human action or volition but from God's favor."

"popular religion." In practice, the popular veneration of Sufi saints challenges this scriptural orthodoxy in treating the bodies of the saints as a valid source for communication with the divine. The shrine is a mark of distinction bestowed on the saints, and the ritual of visiting them because of their continuing presence at their shrines is experienced as the disclosure of the divine in this world. Popular devotion as a mode of *ḥāẓirī* can be seen as the realm of theopoetics, in which the religious consciousness is being opened continuously to new horizons of understanding. More than anything, the saint's body signifies the bond of *'ishq* between human beings and God, a fulfillment of the primordial covenant (*alastū rabbikum*), and "death" brings forth this final union (*wisāl*). In its devotional, performative function, the mystical verses of the Indus saints affect the listener because they are treated as divine utterances delivered in and indeed made possible by the presence of the body of the saint. *Ḥāẓirī* means entering into the living presence of the saints, and listening to the sacred verses in the vicinity of the shrine becomes a form of communication with the Divine beyond discursive speech. The intermediary function of the saints' bodies as microcosms of the divine cosmos belongs to the realm of devotion and love that makes all things whole and holy in their presence.

The formal recitation of the *fātiḥa* is an intervention of the "text" that can potentially rupture the embodied relationship between the devotees and the living presence of the saint. *Fātiḥa* as a supplication (*du'ā*) at the grave of the saint is an ideological signifier for reversing the traditional Sufi cosmology of the saints and their bodily presence in the *dargāh*s through legally proper worship. It "textualizes" the body of the saint as a pious but essentially "dead" body, as is the case with other pious dead Muslims, and marks a clear distinction between the living and the dead. In doing so, it also preserves the orthodox version of eschatology that sees the resurrection of the dead as only possible on the Day of Judgment. The *ulamā' or fuqahā* see the belief in the living presence of the saint as a threat to their role as guardians of the moral economy of Islam, which is the concomitant of their special status as the legitimate inheritors of prophetic knowledge. The popular devotion to the immortal saint overturns their religious ideology of death and moves beyond the dominant religious "discourse" and logo-centrism of the Islamic textual tradition. As this dissertation aims to demonstrate, there is more to ritual practice than what can be learned only through texts or seen as "discourse." The question of ritual gestures being performed "correctly" or "incorrectly" in the case examined here makes a difference. In the process, I question the reliability of studying religious practice as a "discourse" that privileges language over the body's perceptions in interpreting religious experience.

Conclusion

According to Islamic law, all visits undertaken to graveyards and cemeteries must be for the purpose of sending the gift of pious merit (*īṣāl-i ṣawāb*) to the dead. In the legal interpretation or textual Islam of the *fuqaha*, the ritual logic of *īṣāl-i ṣawāb* is applicable in the case of visits (*ziyāra*) made to the graves of saints (*awliyā' allāh*). The act of "sending merit," most commonly through the recitation of *sura al-fātiḥa* (the opening chapter of the Qur'ān), by the living to the dead is a ritual mechanism to pray or supplicate (*du'ā*) for the souls of the dead. Especially in the case of Indus Sufism, however, in the eyes of the faithful, the saints such as Bulleh Shāh of Qasūr should not be considered "dead"; instead, their death anniversaries are festive occasions, ritually celebrated as a *melā* (*'urs*). The prevailing belief is that saints are *resurrected* upon their deaths, and therefore ordinary people visit them daily at their *dargāh*s to perform *ḥāẓirī*. In the metaphysics of *insān-i kamil*, and reflected in popular devotion, the sacred bodies of the saints are held to be absorbed in God upon their deaths. Since *ḥāẓirī* is an act of communion with the dead saints and the Divine in the lived practice of ordinary folk, the *dargāh* is effectively a place of worship (*'ibāda*). For this reason, in "official" Islam, this ritual practice is declared idolatry (*shirk*). Since the 19th century and anticipated by earlier scholars like Sirhindī and Walī Allāh, whose intellectual contributions continue to inform the religious thought of Muslim scholars, Islamic reformers have been most concerned with reforming religious rituals and practices associated with the pilgrimage to and devotional activities at saints' shrines.

In this thesis, my goal has been to exegete the nature of this *incongruity* between the conceptualization of pilgrimage to shrines of saints in textual Islam (specifically in Sunni orthodoxy) and its practice in the lived experience of devotees in Indus Sufism. My first challenge, therefore, has been to explain this discrepancy and illumine its significance in the current situation of the Sufi shrine—a situation that is currently fraught with tension throughout Pakistan. What kind of threat does the Sufi shrine pose to Sunni religious orthodoxy? According to reformer figures like Sayyid Ahmad Barelwī, such "false" rituals of worship are to be seen as responsible for enervating the "true" spirit of Islam and as the cause of the political decline of Islam. As a result of the loss of political power in the Mughal dynasty, as I have discussed in chapter three, the main aim of the *Ṭariqa-i Muḥammadī* was to convert or to re-channel the devotional culture of Sufi shrines away from the preoccupation with finding God—as a lover seeks the beloved—through mystical contemplation, insight, and ritual practices at the shrines, and to turn that culture towards the martial arts of

https://doi.org/10.1515/9783110781496-009

war, to find martyrdom through the ecstasy of religious violence on the battlefield: in other words away from internal religious experience and towards external state-building.

Modern religious reform and revivalism have been centrally preoccupied with returning to the "authentic" Islam preached by the Prophet Muḥammad and followed by his Companions. After the Partition of India, the Pakistani state emerged as a political entity in which its "Islamic" identity is considered fundamental to its ideological and sovereign existence (Pakistan means "land of the pure"). I have argued that these reformist narratives have been deployed to frame Pakistani identity through the ideology of an "Islamic state." The religious ideologues represent this as *nizām-i mustafa* (the system or order of the Prophet). In this formulation, the state represents the political body of the Prophet; a configuration achieved because of the *siyāsa-sharīʿa* model of governance. Therefore, the bodies of its citizens are to be seen as resources for state-building rather than individual loci for seeking union with God through the miraculous bodies of His saints.

Through my research, I have sought to show that revivalists, *ʿulamā* hold the devotional practices at Sufi shrines and their political partners to undermine the identity of Islam as a religion that encompasses the spheres of religion and politics. In becoming popular pilgrimage sites, the *dargāh*s, where the saints are accessible through their living presence, subvert the religious authority of the *ʿulamā* in their claims to be "heirs of the prophets." Historically, this privileged status has allowed the *ʿulamā* to determine the political legitimacy of rulers in exchange for promising to uphold the *sharīʿa*. In principle or theory, such an arrangement has been essential to preserving the marriage between religion and state power. Most importantly, the arrangement is meant to ensure, again in theory, the dominance of Islamic law or textual Islam (and scripture) over the democratic political structure of the state. The state Ministry of *Awqāf* directly regulates the ritual space of Sufi shrines of the Indus saints; this is a traditional mechanism by which the Pakistani state attempts to co-opt or appropriate the Sufi shrine for its political agenda. As discussed in chapter three, various governments since 1947 have employed the idiom of Islam in highly essentialized ways to gain legitimacy for their particular political projects. Despite the persistence of interpretive disagreements between the *ʿulamā* and the modernists, the religious identity of Pakistan as an "Islamic" state affords a potent logic for sanctioning, and even for glorifying, religious violence against allegedly heterodox forms of Islam.

In Chapter 2, I have taken issue with how (Foucauldian) scholarship on Sufism has failed to study ritual practices insofar as these practices can tell us about the *dargāh*—Sufi shrine. As a direct consequence of the theoretical assumptions,

I have argued that some of the model scholarship—such as the earlier work done by Mark Woodward and more recent scholars like Omid Safi and Pnina Werbner —on the subject privileges textual Islam over *lived* experience. So far, scholars have not paid attention to the role of the living presence of the saints in their analysis of Sufism, especially how such a belief in the power of saints' bodies, invested in them by God, informs the religious experience of *ḥāẓirī*. Hence, ironically, despite the strong interest in understanding the historical and sociological significance of the Sufi shrine, it remains by and large peripheral to the concerns of scholarship. As such, the Chapter has discussed the limitations of "discourse analysis" concerning the study of Sufism.

Chapter four focused on Bulleh Shāh's mysticism and defiance of prescriptive Islam. In Bulleh Shāh's poetry, we find an exaltation of direct experience in gaining knowledge of God and love *('ishq)*. In this quest, his verses express the need to purify the "heart" to overcome the limitations of knowledge obtained solely from books or texts. In his search, Bulleh Shāh found his Beloved through incurring the blame of the orthodox clergy *('ulamā)*; as he took to the streets to dance with *ghungru* on his legs, a Sayyid turned *mirasī*. Thus, the Chapter places the ritual body of Bulleh Shāh at the center of the inquiry in understanding the "metaphysics of presence" at work in the case of this paradigmatic saint of Indus Sufism.

The festival of *'Urs* is one of the most distinctive features of Indus Sufism, a celebration of the saint's special immortality or "resurrection" (*qiyāma*). Chapter 5 deals with understanding the theopoetic meaning of this ritual celebration as a *mela* (literally, a festival). I examine *kāfīs*, the spoken word of saints, to engage with the meaning of the festival as it is celebrated in Indus Sufism. In this function of collective joy, the *'Urs* brings forth an occasion for different religious communities to gather in a sacred space to participate without concern over the religious identity of the pilgrims. The *mela* of saints is a celebration of religious pluralism and a Divine manifestation of the multi-hued nature of human existence.

The last Chapter (6) looked at the relationship between the ritual of *ḥāẓirī* and *samā'* performed in the vicinity of the Sufi *dargāh*. I argued, drawing primarily on the work of William A. Graham, that the devotees' emotional relationship (*nisbat*) with their saint is pivotal to how the poetry is experienced and, as a consequence of its ritual aesthetics, functions as equivalent to the Divine word of God. The primary channel for this scriptural equivalence is the sacred body of the saint. Lastly, I offer a phenomenological analysis of *samā'*, particularly in how the body of a devotee experiences ecstasy during its performance. The main aim of the Chapter was to show how the ritual space of the *dargāh* owes its sacred character to the living presence of the saint. Listening to Sufi *kalām* is a form of *'ibāda* for the devotees and a means for experiencing Divine Presence, as if it were, through the body of the saint, a fulfillment of the Primordial Covenant in this ritual of *samā'*

Such devotional realities as those above, all associated with Sufism and worship at saints' tombs, have been the targets of the attempts by the state to regulate saint-tomb visitation as a means to control the devotional culture of these sacred spaces. To ensure its total control, the state entered into the ritual space of the *dargāh* in the 1960s by supervising ritual activities. I have argued that the state has attempted to substitute the prayer for the dead based on the scripturally-based *al-fātiḥa* for the centuries-old tradition of *ḥāẓirī* as a means to reverse the *lived* metaphysics of presence. Through this substitution, the state sought to normalize and disempower the "resurrected" bodies of saints by recasting them as "dead" bodies. My view, for the reasons I have tried to elucidate in this work, is that this project has been a theological failure since Sufi saints of the Indus continue to command the allegiance and veneration of millions who still come for *ḥāẓirī*, the experience of the power of Sufi saints' "presence' at their shrines.

Bibliography

Sources in Arabic, Persian, Urdu, and Punjabi

ʻAbd al-Samī Sahāranpurī, Muḥammad. *Anwār Sāṭiʻah dar Bayān Mawlūd wa Fātiha*. Edited by Muḥammad Afroz Qādirī Chariyākutī. Reprint, Lahore: Faiẓ Ganj-i Baksh, 2012.

Aḥmad, Sayyid Nāzīr, trans. *Kalām-i Bulleh Shāh*. Lahore: Packages Limited, 1976.

Ājiz, AbdurRahman. *Reality of Death*. Trans. Ghani al-Akram Sabzwari. Karachi: Library Promotion Bureau, 2002.

Allama Jalāl al-Dīn Suyūtī. *Tazkira al-Qabr*. Ed. and trans. Mawlana Muhammad Sharīf Naqshbandī. Lahore: Ziya al-Qur'ān Publications, 2001.

Anbaṭawī, Khalīl Aḥmad. *Barāhīn-i Qāṭiʻah*. Deoband: Kutub-khāna Imdādīyah, n.d.

Bhattī, Amjad ʻAlī. *Bulleh Shāh: Muntakhab Kalām Maʻ Urdu Tarjumah*. Islamabad: National Book Foundation, 2014.

Fārūq, Sayyid Muḥammad. *Rasā'il Shāh Walīullāh*. Gujaranwalla: Awaisī Būk Sṭāl, n.d.

Faqīr, Muḥammad Faqīr. *Kulliyāt Bulleh Shāh*. Lahore: Al-Faisal Nashran, 2012.

Hujwīrī, ʻAlī Ibn ʻUthmān. *Kashf al-Maḥjūb*. Translation by Reynold Nicholson. Lahore: Sang-e Meel Publications, 2007.

Al-Jawzīyah, Ibn Qayyim. *The Soul: the Condition of the Souls of the Dead and the Living According to the Book, the Sunnah, the Traditions and the Sayings of the Scholars, al-Ruh*. Translation by Ismail Abdus Salam. Beriut: Dar al-Kutib al-ilmiyah, 2008.

Ishāq, Pīr Muḥammad. *Taḥqiq Lājawāb fī Masla Īṣāl-i Sawāb* (Faisalabad: Maktaba Nūriya Riẓviyya, 2007).

Ismāʻīl, Mawlānā Qāẓī Sayyid. *Fātiha kā Sahih Ṭarīqa*. Raiwind: Maktaba Anwāriya, 1981.

Kāndhalawi, Nūr al-Ḥasan Rashīd. "*Taqwiyat al-Īmān aur Shāh Ismāʻīl Shahīd ke Khilāf bar-pā Shūrish Tārīkh aur Ḥaqiqat ke Ā'īna meñ*," *al-Furqān* lxi, no. 7 (1991) to lxi, no. 10 – 12 (1994): 170 – 196.

Khan, Muḥammad Naʻīmullah. *Maḥfil-i Īṣāl-i Sawāb par Aʻitarāẓāt kā ʻAmalī Mahāsaba*. Lahore: Qadirī Riẓvī Kutub-khāna, 2010.

Khan, Muḥammad Āsif. *Ākhayā Bulleh Shāh Ney*. Lahore: Pakistan Punjabī Adabī Board, 1992.

Maqbūl, Mian Ẓafar. *Terey ʻIshq Nachāyā: Sharḥ Kalām Bulleh Shāh*. Lahore: Maktaba Dāniyāl, n.d.

Naʻimī, Aḥmad Yār Khān. *Wasīla-i Auliyā Allāh*. Lahore: Maktaba Islamīyya, n.d.

Noorani, Okarvi Kaukab. *Vālidain-i Risālat Ma'āb: Ma'Risālah-yi Qabr ke Ahkām o Ādāb*. Lahore: Ziya al-Qur'ān Publications, 2004.

Puri, J. R. & T. R. Shangari. *Bulleh Shah*. Dera Baba Jaimal Singh, Amritsar: Radha Soami Satsang Beas, 1986.

Qādirī, Muḥammad Ṭāhir. *Īṣāl-i Sawāb kī Shariʻī Ḥīsīyyat*. Lahore: Mihāj al-Qur'ān Publications, 2003.

Qaddūsī, Mawlānā Aʻijāz al-Ḥaq. *Tazkira Auliyā-i Panjāb*. Lahore: Kutub-khāna Shān-i Islām, n.d.

Qurtubī, Muhammad ibn Aḥmad. *Al-Tazkira fai Aḥvāl al-Mawta Va Umūr al-Akḥirat: Safar-i Akḥirat Kī Manāzil 2 Vols*. Trans. Ghulām Naṣīruddīn. Lahore: Farīd Buk Sṭāl, 2008.

https://doi.org/10.1515/9783110781496-010

Al-Qushayri, Abu'l-Qasim. *Al-Risāla al-Qushayriyya fi 'ilm al-taṣawwuf. Al-Qushayri's Epistle on Sufism*. Trans. Alexander Knysh. Lahore: Suhail Academy, 2011.
Rafīq, Abu Hamza Muḥammad. *Iṣāl-i Sawāb kī Shari'ī Ḥisiyyat*. Lahore: Jāmi'a Faiẓ al-Islām, 2014.
Rana, Aslam. *Ramz Rawāyat*. Lahore: Aziz Publishers, 1990.
Rizqī, Muhammad. *Haqīqat al-Mawt 'inda al-Ṣufīyah*. Sīlyānah, Tunīs: Dār Sanābil lil Nashr was al-Tawzī, 2007.
Sābir, Muḥammad Sharīf (ed.), *Bulleh Shāh: Mukammal Kāfīāñ*. Lahore: Sayyid Ajmal Ḥussaīn Memorial Society, 1991.
Shackle, Christopher (ed. and trans.). *Bullhe Shah: Sufi Lyrics*. Cambridge: Harvard University Press, 2015.
Shikūh, Dārā. *Majma ul-Baḥrain or The Mingling of the Two Oceans*." Translated by M. Mahfuz ul-Haq. Karachi: Royal Book Company, 1990; first published in Calcutta in 1929.
Taymiyya, Taqī al-Dīn Aḥmad. *Kitāb al-Wasīla, Urdu*. Lahore: Islāmī Academy, 1979.
Taymiyya, Taqī al-Dīn Aḥmad. *Epistle on Worship: Risālat Al-'Ubūdiyya*. Trans. James Pavlin. Cambridge: Islamic Texts Society, 2015.
Thānwī, Imdādullāh. *Faiṣalah-i Haft Masalah*. Edited by Muḥammad Khalīl Khān Qādirī. Lahore: Farīd Buk Sṭāl, 2000.
Turkī, Ibrāhīm Muhammad. *Falsafat al-Mawt 'inda al-Sufiya*. Al-Iskandarīyah: Dār al-Wafā Li-Dunyā al-Tibā'ah wa al-Nashr, 2003.
Al-Wahhāb, Muhammad ibn. *Qabron ke Halāt: Ahādīs kī Raushanī mein*. Translated by Shafqāt Aḥmad Mujaddidī. Lahore: Qadirī Razvī Kutub Khānah, 2006.
Yūsuf Ludhiyānavī, Muhammad. *Jab Ankhen Khulen gī: Maut aur Maut Ke Ba'd kī Zindagī key Tamām Marāḥil*. Edited by Muḥammad Zubair Tahīr. Karachi: Maktabah-i Ludhiyānvī, 2012.

Sources in English

Abbas, Shemeem Burney. The Female Voice in Sufi Ritual: Devotional Practices of Pakistan and India. University of Texas Press, 2002.
Abbott, Freeland. "The Decline of the Mughal Empire and Shah Waliullah." Muslim World 52 (1962): 114 – 123.
Abbott, Freeland. "The Jihād of Sayyid Ahmad Shahīd." Muslim World 52 (1962): 216 – 22.
Abbott, Freeland. "The Transformation of the Jihād Movement." Muslim World 52 (1962): 288 – 95.
Abrahamov, Binyamin. Anthropomorphism and Interpretation of the Qur'an in the Theology of al-Qasim ibn Ibrahim. Leiden: Brill, 1996.
Abrahamov, Binyamin. Divine Love in Islamic Mysticism: The Teachings of al-Ghazālī and al-Dabbāgh. New York: Routledge, 2003.
Ahmad, Imtiaz (ed.). Ritual and Religion among Muslims in India. New Delhi: Manohar, 1981.
Ahmad, Irfan. "Genealogy of the Islamic State: Reflections on Maududi's Political Thought and Islamism." The Journal of the Royal Anthropological Institute 15 (2009): S145-S162.
Ahmad, Qeyamuddin. The Wahhabi Movement in India. Calcutta: Firma K.L. Mukhopadhyay, 1966.

Ahmed, Aziz. Studies in Islamic Culture in the Indian Environment. Oxford: Oxford University Press, 1964.
Ahmed, Aziz. "Cultural and Intellectual Trends in Pakistan." Middle East Journal 19.1 (1965): 35–44.
Ahmed, Manzooruddin. Pakistan: The Emerging Islamic State. Karachi: The Allies Book Corporation, 1966.
Ahmed, Manzooruddin. "Sovereignty of God in the Constitution of Pakistan: A Study in the Conflict of Traditionalism and Modernism." Islamic Studies 4.2 (1965): 201–212.
Ahmed, Rumee. Narratives of Islamic Legal Theory. Oxford: Oxford University Press, 2012.
Ahmed, Safdar. Modernity in Islam: The Philosophical, Cultural and Political Discourses among Muslim Reformers. London: I. B. Tauris, 2013.
Alam Muzaffar. The Languages of Political Islam in India, c. 1200–1800. New Delhi: Permanent Black, 2004.
Alam, M., and F. N. Delvoye et al. (eds.). The Making of Indo-Persian Culture: Indian and French Studies. New Delhi: Manohar, 2000.
Al-Bayrak, Ahmet. "Insan/Nas." In Oliver Leaman (ed.). The Qur'an: an encyclopedia. New York: Routledge, 2006.
Algar, Hamid. "The Naqshbandī Order: A Preliminary Survey of Its History and Significance." Studia Islamica 44 (1976): 123–152.
Al Huda, Qamar. "Celebrating Death and Engaging in Texts at Dātā Ganj Bakhsh's 'Urs." The Muslim World 90 3/4 (2000), 377–394.
Al-Qadi, Wadad Kadi. "The Primordial Covenant and Human History in the Qur'ān." Proceedings of the American Philosophical Society 147.4 (2003): 332–338.
Ansari, Abdul Haq. "Shah Waliy Allah Attempts to Revise 'Waḥdat al-Wujūd.' Arabica 35.2 (1988): 197–213.
Ansari, Sarah F. D. Sufi Saints and State Power: The Pīrs of Sind, 1843–1947. Cambridge: Cambridge University Press, 1992.
Arjomand, Said Amir. "The Law, Agency, and Policy in Medieval Islamic Society: development of the Institutions of Learning from the Tenth to the Fifteenth Century." Comparative Studies in Society and History 41.2 (1999): 263–293.
Arjomand, Said Amir. "The Constitution of Medina: A Sociolegal Interpretation of Muhammad's Acts of Foundation of the "Umma". International Journal of Middle East Studies 41.4 (2009): 555–575.
Arkoun, Muhammad. Islam: To Reform or To Subvert? London: Saqi Books, 2006.
Asad, Talal. "Anthropology and the Analysis of Ideology." Man 14.4 (1979): 607–627.
Asad, Talal. "The Idea of an Anthropology of Islam." Washington, DC: Georgetown University Center for Contemporary Arab Studies, 1986.
Asad, Talal. "Religion and Politics: An Introduction." Social Research 59.1 (1992): 3–16.
Asad, Talal. Genealogies of Religion: Discipline and Reasons of Power in Christianity and Islam. Baltimore: John Hopkins University Press, 2009.
Asani, Ali A. "Sufi Poetry in the Folk Tradition of Indo-Pakistan." Literature and Religion 20.1 (1988): 81–94.
Asani, Ali S. Ecstasy and Enlightenment: The Ismaili Literature of South Asia. London: I. B. Tauris (in association with The Institute of Ismaili Studies), 2002.
Asani, Ali S. "Sindhi Literary Culture." In Sheldon Pollock (ed.), Literary Cultures in History: Reconstructions from South Asia. Berkeley: University of California Press, 2003.

Atiyeh, George N. The Book in the Islamic World: The Written Word and Communication in the Middle East. Albany: State University of New York, 1995.
Auer, Blain H. Symbols of Authority in Medieval Islam: History, religion, and Muslim Legitimacy in the Delhi Sultanate. New York: I. B. Tauris, 2012.
Austin, J. L. How to Do Things with Words. Cambridge: Harvard University Press, 1962.
Awn, Peter J. "The Ethical Concerns of Classical Sufism." The Journal of Religious Ethics 11.2 (1983): 240–263.
Ayoub, Mahmoud M. "The Prayer of Islam: A Presentation of the Surat al-Fātiḥa in Muslim Exegesis," Journal of the American Academy of Religion, Thematic Issue S, XLVII 4 (1980): 636–47.
Baljon, J. M. S. The Reforms and Religious Ideas of Sir Sayyid Ahmad Khan. 3rd revised edition, Lahore: Sh. Muhammad Ashraf, 1964.
Baljon, J. M. S. Modern Muslim Koran Interpretation (1880–1960). Leiden: E. J. Brill, 1968.
Baljon, J. M. S. "Prophetology of Shāh Walī Allāh." Islamic Studies 9.1 (1970): 69–79.
Bashir, Shahzad. Sufi Bodies: Religion and Society in Medieval Islam. New York: Columbia University Press, 2011.
Bauman, Richard (ed.). Folklore, Cultural Performances, and Popular Entertainments. A Communications-centered Handbook. Oxford: Oxford University Press, 1992.
Behl, Aditya. Love's Subtle Magic: An Indian Islamic Literary Tradition, 1379–1545. Oxford: Oxford University Press, 2012.
Bell, Catherine. Ritual Theory, Ritual Practice. Oxford: Oxford University Press, 1992.
Bell, Catherine. Ritual: Perspectives and Dimensions. Oxford: Oxford University Press, 2009.
Bell, Joseph Norment. Love Theory in Later Ḥanbalite Islam. Albany: State University of New York Press, 1979.
Bennett, Clinton and Charles M. Ramsey (eds.). South Asian Sufis: Devotion, Deviation, and Destiny. London: Bloomsbury, 2012.
Berkey, Jonathan P. Popular Preaching & Religious Authority in the Medieval Islamic Near East. Seattle: University of Washington Press, 2001.
Bigelow, Anna. Sharing the Sacred: Practicing Pluralism in Muslim North India. Oxford: Oxford University Press, 2010.
Binder, Leonard. "Pakistan and Modern Islamic-Nationalist Theory." Middle East Journal 11.4 (1957): 382–396.
Binder, Leonard. "Problems of Islamic Political Thought in the Light of Recent Developments in Pakistan." The Journal of Politics 20.4 (1958): 655–675.
Binder, Leonard. "Pakistan and Modern Islamic-Nationalist Theory. Part II." Middle East Journal 12.1 (1958): 45–56.
Binder, Leonard. Religion and Politics in Pakistan. Berkeley: University of California Press, 1963.
Binder, Leonard. Islamic Liberalism: A Critique of Development Ideologies. Chicago: The University of Chicago Press, 1988.
Blake, Stephen P. "The Patrimonial-Bureaucratic Empire of the Mughals." The Journal of Asian Studies 39.1 (1979): 77–94.
Bowen, John R. "Salat in Indonesia: The Social Meanings of an Islamic Ritual," Man 24.4 (1989): 600–619.
Bowen, John R. Muslims Through Discourse. Religion and Ritual in Gayo Society. Princeton: Princeton University Press, 1993.

Bowen, John R. A New Anthropology of Islam. Cambridge: Cambridge University Press, 2012.
Böwering, Gerhard. The Mystical Vision of Existence in Classical Islam: The Qur'ānic Hermeneutics of the Sufi Sahl al-Tustari. Walter de Gruyter, 1979.
Böwering, Gerhard. "Time in Persian Sufism." Iran vol. 30 (1992), 77–79.
Böwering, Gerhard. "The Light Verse: Qur'ānic Text and Ṣūfī Interpretation." Oriens 36 Journal of the International Society for Oriental Research (Leiden: Brill, 2001): 113–144.
Bowker, J. W. "Intercession in the Qur'ān and the Jewish Tradition." Journal of Semitic Studies 11 (1966): 69–84.
Brockopp, Jonathan E. (ed.). The Cambridge Companion to Muḥammad. Cambridge: Cambridge University Press, 2010.
Brown, Nathan. "Sharia and State in the Modern Middle East." International Journal of Middle East Studies 29.3 (1997): 359–376.
Brown, Norman O. "The Apocalypse of Islam." Social Text 8 (1983–1984): 155–171.
Brown, Norman O. "The Prophetic Tradition." Studies in Romanticism 21.3 (1982): 367–386.
Buehler, Arthur. "Translation Issues in Aḥmad Sirhindī's "Collected Letters: Why Šarī'ah is Lot More Than Just "Islamic Law." Oriente Moderno 92.2 (2012): 311–321.
Buehler, Arthur F. Recognizing Sufism: Contemplation in the Islamic Tradition. London: I.B. Tauris, 2016.
Bukharī, Tanwīr. Panjābī Urdū Lughat. Lāhaur: Urdū Sā'ins Būrḍ, 1989.
Butler, Judith, Jürgen Habermas, Charles Taylor, and Cornel West. The Power of Religion in the Public Sphere. Ed. and Introduced by Eduardo Mendieta with afterword by Craig Calhoun. New York: Columbia University Press, 2011.
Bynum, Carolyn. Fragmentation and Redemption: Essays on Gender and the Human Body in Medieval Religion. New York: Zone, 1991.
Calder, Norman. "Friday Prayer and the Juristic Theory of Government: Sarakhsī, Shīrāzī, Māwardi." Bulletin of the School of Oriental and African Studies 49 (1986): 35–47.
Cannell, Fenella (ed.). The Anthropology of Christianity. Durham: Duke University Press, 2006.
Chaudhuri, A. B. Peasant History of Late Pre-Colonial and Colonial India. New Delhi: Pearson Longman, 2008.
Chauhan, Abha (ed.). Understanding Culture and Society in India. A Study of Sufis, Saints and Deities in Jammu Region. Singapore: Springer, 2021.
Chipman, Leigh N. B. "Adam and the Angels: An Examination of the Mythic Elements in Islamic Sources." Arabica 49.4 (2002): 429–455.
Chittick, William. "The Perfect Man as the Prototype of the Self in the Sufism of Jami." Studia Islamica 49 (1979): 71–101.
Chittick, William. Divine Love: Islamic Literature and the Path to God. New Haven: Yale University Press, 2013.
Chodkiewicz, Michel. Seal of the Saints: Prophethood and Sainthood in the Doctrine of Ibn 'Arabī. Trans. Liadain Sherrard. Cambridge: The Islamic Texts Society, 1993.
Coakley, Sarah, (ed.). Religion and the Body. Cambridge Studies in Religious Traditions 8. Cambridge: Cambridge University Press, 1997.
Cook, David. Martyrdom in Islam. Cambridge: Cambridge University Press, 2007.
Cornell, Vincent J. Realm of the Saint: Power and Authority in Moroccan Sufism. Austin: University of Texas Press, 1998.

Cornell, Vincent J. "Fruit of the Tree of Knowledge: The Relationship Between faith and Practice in Islam." In John L. Esposito, ed., The Oxford History of Islam. Oxford: Oxford University Press, 1999.
Currie, P. M. The Shrine and Cult of Mu'īn al-Dīn Chishtī of Ajmer. Delhi: Oxford University Press, 1989.
Curry, John J., and Erik S. Ohlander (eds.). Sufism and Society: Arrangements of the Mystical in the Muslim World, 1200–1800. New York: Routledge, 2012.
Cook, David. "Moral Apocalyptic in Islam." Studia Islamica 86 (1997): 37–69.
Cornell, Vincent J. Realm of the Saint: Power and Authority in Moroccan Sufism. Austin: University of Texas Press, 1998.
Coulson, Noel James. "Doctrine and Practice in Islamic Law: One Aspect of the Problem." Bulletin of the School of Oriental and African Studies 18.2 (1956): 211–226.
Coulson, Noel James. "The State and the Individual in Islamic Law. International and Comparative Law Quarterly 5.1 (1957): 49–60.
Coulson, Noel James. A History of Islamic Law. Edinburgh: Edinburgh University Press, 1964.
Darling, Linda T. "The Vicegerent of God, from Him We Expect Rain": The Incorporation of the Pre-Islamic State in Early Islamic Political Culture." Journal of the American Oriental Society 134.3 (2014): 407–429.
De Jong, Frederick and Bernd Radtke (eds.). Islamic Mysticism Contested: Thirteen Centuries of Controversies and Polemics. Leiden: Brill, 1999.
De Jong, F. "Cairene Ziyāra-Days: A Contribution to the Study of Saint Veneration in Islam," Die Welt des Islams 17.1/4 (1976–1977): 26–43.
Diem, Werner and Marco Scholler. The Living and the Dead in Islam. Studies in Arabic Epitaphs. Wiesbaden: Harrassowitz Verlag, 2004.
Digby, Simon. "The Sufi Shaykh and the Sultan: A Conflict of Claims to Authority in Medieval India. Iran 28 (1990): 71–81.
Douglas, Mary. Purity and Danger: An Analysis of Concepts of Pollution and Taboo. New York: Routledge, 1966.
Douglas, Mary. Natural Symbols: Explorations in Cosmology. New York: Pantheon Books, 1973.
Duderija, Adis. "A Paradigm Shift in Assessing/Evaluating the Value and Significance of Ḥadīth in Islamic Thought: From 'ulūmu-l-isnād/rijāl to 'usūl-l-fiqh." Arab Law Quarterly 23.2 (2009): 195–206.
Duderija, Adis. "Evolution in the Canonical Sunni Ḥadith Body of Literature and the Concept of an Authentic Ḥadith During the Formative Period of Islamic Thought as Based on Recent Western Scholarship." Arab Law Quarterly 23 (2009): 389–415.
During, Jean et al. "Hearing and Understanding in Islamic Gnosis." The World of Music 39.2 (1997): 127–137.
Eaton, Richard M. "The Court and the Dargah in the Seventeenth Century Deccan." Indian Economic and Social History Review 10 (1973): 50–63.
Eaton, Richard M. "Sufi Folk Literature and the Expansion of Indian Islam." History of Religions 14 (1974): 117–27.
Eaton, Richard M. Sufis of Bijapur: 1300–1700: Social Roles of Sufis in Medieval India. Princeton: Princeton University Press, 1978.
Eck, Diana L. Darsan: Seeing the Divine Image in India. New York: Columbia University Press, 1998.

Eickelman, Dale. Moroccan Islam: Tradition and Pilgrimage in a Traditional Society. Austin: University of Texas Press, 1976.
El-Fadl, Abou. "The Centrality of Sharī'ah to Government and Constitutionalism in Islam." Rainer Grote and Tilmann Röder, Constitutionalism in Islamic Countries: Between Upheaval and Continuity. Oxford: Oxford University Press, 2012.
El-Rouayheb, Khaled. "Heresy and Sufism in the Arabo-Islamic World, 1550–1750: Some Preliminary Observations." Bulletin of the School of Oriental and African Studies 73.3 (2010): 357–380.
El-Rouayheb, Khalid. Islamic Intellectual History in Seventeenth Century: Scholarly Currents in the Ottoman Empire and the Maghreb. Cambridge: Cambridge University Press, 2015.
Elias, Jamal (ed.). Key Themes for the Study of Islam. Oxford: Oneworld, 2010.
Ephrat, Daphna. "Spiritual Heirs of the Prophet: Sufi Masters in a Period of Sunni Revitalization." In Daphna Ephrat and Meir Hatina, (eds.). Religious Knowledge, Authority, and Charisma: Islamic and Jewish Perspectives. Utah: University of Utah Press, 2013.
Ernst, Carl W. Eternal Garden: Mysticism, History, and Politics at a South Asian Sufi Center. Albany: State University of New York Press, 1992.
Ernst, Carl W and Bruce B. Lawrence. Sufi Martyrs of Love: The Chishti Order in South Asia and Beyond. London: Palgrave Macmillan, 2002.
Ernst, Carl W. "An Indo-Persian Guide to Sufi Shrine Pilgrimage," in It's Not Just Academic!: Essays on Sufism and Islamic Studies. New Delhi: Sage, 2018.
Erzen, Jale Nejdet. "Reading Mosques: Meaning and Architecture in Islam." The Journal of Aesthetics and Art Criticism 69.1 (2011): 125–131.
Ewing, Katherine P. Arguing Sainthood: Modernity, Psychoanalysis and Islam. Durham: Duke University Press, 1997.
Ewing, Katherine P. "Devotional Service at Sufi Shrines: A Punjabi Charagwala and his Votive Offering of Light." Journal of the History of Sufism 4 (2003–2004): 255–61.
Fakhry, Majid. "Three Varieties of Mysticism in Islam." International Journal of Philosophy vol. 2.4 (1971): 193–207.
Faruqi, Burham Ahmad. The Mujaddid's Conception of Tawhid. Lahore: Sh. Muhammad Ashraf, 1970; reprint edition.
Ferrer, Jorge N. and Jacob H. Sherman (eds.). The Participatory Turn: Spirituality, Mysticism, Religious Studies. Albany: State University of New York Press,2008.
Friedmann, Yohanan. Shaykh Ahmad Sirhindi: An Outline of his Thought and a Study of His Image in the Eyes of Posterity. Montreal: McGill University Islamic Studies II, 1971.
Foucault, Michel. The Archaeology of Knowledge. New York: Routledge Classics, 2002; first published 1969.
Frembgen, Jurgen Wasim. "Dhamāl and the Performing Body: Trance Dance in the Devotional Sufi Practice of Pakistan." Journal of Sufi Studies 1 (2012): 77–113.
Friedmann, Yohanan. "The Attitude of the Jamiyy'at-i Ulama-i Hind to the Indian National Movement and to the Establishment of Pakistan." Asian and African Studies 7. (1971): 157–180.
Frishkopf, Michael. "Authorship in Sufi Poetry." Alif: Journal of Comparative Poetics 23 (2003): 78–108.
Fueck, J. "The Role of Traditionalism in Islam." In Merlin L. Swartz (ed.), Studies on Islam. Oxford: Oxford University Press, 1981.

Furey, Constance M. "Body, Society, and Subjectivity in Religious Studies." Journal of the American Academy of Religion 80.1 (2012): 7–33.
Fusfeld, Warren Edward. "The Shaping of Sufi Leadership in Delhi: The Naqshbandi Mujaddidiyya, 1750 to 1920." Ph.D. dissertation, University of Pennsylvania, 1981.
Fusfeld, Warrem Edward. "Naqshbandi Sufism and Reformist Islam." Journal of Asian and African Studies XVIII.3–4 (1983) 241–262.
Gaborieau, Marc. "Late Persian, Early Urdu: The Case of "Wahhabi" Literature" (1818–1857). In Francoise 'Nalini' Delvoye (ed.). Confluence of Cultures: French Contributions to Indo-Persian Studies. New Delhi: Manohar, 1994.
Gaborieau, Marc. "Sufism in the First Indian Wahhabi Manifesto: Sirātu'l mustaqīm by Ismail Shahid and 'Abdu'l Hayy." In Muzaffar Alam et al., eds., The Making of Indo-Persian Culture: Indian and French Studies. New Delhi: Manohar, 2000.
Garden, Kenneth. The First Islamic Reviver: Abū Ḥamīd al-Ghazālī and his Revival of the Religious Sciences. Oxford: Oxford University Press, 2014.
Gauvain, Richard. "Ritual Rewards: A Consideration of Three Recent Approaches to Sunni Purity Law." Islamic Law and Society 12.3 (2005), 333–393.
Gearhart, Suzanne. "The Taming of Michel Foucault: New Historicism, Psychoanalysis, and the Subversion of Power." New Literary History 28.3 (1997): 457–480.
Geertz, Clifford. The Religion of Java. Chicago: The University of Chicago Press, 1976.
Geertz, Clifford. Islam Observed: Religious Development in Morocco and Indonesia. Chicago: The University of Chicago, 1971 edition.
Gilmartin, David. "Partition, Pakistan, and South Asian History: In Search of a Narrative." The Journal of Asian Studies 57.4 (1998): 1068–1095.
Gilmartin, David, and Bruce B. Lawrence. Beyond Turk and Hindu: Rethinking Religious Identities in Islamicate South Asia. New Delhi" India Research Press, 2002.
Gilmartin, David. "Rethinking the Public Through the Lens of Sovereignty." South Asia: Journal of South Asian Studies 38.3 (2015): 371–386.
Gilsenan, Michael. Saint and Sufi in Modern Egypt: An Essay in Sociology of Religion. Oxford: Clarendon Press, 1973.
Gilsenan, Michael. Recognizing Islam: Religion and Society in the Modern Middle East. New York: I. B. Tauris, 2005.
Gleave, Robert. Islam and Literalism: Literal Meaning and Interpretation in Islamic Legal Theory. Edinburgh: Edinburgh University Press, 2012.
Goldziher, Ignaz. Muslim Studies. S. M. Stern (ed.) and trans. 2 vols. London: George Allen & Unwin, 1971.
Goodman, Lenn Evan. Jewish and Islamic Philosophy: Crosspollinations in the Classic Age. Edinburgh: Edinburgh University Press, 1999.
Graham, William A. Divine Word and Prophetic Word in Early Islam. Mouton: Walter de Gruyter, 1977.
Graham, William. Beyond the Written Word: Oral Aspects of Scripture in the History of Religion. Cambridge: Cambridge University Press, 1987.
Graham, William. Islamic and Comparative Religious Studies. Burlington: Ashgate, 2010.
Green, Nile. Sufism: A Global History. Chichester: Wiley-Blackwell, 2012.
Gupta, Shakti M. Festivals, Fairs and Fasts of India. Delhi: Clarion Books, 1991.
Haj, Samira. Reconfiguring Islamic Tradition: Reform, Rationality, and Modernity. Oxford: Oxford University Press, 2008.

Halevi, Leor. Muhammad's Grave: Death Rites and the Making of Islamic Society. New York: Columbia University Press, 2007.
Halevi, Leor. "The Paradox of Islamization: Tombstone Inscriptions, Qur'ānic Recitations, and the Problem of Religious Change." History of Religions 44.2 (2004): 120–152.
Hallaq, Wael B. "On the Authoritativeness of Sunni Consensus." IJMES 18.4 (1986): 427–454
Hallaq, Wael B. "Juristic Authority vs. State Power: The Legal Crisis of Modern Islam." Journal of Law and Religion 19.2 (2003–2004): 243–258.
Hallaq, Wael B. The Impossible State: Islam, Politics, and Modernity's Moral Predicament. New York: Columbia University Press, 2013.
Hamza, Feras, Sajjad Rizvi, and Farhana Mayer (eds.). An Anthology of Qur'anic Commentaries, Volume I, On the Nature of the Divine. Oxford: Oxford University Press, 2008.
Hannoum, Abdelmajid (ed.). Practicing Sufism. Sufi Politics and Performance in Africa. New York: Routledge, 2016.
Hardy, Peter. The Muslims of British India. Cambridge: Cambridge University Press, 1972.
Hasan, Mushir-ul (ed.). Knowledge, Power & Politics: Educational Institutions in India. New Delhi: Lotus Collection, 1998.
Hawting, Gerald (ed.). The Development of Islamic Ritual. Burlington: Ashgate Variorum, 2006.
Heck, Paul L. "The Hierarchy of Knowledge in Islamic Civilization." Arabica 49.1 (2002): 27–54.
Heck, Paul L. "Mysticism as Morality: The Case of Sufism." The Journal of Religious Ethics 34.2 (2006): 253–286.
Heck, Paul L. Sufism and Politics: The Power of Spirituality. Princeton: Markus Wiener Publishers, 2007.
Heck, Paul L. "Conscience Across Cultures: The Case of Islam." The Journal of Religion 94.3 (2014): 292–324.
Hefner, Robert W. (ed.). Shari'a Politics: Islamic Law and Society in the Modern World. Bloomington: Indiana University Press, 2011.
Hennigan, Peter C. The Birth of a Legal Institution: The Formation of the Waqf in Thrid-Century A.H. Ḥanafī Legal Discourse. Leiden: Brill, 2004.
Hermansen, Marcia K. "Shāh Walī Allāh's Theory of the Subtle Spiritual Centers (Ḷatā'if): A Sufi Model of Personhood and Self-Transformation." Journal of Near Eastern Studies 47.1 (1988): 1–25.
Hirschkind, Charles. The Ethical Soundscape: Cassette Sermons and Islamic Counter Publics. New York: Columbia University Press, 2006.
Hocart, A. M. "The Origin of Monotheism." Folklore 33.3 (1922): 282–293.
Hoexeter, Miriam, Shmuel N. Eisenstadt, and Nehemia Levtzion (eds.). The Public Sphere in Muslim Societies. Albany: State University of New York Press, 2002.
Hoexeter, Miriam. "Waqf Studies in the Twentieth Century: The State of the Art." Journal of Economic and Social History of the Orient 41.4 (1998): 474–495.
Hofer, Nathan. The Popularization of Sufism in Ayyubid and Mamluk Egypt, 1173–1325. Edinburgh: Edinburgh University Press, 2015.
Hoffman-Ladd, Valerie J. "Devotion to the Prophet and His Family in Egyptian Sufism," IJMES 24 (1992): 615–637.

Ho, Enseng. Graves of Tarim: Genealogy and Mobility across the Indian Ocean. Berkeley: University of California, 2006.
Homerin, Emil Th. From Arab Poet to Muslim Saint: Ibn al-Fāriḍ, His Verse, and His Shrine. Columbia: University of South Carolina Press, 1994.
Hourani, George F. "The Basis of Authority of Consensus of Sunnite Islam." Studia Islamica 21 (1964), 13–60.
Howes, David. "Controlling Textuality: A Call for a Return to the Senses." Anthropologica 32.1 (1990), 55–73.
Howes, David (ed.). The Varieties of Sensory Experience: A Sourcebook in the Anthropology of the Senses. Toronto: University of Toronto Press, 1991.
Hurvitz, Nimrod. "Biographies and Mild Asceticism: A Study of Islamic Moral Imagination." Studia Islamica 85 (1997): 41–65.
Ingram, Brannon. "Crisis of the Public in Muslim India: Critiquing 'Custom' at Aligarh and Deoband." South Asia: Journal of South Asian Studies 38.3 (2015): 403–418
Iqtidar, Humeira, and David Gilmartin. "Secularism and the State in Pakistan." Modern Asian Studies 45.3 (2011): 491–499.
Irani, Ayesha. "The Prophetic Principle of Light and Love: Nūr Muḥammad in Early Modern Bengali Literature." History of Religions 55.4 (2016): 391–428.
Izutsu, Toshihiko. Sufism and Daoism: A Comparative Study of Key Philosophical Concepts. Berkeley: University of California Press, 1983.
Jackson, Michael. The Palm at the End of the Mind: Relatedness Religiosity, and the Real. Durham: Duke University Press, 1989.
Jackson, Michael and Albert Piette. What is Existential Anthropology? Oxford: Berghahn Books, 2017.
Jaffee Martin S. "One God, One Revelation, One People: On the Symbolic Structure of Elective Monotheism." Journal of the American Academy of Religion 69.4 (2001): 753–775.
Jalal, Ayesha. "Conjuring Pakistan: History as Official Imagining." International Journal of Middle East Studies 27.1 (1995): 73–89.
James, William. The Varieties of Religious Experience: A Study in Human Nature. New York: Penguin Books, 1985 reprint.
Jones, Kenneth W. Socio-Religious Reform Movements in British India. Cambridge: Cambridge University Press, 1989.
Joppke, Christian. "Islam and the Legal Enforcement of Morality." Theory and Society 43.6 (2014): 589–615.
Joseph, Roger. "The Semiotics of the Islamic Mosque." Arab Studies Quarterly 3.3 (1981): 285–295.
Kadivar, Mohsen. "From Traditional Islam to Islam as an End in Itself." Die Welt des Islams 51.3/4 (2011): 459–484.
Kamran, Tahir. "The Pre-History of Religious Exclusionism in Contemporary Pakistan: Khatam-e-Nubuwwat 1889–1953." Modern Asian Studies 49.6 (2015): 1840–1874.
Karamustafa, Ahmet T. God's Unruly Friends: Dervesh Groups in the Later Middle Period 1200–1550. Salt Lake City: University of Utah Press, 1994.
Karawan, Ibrahim A. "Monarchs, Mullas, and Marshals: Islamic Regimes?" The Annals of the American Academy of Political and Social Science 524 (1992): 103–119.
Katz, Marion. "The Ḥajj and the Study of Islamic Ritual," Studia Islamica 98/99 (2004), 95–129.

Katz, Marion Holmes. The Birth of the Prophet Muḥammad: Devotional Piety and Sunni Islam. New York: Routledge, 2007.
Kazuo, Morimoto. Sayyids and Sharifs in Muslim Societies: The Living Links to the Prophet. New York: Routledge, 2012.
Keddie, Nikki R. "Symbol and Sincerity in Islam." Studia Islamica 19 (1963): 27–63.
Khalid, Detlev H. "Theocracy and the Location of Sovereignty." Islamic Studies 11.3 (1972): 187–209.
Khieckhefer, R., and B. Radtke (eds.). Sainthood: Its Manifestations in World Religions. Berkeley: University of California Press, 2007.
Khodamoradi, Soraya. "Ṭarīqah Muḥammadiyyah" as "Ṭarīqah Jāmi'ah": Khawājah Mīr Dard's Experience Beyond "Jamal" and "Jalāl." Islamic Studies 51.4 (2012): 367–402.
Khosronejad, Pedram. Saints and their Pilgrims in Iran and Neighbouring Countries. UK: Sean Kingston Publishing, 2012.
Kugle, Scott. Sufis & Saints' Bodies: Mysticism, Corporeality, & Sacred Power in Islam. Chapel Hill: The University of North Carolina Press, 2007.
Kurin, Richard. "The Structure of Blessedness at a Muslim Shrine in Pakistan." Middle Eastern Studies 19 (1983): 312–25.
Kynsh, Alexander D. Ibn 'Arabi in the Later Islamic Tradition: The Making of a Polemical Image in Medieval Islam. Albany: State University of New York Press, 1999.
Landau-Tasseron, Ella. "The "Cyclical Reform": A Study of the mujaddid Tradition." Studia Islamica 70 (1989): 79–117.
Lane, Edward William. An Account of the Manners and Customs of the Modern Egyptians: Written in Egypt During the Years 1833–1835. London: John Murray, 1836.
Lange, Christian. "Barzakh." In Kate Fleet, Gudrun Krämer, Denis Matringe, John Nawas, Everett Rowson (eds.), Encyclopedia of Islam, Three. Leiden: Brill, 2011.
Lapidus, Ira M. "State and Religion in Islamic Societies." Past & Present 151 (1996): 3–27.
Lapidus, Ira M. "Islamic Revival and Modernity: The Contemporary Movements and the Historical Paradigms." Journal of the Economic and Social History of the Orient 40.4 (1997): 444–460.
Laude, Patrick. Singing the Way: Insights in Poetry and Spiritual Transformation. Bloomington: World Wisdom, 2005.
Layish, Aharon. " 'Waqf' and Ṣūfī Monasteries in the Ottoman Policy of Colonization: Sulṭan Selīm I's 'waqf' of 1516 in Favour of Dayr al-Asad." Bulletin of the School or Oriental and African Studies 50.1 (1987): 61–89.
Leeuwen, Richard Van. Waqfs and Urban Structures: The Case of Ottoman Damascus. Leiden: Brill, 1999.
Lewisohn, Leonard (ed.). Classical Persian Sufism: From its Origins to Rumi. London: Khaniqahi Nimatullah Publications, 1992.
Lewisohn, Leonard (ed.). Hafiz and the Religion of Love in Classical Persian Poetry. New York: I. B. Taurus, 2010.
Librande, Leonard. "Ibn Abī al-Dunyā: Certainty and Morality." Studia Islamica 100/101 (2005): 5–42.
Liebeskind, Claudia. Piety on Its Knees: Three Sufi Traditions in South Asia in Modern Times. Delhi: Oxford University Press, 1998.
Liew, Han Hsein. "The Caliphate of Adam: Theological Politics of the Qur'ānic Term Ḥalīfa. Arabica 63 (2016): 1–29.

Lings, Martin. Symbol and Archetype: A Study of the Meaning of Existence. Lahore: Suhail Academy, 2000.
Makdisi, George. "Muslim Institutions of Learning in Eleventh-Century Bagdad." Bulletin of the School of Oriental and African Studies 24.1 (1961): 1–56.
Malik, Jamil S. "The Luminous Nurani: Charisma and Political Mobilization among the Barelwis in Pakistan." Social Analysis: The International Journal of Social and Cultural Practice 28 (1990): 38–50.
Malik, Jamal S. Colonization of Islam: Dissolution of Traditional Institutions in Pakistan. New Delhi: Manohar, 1998.
Mandair, Arvind-Pal S. Religion and the Specter of the West: Sikhism, India, Postcoloniality, and the Politics of Translation. New York: Columbia University Press, 2009.
March, Andrew F. "Genealogies of Sovereignty in Islamic Political Theology." Social Research 80.1 (2013): 293–320.
Margalit, Avishai. "Revisiting God's Authority." Social Research 80.1 (2013): 77–100.
Marshall, David and Lucinda Mosher (eds.). Death, Resurrection, and Human Destiny. Georgetown: Georgetown University Press, 2014.
Martin, Richard C., Mark R. Woodward, and Dwi S. Atmaja. Defenders of Reason: Mu'tazilism from Medieval School to Modern Symbol. London: Oneworld Publications, 1997.
Mason, David R. "An Examination of "Worship" as a Key for Re-Examining the God Problem." The Journal of Religion 55.1 (1975): 76–94.
Mathur, Y. B. "The Mujahiddin Movement in Modern India." Studies in Islam 6 (April-October, 1969), pp. 193–217.
Matringe, Denis. "Kṛṣṇaite and Nāth Elements in the Poetry of the Eighteenth-Century Punjabī Sūfi Bullhe Sāh." In R. S. McGregror. ed., Devotional Literature in South Asia. Cambridge: Cambridge University Press, 1999.
McDonough, Sheila. "Some Leading Ideas Constitutive of Pakistan's Nationhood." Islamic Studies 7.1 (1968): 9–31.
Meri, Josef W. The Cult of Saints Among Muslims and Jews in Medieval Syria. Oxford: Oxford University Press, 2002.
Metcalf, Barbara. Islamic Revival in British India: Deoband, 1860–1900. Princeton: Princeton University Press, 1982.
Melchert, Christopher. "Traditionist-Jurisprudents and the Framing of Islamic Law." Islamic and Society 8.3 (2001): 383–406.
Melchert, Christopher. "The Piety of the Hadith Folk." International Journal of Middle East Studies 34.3 (2002): 425–439.
Melchert, Christopher. " 'God Created Adam in His Image'." Journal of Qur'anic Studies 13.1 (2011): 113–124.
Memon, Muhammad Umar. Ibn Taymīya's Struggle against Popular Religion. Paris: Mouton & Co., 1976.
Merleau-Ponty, Maurice. Phenomenology of Perception, translated by Colin Smith. New Jersey: The Humanities Press, 1962.
Mills, Sara. Discourse. London: Routledge, 1997.
Mir, Farina. The Social Space of Language: Vernacular Culture in British Colonial Punjab. Berkeley: University of California Press, 2010.
Mir, Mustansir. "Teachings of the Two Punjabi Poets." In Donald S. Lopez, ed., Religions of India in Practice. Princeton: Princeton University Press, 1995.

Michell, George (ed.). Architecture of the Islamic World: Its History and Social Meaning. London: Thames and Hudson Ltd., 1978.
Moad, Edward Omar. "A Path to the Oasis: ""Sharī'ah" and Reason in Islamic Moral Epistemology." International Journal for Philosophy of Religion 62.3 (2007): 135–148.
Moaddel, Mansoor. "Conditions for Ideological Production: The Origins of Islamic Modernism in India, Egypt, and Iran." Theory and Society 30.5 (2001): 669–731.
Moini, S. Liyaqat H. The Chishti Shrine of Ajmer, Pirs, Pilgrims, Practices. Jaipur: Publication Scheme, 2004.
Morgan, Claire. "Islam and Civil Society: The Waqf." The Good Society 10.1 (2001): 21–24
Morgan, David. The Sacred Gaze: Religious Visual Culture in Theory and Practice. Berkeley: University of California Press, 2005.
Nasr, Seyyed Hossein. "Shi'ism and Sufism" Their Relationship in Essence and in History." Religious Studies 6.3 (1970): 229–242.
Nasr, Seyyed Hossein. An Introduction to Islamic Cosmological Doctrines: Conceptions of Nature and Methods Used for Its Study by the Ikhwān al-Ṣafā, al-Bīrūnī, and And Ibn Sīnā. Albany: State University of New York Press, 1993.
Nasr, Seyyed Hossein. Religion & The Order of Nature. The 1994 Cadbury Lectures at the University of Birmingham. New York: Oxford University Press, 1996.
Nasr, Seyyed Hossein. The Garden of Truth: The Vision and Promise of Sufism, Islam's Mystical Tradition. New York: Harper Collins, 2007.
Nizami, K. A. "Shah Wali-Ullah Dehlavi and Indian Politics in the 18th century." Islamic Culture 25 (1951), pp. 133–45.
Nizami, K. A. Royalty in Medieval India. New Delhi: Munshiram Manoharlal Publishers Pvt. Ltd., 1997.
O'Fahey, R. S., and Bernd Radtke. "Neo-Sufism Reconsidered." Der Islam 70 (1993) 52–87.
Omar, Irfan A. "Kiźr-i Rāh: The Pre-Eminent Guide to Action in Muhammad Iqbal's Thought." Islamic Studies 43.1 (2004): 39–50.
Ortner, Sherry B. Anthropology and Social Theory: Culture, Power, and the Acting Subject. Durham: Duke University Press, 2006.
Osella, Filippo and Caroline Osella. "Introduction: Islamic Reformism in South Asia." Modern Asian Studies 42.2/3 (2008): 247–257.
O'Shaughnessy, Thomas J. Muhammad's Thoughts on Death. Leiden: E. J. Brill, 1969.
O'Shaughnessy, Thomas J. "Islam—Surrender to God." Philippine Studies 15.1 (1967): 108–129.
O'Shaughnessy, Thomas J. "God's Throne and Biblical Symbolism of the Qur'ān," Numen 20.3 (1973): 202–221.
Padwick, Constance. Muslim Devotions. London: S.P.C.K, 1961.
Pagani, Samuela. "Ibn 'Arabī, Ibn Qayyim al-Jawziyya, and the Political Functions of Punishment in the Islamic Hell." In Christian Lange, ed., Locating Hell in Islamic Traditions. Leiden: Brill, 2016.
Parodi, Laura E. "Soloman, the Messenger and the Throne Themes from a Mughal Tomb." East and West 51.1/2 (2001): 127–142.
Patton, Kimberley C. and Benjamin C. Ray (eds.). A Magic Still Dwells: Comparative Religion in the Postmodern Age. Berkeley: University of California Press, 2000.
Patton, Kimberley C. Religion of the Gods: Ritual, Paradox and Reflexivity. Oxford: Oxford University Press, 2009.

Pearson, Harlan Otto. Islamic Reform and Revival in Nineteenth-century India: Tarīqah-i-Muhammadīyah. New Delhi: Yoda Press, 2008.
Peters, Rudolph. "Idjtihād and Taqlīd in 18th and 19th Century Islam." Die Welt des Islams 20.3/4 (1980): 131–145.
Philippon, Alix. "The 'urs of the Patron of Lahore: National Popular Festival and Sacred Union Between the Pakistani State and Society? Social Compass 59.3 (2012): 289–297.
Philippon, Alix. "A Sublime, Yet Disputed, Object of Political Ideology? Sufism in Pakistan at the Crossroads." Commonwealth & Comparative Politics, 52.2 (2014): 271–292.
Pinto, Paulo G. "The Anthropologist and the Initiated: Reflections on the Ethnography of Mystical Experience among the Sufis of Aleppo, Syria." Social Compass 57.4 (2010): 464–478.
Pirbhai, M. Reza. Reconsidering Islam in South Asian Context. Leiden: Brill, 2009.
Platts, John T. A Dictionary of Urdu, Classical Hindi, and English. New Delhi: Munshiram Manoharlal Publishers, 2000.
Porta, Sergio La and David Dean Shulman (eds.). The Poetics of Grammar and the Metaphysics of Sound and Sign. Leiden: Brill, 2007.
Powers, Paul R. "Interiors, Intentions, and the "Spirituality" of Islamic Ritual Practice." Journal of the American Academy of Religion 72.2 (2004): 425–459.
Puri, J. R. & T. R. Shangari. Bulleh Shah: The Love-intoxicated Iconoclast. Dera Baba Jaimal Singh, Dist. Amritsar: Radha Soami Satsang Beas, 1986.
Qasmi, Ali Usmani. Questioning the Authority of the Past: The Ahl al-Qur'an Movements in Punjab. Karachi: Oxford University Press, 2011.
Quraeshi, Samina. Sacred Spaces: A Journey with the Sufis of the Indus. Cambridge: Peabody Museum Press, 2009.
Rahman, Fazlur. "Implementation of the Islamic Concept of State in the Pakistani Milieu." Islamic Studies 6.3 (1967): 205–223.
Rahman, Fazlur. "Islam and the Constitutional Problem of Pakistan." Studia Islamica 32 (1970): 275–287.
Raudvere, Catharina and Leif Stenberg (eds.). Sufism Today: Heritage and Tradition in the Global Community. London: I.B. Tauris, 2009.
Reetz, Dietrich. Islam in the Public Sphere: Religious Groups in India, 1900–1947. New York: Oxford University Press, 2006.
Reinhart, Kevin. Before Revelation: The Boundaries of Muslim Moral Thought. Albany: State University of New York Press, 1995.
Renard, John. All the King's Falcons: Rumi on Prophets and Revelation. Albany: SUNY Press, 1994.
Renard, John. Historical Dictionary of Sufism. Oxford: The Scarecrow Press, Inc., 2005.
Renard, John. Friends of God: Islamic Images of Piety, Commitment, and Servanthood. Berkeley: University of California Press, 2008.
Ridgeon, Lloyd. " 'Azīz Nasafī's Six Ontological Faces (Whenever You Turn There Is The Face of God)." Iran: Journal of Persian Studies 34 (1996): 85–99.
Riexinger, Martin. "Rendering Muḥammad Human Again: The Prophetology of Muḥammad b. 'Abd al-Wahhāb (1703–1792)." Numen 60 (2013): 103–118.
Rinehart, Robin. "The Portability of Bullhe Shah: Biography, Categorization, and Authorship in the Study of Punjabi Sufi Poetry." Numen 46.1 (1999): 53–87.
Rippin, Andrew (ed.). Defining Islam: A Reader. New York: Routledge, 2014.

Rippin, Andrew and Roberto Tottoli (eds.). Books and Written Culture of Islamic World Studies Presented to Claude Gilliot on the Occasion of his 75th Birthday. Leiden: Brill, 2014.

Rizvi, Saiyid Athar Abbas. Muslim Revivalist Movements in Northern India in the Sixteenth and Seventeenth Centuries. Agra: Balkrishna Book Co., 1965.

Rizvi, Saiyid Athar Abbas. "The Breakdown of Traditional Society." In P. M. Holt, ed., Cambridge History of Islam. Cambridge: Cambridge University Press, 1971.

Rizvi, Saiyid Athar Abbas. The Wonder That Was India, Volume II, A Survey of the History of the Indian Sub-Continent From the Coming of the Muslims to the British Conquest. London: Sidgwick & Jackson, 1987.

Robison, Francis. "Crisis of Authority: Crisis of Islam? Journal of the Royal Asiatic Society 19.3 (2009): 339–354.

Rosenthal, Erwin I. J., and E. J. J. Rosenthal. "Some Reflections on the Separation of Religion and Politics in Modern Islam." Islamic Studies 3.3 (1964): 249–284.

Robison, Francis. "Islamic Reform and Modernities in South Asia." Modern Asian Studies 42.2–3 (2008): 259–281.

Rogers, Michael. "Waqf and Patronage in Seljuk Anatolia: The Epigraphic Evidence." Anatolian Studies 26 (1976): 69–103.

Rogers, Edward Thomas, Mary Eliza Rogers, and George L. Seymour. "Cemeteries and Mosque Tombs, Cairo." The Art Journal (1875–1887) 6 (1880): 201–204.

Rubin, Uri. "Pre-Existence and Light: Aspects of the Concept of Nūr Muḥammad." Israel Oriental Studies 5 (1975): 62–119.

Rudbeck, Jens. "Popular Sovereignty and the Historical Origin of the Social Movement." Theory and Society 41.6 (2012): 581–601.

Safi, Omid. "Bargaining with Baraka: Persian Sufism, "Mysticism," and Premodern Politics." The Muslim World 90 (2000): 259–87.

Safi Omid. The Politics of Knowledge in Pre-modern Islam: Negotiating Ideology and Religious Inquiry. Chapel Hill: University of North Carolina Press, 2006.

Sanford, A. Whitney. Sacred Play: Ritual Levity and Humor in South Asian Religions. Albany: State University of New York Press, 2010.

Sanyal, Usha. Devotional Islam & Politics in British India: Ahmad Riza Khan and His Movement, 1870–1920. Oxford: Oxford University Press, 1996.

Sanyal, Usha. Ahmad Riza Khan Barelwi: In the Path of the Prophet. Oxford: Oneworld, 2005.

Salomon, Noah. For the Love of the Prophet: An Ethnography of Sudan's Islamic State. Princeton: Princeton University Press, 2016.

Salvatore, Armando and Dale F. Eickelman. Public Islam and the Common Good. Leiden: Brill, 2004.

Salvatore, Armando. The Public Sphere: Liberal Modernity, Catholicism, Islam. New York: Palgrave Macmillan, 2007.

Sarrio, Diego R. "Spiritual Anti-elitism: Ibn Taymiyya's Doctrine of Sainthood (walāya)." Islam and Christian-Muslim Relations 22.3 (2011): 275–291.

Scharbrodt, Oliver. "The Salafiyya and Sufism: Muḥammad 'Abduh and his Risālat al-Warīdāt (Treatise on Mystical Inspirations)." Bulletin of the School of Oriental and African Studies 70.1 (2007): 89–115.

Scheilke, Samuli J. The Perils of Joy: Contesting Mulid Festivals in Contemporary Egypt. Syracuse: Syracuse University Press, 2012.

Schimmel, Annemarie. "The Place of the Prophet of Islam in Iqbāl's Thought." Islamic Studies 1.4 (1962): 111–130.
Schimmel, Annemarie. The Triumphal Sun: A Study of the Works of Jalāloddin Rumi. Albany: State University of New York Press, 1978.
Schimmel, Annemarie. Islam in the Indian Subcontinent. Leiden: E.J. Brill, 1980.
Schimmel, Annemarie. As Through a Veil: Mystical Poetry in Islam. New York: Columbia University Press, 1982.
Schimmel, Annemarie. Mystical Dimensions of Islam. Chapel Hill: University of North Carolina Press, 1975.
Schimmel, Annemarie. And Muhammad is His Messenger: The Veneration of the Prophet in Islamic Piety. Chapel Hill: The University of North Carolina Press, 1985.
Schimmel, Annemarie. Deciphering the Signs of God: a Phenomenological Approach to Islam. Albany: State University of New York, 1994.
Schoeler, Gregor. The Oral and the Written in Early Islam. Trans. Uwe Vagelpohl and ed. James E. Montgomery. New York: Routledge, 2006.
Seferta, Yusuf H. R. "The Doctrine of Prophethood in the Writings of Muḥammad 'Abduh and Rashīd Riḍā." Islamic Studies 24.2 (1985): 139–165.
Sells, Michael. Mystical Languages of Unsaying. Chicago: The University of Chicago Press, 1994.
Shackle, Christopher. "Early Vernacular Poetry in the Indus Valley: Its Contents and Character." In Anna Libera Dallapiccola and Stephanie Zingel-Ave Lallemant (eds.), Islam and Indian Regions. Stuttgart: Steiner, 1993.
Shaikh, Sa'diyya. "In Search of "Al-Insān": Sufism, Islamic Law, and Gender." Journal of the American Academy 77.4 (2009): 781–822.
Shahzad, Qaiser. "Ibn 'Arabī's Contribution to the Ethics of Divine Names." Islamic Studies 43.1 (2004): 5–38.
Shahzad, Qaiser. "Ibn 'Arabī's Metaphysics of the Human Body." Islamic Studies 46.4 (2007): 499–525.
Shapiro, Sheldon. "Morality in Religious Reformations." Comparative Studies in Society and History 18.4 (1976): 438–457.
Shepard, William E. "Islam and Ideology: Towards a Typology." International Journal of Middle East Studies 19.3 (1987): 307–335.
Silverstein, Brian. "Sufism and Governmentality in the Late Ottoman Empire." Comparative Studies of South Asia, Africa and the Middle East 29.2 (2009): 171–185.
Simpson, Gary M. "Whither Wolfhart Pannenberg? Reciprocity and Political Theology." The Journal of Religion 67.1 (1987): 33–49.
Singer, Amy. Charity in Islamic Societies. Cambridge: Cambridge University Press, 2008.
Singer, Milton (ed.). Krishna: Myths, Rites and Attitudes. Chicago: University of Chicago Press, 1966; first published by East-West Center, University of Hawaii.
Singh, Surinder and Ishwar Dayal Gaur (eds.). Sufism in Punjab: Mystics, Literature and Shrines. Delhi: Aakar Books, 2009.
Sirriyeh, Elizabeth. Sufis and Anti-Sufis: The Defense, Rethinking and Rejection of Sufism in the Modern World. Surrey: Curzon, 1999.
Smith, Grace Martin and Carl W. Ernst (eds.), Manifestations of Sainthood in Islam. Istanbul: Isis, 1993.
Smith, Jonathan Z. Map is Not a Territory. Chicago: The University of Chicago Press, 1993.

Smith, Jane I. "The Understanding of Nafs and Ruh in Contemporary Muslim Considerations of the Nature of Sleep and Death." Muslim World 69 (1979): 151–62.
Smith, Jane I., and Yvonne Haddad. The Islamic Understanding of Death and Resurrection. Oxford: Oxford University Press, 2002.
Smith, Wilfred Cantwell. The Meaning and End of Religion. Minneapolis: Fortress Press, 1991.
Solomon, Norman, Richard Harries and Tim Winter (eds.). Abraham's Children: Jews, Christians and Muslims in Conversation. New York: T & T Clark, 2005.
Stauth, Georg. "Leonard Binder and the Hermeneutic of Authenticity: Critical Note." Arabica 40.1 (1993): 62–83.
Stauth, George, and Armando Salvatore (eds.). On Archaeology of Sainthood and Local Spirituality of Islam. Yearbook of the Sociology of Islam, vol. 5. Bielefeld: Transcript, 2005.
Stauth, Georg. "Afterword: Holism, Individualism, Secularism." Asian Journal of Social Science 33.3 (2005): 529–543.
Stepputat, Finn, and Thomas Blom Hansen. "Sovereignty Revisited." Annual Review of Anthropology 35 (2006): 295–315.
Steppat, Fritz. "God's Deputy: Materials on Islam's Image of Man." Arabica (1989): 163–172.
Stewart, Tony. "In Search of Equivalence: Conceiving the Muslim-Hindu Encounter Through Translation Theory." History of Religions 40.3 (2001): 260–87.
Strothmann, Linus. "Giving Comfort, Dispelling Fear: Social Welfare at the Shrine of Data Ganj Bukhsh in Lahore, Pakistan." Erdkunde 67.1 (2013): 49–61.
Swartz, David. "Bridging the Study of Culture and Religion: Pierre Bourdieu's Political Economy of Symbolic Power." Sociology of Religion 57 1 (1996), 71–85
Swartz, Merlin L. A Medieval Critique of Anthropomorphism: Ibn al-Jawzī's Kitāb Akhbār aṣ-ṣifāt: A Critical Edition of the Arabic Text with Translation, Introduction and Notes. Leiden: Brill, 2002.
Synnott, Anthony. "Tomb, Temple, Machine and Self: The Social Construction of the Body." The British Journal of Sociology 43.1 (1992): 79–110.
Talmon-Heller, Daniella. Piety in Medieval Syria: Mosques, Cemeteries and Sermons under the Zangids and Ayyūbids. Leiden: Brill, 2007.
Tambar, Kabir. "Islamic Reflexivity and the Uncritical Subject." The Journal of the Royal Anthropological Institute 18.3 (2012): 652–672.
Tapper, Nancy and Richard Tapper. "The Birth of the Prophet: Ritual and Gender in Turkish Islam." Journal of the Royal Anthropological Society 22 (1987): 62–79.
Tareen, Sherali. "Narratives of Emancipation in Modern Islam: Temporality, Hermeneutics, and Sovereignty." Islamic Studies 52.1 (2013): 5–28.
Taylor, Christopher S. In the Vicinity of the Righteous: Ziyāra and the Veneration of the Muslim Saints in Late Medieval Egypt. Leiden: Brill, 1999.
Tayyara, Abdel el-Rahman. "Prophethood and Kingship in Early Islamic Historical Thought." Der Islam 84.1 (2007): 73–102.
Tayob, Abdelkader I. "Ṭabarī on the Companions of the Prophet: Moral and Political Contours in Islamic Historical Writing." Journal of the American Oriental Society 119.2 (1999): 203–210.
Terzioğlu, Derin. "Man in the Image of God in the Image of Times: Self-Narratives and Diary of Niyāzī-i Miṣri (1618–94)." Studia Islamica 94 (2002): 139–165.

Tibi, Bassam. "The Return of the Sacred to Politics as a Constitutional Law: The Case of the of the Shari'atization of Politics in Islamic Civilization." Theoria: A Journal of Social and Political Theory 115 (2008): 91–119.
Tottoli, Roberto. "Muslim Attitudes Towards Prostration (sujūd): I. Arabs and Prostration at the Beginning of Islam and in the Qur'ān." Studia Islamica 88 (1988): 5–34.
Trimingham, Spencer J. S. The Sufi Orders of Islam. Oxford: Oxford University Press, 1998.
Troll, Christian W. (ed.). Muslim Shrines in India: Their Character, History, and Significance. New Delhi: Oxford University Press, 1989.
Turner, Colin (ed.). The Koran: Critical Concepts in Islamic Studies. New York: Routledge Curzon, 2004.
Turner, Victor. The Ritual Process: Structure and Anti-Structure. New Brunswick: Aldine Transaction, 1969.
Tweed, Thomas A. Crossings and Dwellings: A Theory of Religion. Cambridge: Cambridge University Press, 2006.
Udo, Simon, and Robert Langer. "The Dynamics of Orthodoxy and Heterodoxy. Dealing with Divergence in Muslim Discourses and Islamic Studies." Die Welt des Islams (2008): 273–288.
Uzma, Rehman. "Religion, Politics and Holy Shrines in Pakistan." Nordic Journal of Religion and Society 19 (2006): 17–28.
Van der Veer, Peter. "Playing or Praying: A Sufi Saint's Day in Surat." Journal of Asian Studies 51.3 (1992): 545–64.
Viitamäki, Mikko. "Mystical Gazal in Sufi Practice. Poetics, Music and Metaphysics in the Verses of Sāh Niyāz Aḥmad." Oriente Moderno 92.3 (2012): 519–536.
Vikor, Knut S. Between God and the Sultan: A History of Islamic Law. Oxford: Oxford University Press, 2005.
Voll, John O. "Hadith Scholars and Tariqahs: An Ulama Group in the 18th Century Haramayn and their Impact in the Islamic World." Journal of Asian and African Studies XV.3–4 (1980): 264–273.
Voll, John O. "Wahhabism and Mahdism: Alternative Styles of Islamic Renewals." Arab Studies Quarterly 4.1/2 (1982): 110–126.
Wade, Bonnie C. (ed.). Performing Arts in India: Essays on Music, Dance, and Drama. Berkeley: University of California, 1983.
Walls, Jerry D. (ed.). Oxford Handbook on Eschatology. Oxford: Oxford University Press, 2007.
Walker, Bethany J. "Commemorating the Sacred Spaces of the Past: The Mamluks and the Umayyad Mosque at Damascus." Near Eastern Archaeology 67.1 (2004): 26–39.
Waugh, Earle H. "Images of Muḥammad in the Work of Iqbal: Tradition and Alterations." History of Religions 23.2 (1983): 156–168.
Weiss, Bernard. The Spirit of Islamic Law. Athens: The University of Georgia Press, 2006.
Wheeler, Brannon. "Gift of the Body in Islam: The Prophet Muhammad's Camel Sacrifice and Distribution of Hair and Nails at his Farewell Pilgrimage." Numen 57.3/4 (2010): 341–388.
Whelch, Alford T. "Death and Dying in the Qur'an." In Frank E. Reynolds and Earle W. Waugh (eds.), Religious Encounters with Death. University Park: The Pennsylvania State Press, 1977.
Werbner, Pnina and Helene Basu (eds.). Embodying Charisma: Modernity, Locality and the Performance of Emotion in Sufi Cults. London: Routledge, 1998.

Werbner, Pnina. Pilgrims of Love: The Anthropology of a Global Sufi Cult. Oxford: Oxford University Press, 2003.
Williams, Wesley. "Aspects of the Creed of Iman Ahmad Ibn Hanbal: A Study of Anthropomorphism in Early Islam." IJMES 34.3 (2002): 441–463.
Wilson, John (ed.). Pakistan: The Struggle Within. India: Dorling Kindersley Pvt. Ltd., 2009.
Wilson, Stephen. Saints and their Cults: Studies in Religious Sociology, Folklore and History. Cambridge: Cambridge University Press, 1983.
Wolf, Richard K. "The Poetics of "Sufi" Practice: Drumming, Dancing, and Complex Agency of Madho Lal Husain (and Beyond)." American Ethnologist 33 (2006): 246–68.
Wolper, Ethel Sara. "The Politics of Patronage: Political Change and the Construction of Dervish Lodges in Sivas." Muqarnas 12 (1995): 39–47.
Wright, Zachary Valentine. Living Knowledge in West African Islam. The Sufi Community of Ibrāhīm Niasse. Leiden: Brill, 2015.
Yamba, Bawa C. Permanent Pilgrims: The Role of Pilgrimage in the Lives of West African Muslims in Sudan. Washington D. C.: Smithsonian Institution, 1995.
Yogish, Snehi. Spatializing Popular Sufi Shrines in Punjab: Dreams, Memories, Territoriality. New York: Routledge, 2019.
Zaman, Muhammad Qasim. "The 'Abbāsid Revolution: A Study of the Nature and Role of Religious Dynamics." Journal of Asian History 21.2 (1987): 119–149.
Zaman, Muhammad Qasim. "The Caliphs, the 'Ulamā', and the Law: Defining the Role and Function of the Caliph in the Early 'Abbāsid Period." Islamic Law and Society 4.1 (1997): 1–36.
Zaman, Muhammad Qasim. "Death, Funeral Processions, and the Articulation of Religious Authority in Early Islam." Studia Islamica 93 (2001): 27–58.
Zaman, Muhammad Qasim. The Ulama in Contemporary Islam: Custodians of Change. Princeton: Princeton University Press, 2002.
Zaman, Muhammad Qasim. "Religious Discourse and the Public Sphere in Contemporary Pakistan." Revue des Mondes Musulmans et de la Méditerranée 123 (2008): 55–73.
Zaman, Muhammad Qasim. "Transmitters of authority and ideas across cultural boundaries, eleventh to eighteenth centuries." In David O. Morgan and Anthony Reid, eds., The New Cambridge History of Islam Volume 3: The Eastern Islamic World, Eleventh to Eighteenth Centuries. Cambridge: Cambridge University Press, 2010.
Zaman, Muhammad Qasim. "The Sovereignty of God in Modern Islamic Thought." Journal of the Royal Asiatic Society 25 (2015): 389–418.
Zeghal, Malika. "Religion and Politics: The Complex Relationship Between Islamism and the State." Middle East Studies Association Bulletin 32.1 (1998): 25–28.
Ziad, Homayra. "Poetry, Music and the Muḥammadī Path: How Khvājah Mīr Dard Brought Three Worlds Together in Eighteenth-Century Delhi." Journal of Islamic Studies 21 3 (2010), 345–376.
Ziring, Lawrence. "From Islamic Republic to Islamic State in Pakistan." Asian Survey 24.9 (1984): 931–946.

Index

https://doi.org/10.1515/9783110781496-011